Atomic and Nuclear Physics

[*For B.Sc. Classes as per UGC Model Syllabus*]

N. SUBRAHMANYAM, M.Sc., Ph.D.

Deptt. of Physics
Kirori Mal College, University of Delhi,
Delhi - 110 007

and

BRIJ LAL, M.Sc.,

Reader in Physics
Hindu College, University of Delhi,
Delhi - 110 007

Revised by

JIVAN SESHAN

Head of Physics Department
Jai Hind College,
Mumbai - 20

S. CHAND PUBLISHING

S Chand And Company Limited

(ISO 9001 Certified Company)

S Chand And Company Limited
(ISO 9001 Certified Company)
Head Office: Block B-1, House No. D-1, Ground Floor, Mohan Co-operative Industrial Estate, New Delhi – 110 044 | Phone: 011-66672000
Registered Office: A-27, 2nd Floor, Mohan Co-operative Industrial Estate, New Delhi – 110 044
Phone: 011-49731800
www.**schandpublishing.com**; e-mail: **info@schandpublishing.com**

Branches

Ahmedabad : Ph: 27542369, 27541965; ahmedabad@schandpublishing.com
Bengaluru : Ph: 22354008, 22268048; bangalore@schandpublishing.com
Bhopal : Ph: 4274723, 4209587; bhopal@schandpublishing.com
Bhubaneshwar : Ph: 2951580; bhubaneshwar@schandpublishing.com
Chennai : Ph: 23632120; chennai@schandpublishing.com
Guwahati : Ph: 2738811, 2735640; guwahati@schandpublishing.com
Hyderabad : Ph: 40186018; hyderabad@schandpublishing.com
Jaipur : Ph: 2291317, 2291318; jaipur@schandpublishing.com
Jalandhar : Ph: 4645630; jalandhar@schandpublishing.com
Kochi : Ph: 2576207, 2576208; cochin@schandpublishing.com
Kolkata : Ph: 23357458, 23353914; kolkata@schandpublishing.com
Lucknow : Ph: 4003633; lucknow@schandpublishing.com
Mumbai : Ph: 25000297; mumbai@schandpublishing.com
Nagpur : Ph: 2250230; nagpur@schandpublishing.com
Patna : Ph: 2260011; patna@schandpublishing.com
Ranchi : Ph: 2361178; ranchi@schandpublishing.com
Sahibabad : Ph: 2771238; info@schandpublishing.com

First Edition 1984
Subsequent Editions and Reprints 1994, 96, 97, 99, 2001, 2003, 2005, 2006, 2008
Revised Edition 2009; Reprints 2010, 2012, 2016, 2019
Reprint 2020

ISBN: 978-81-219-0414-8 **Product Code:** H6MOP68PHYS10ENZX09O

PRINTED IN INDIA

By Vikas Publishing House Private Limited, Plot 20/4, Site-IV, Industrial Area Sahibabad, Ghaziabad – 201 010 and Published by S Chand And Company Limited, A-27, 2nd Floor, Mohan Co-operative Industrial Estate, New Delhi – 110 044.

PREFACE TO THE REVISED EDITION

The present edition of the book is revised as per the UGC syllabus. Questions and problems at the end of each chapter have been up-dated. Many new solved examples are included in this edition.

Certain topics have been added so that students from some universities where the syllabus has been modified and upgraded may benefit.

Besides being a text-book we hope that this will benefit students appearing at the IAS, AMIE and other Competitive Examinations.

We are grateful to the students and teachers who have appreciated the book. Suggestions for further improvement of the book will be highly appreciated. Our grateful thanks are due to the staff of S. Chand and Co. Ltd, Shri Navin Joshi VP (Publishing), Shri Bhagirath Kaushik, General Manager for getting the book printed in-time and Shri D.R. Parab, Branch Manager for the co-ordination work.

UGC MODEL SYLLABUS (Course 7)

[Relativity, Quantum Mechanics, Atomic, Molecular and Nuclear Physics]

1. Relativity (15)

Reference systems, inertial frames, Galilean invariance and conservation laws, propagation of light, Michelson-Morley experiment; search for ether. (5)

Postulates for the special theory of relativity, Lorentz transformations, length contraction, time dilation, velocity addition theorem, variation of mass with velocity, mass-energy equivalence, particle with a zero rest mass, (10)

2. Quantum Mechanics (30)

Origin of the quantum theory: Failure of classical physics to explain the phenomena such as black-body spectrum, photoelectric effect, Ri tz combination principle in spectra, stability of an atom. Planck's radiation law, Einstein's explanation of photoelectric effect, Bonr's quantization of angular momentum and its applications to hydrogen atom, limitations of Bohr's theory (5)

Wave-particle duality and uncertainty principle: de Broglie's hypothesis for matter waves, the concept of wave and group velocities, evidence for diffraction and interference of particles', experimental demonstration of matter waves.

Consequene of de Broglie's concepts; quantisation in hydrogen atom; energies of a particle in a box, wave packets, Heisenberg's uncertainty relation for p and x, its extension to energy and time. (3)

Consequence of the uncertainty relation; gamma ray microscope, diffraction at a slit, particle in a box, position of electron in a Bohar orbit.

Quantum Mechanics: Schrödinger's equation. postulatory basis of quantum mechanics: operators, expectation values, transition probabilities, applications to particle in a one-and three-dimensional boxes, harmonic oscillator, reflection at a step potential, transmission across a potential barrier.

Hydrogen atom: natural occurrence of n, 1 and m quantum numbers, the related physical quantities, comparison with Bohr's theory.

3. Atomic Physics (15)

Spectra of hydrogen, deuteron and alkali atoms spectral terms, doublet fine structure, screening constants for alkali spectra for s, p, d, and f states, selection rules. (6)

Singlet and triplet fine structure in alkaline earth spectra, L-S and J-J couplings. (3)

Week spectra: continuous X-ray spectrum and its dependence on voltage, Duane and Hunt's, law. Characteristics X-rays. Moseley's law, doublet structure of X-ray spectra, X-ray absorption spectra. (3)

4. Molecular Physics (15)

Discrete set of electronic energies of molecules, quantisation of vibrational and rotational energies, determination of internuclear distance, pure rotational and rotation vibration spectra. Dissociation limit for the ground and other electronic states, transition rules for pure vibration and electronic vibration spectra. (7)

Raman effect, Stokes and anti-Stokes lines, complimentary character of Raman and infrared spectra, experimental arrangements for Raman spectroscopy. (3)

Spectroscopic techniques: Sources of excitation, prism. and grating spectrographs for visible, UV and IR, absorption spectroscopy, double beam instruments, different recording systems. (5)

5. Nuclear Physics (15)

Interaction of charged particles and neutrons with matter, working of nuclear detectors, G-M counter, proportional counter and scintillation counter, cloud chambers, spark chamber, emulsions. (5)

Structure of nuclei, basic properties (I, l, Q and binding energy), deuteron binding energy. *p-p* and *n-p* scattering and general concepts of nuclear forces. Beta decay, range of alpha particle Geiger-Nuttal law. Gamow's explanation of beta decay, alpha decay and continuous and discrete spectra. (5)

Nuclear reactions, channels, compound nucleus, direct reaction (concepts) (3)

Shell model; liquid drop model, fission and fusion (concepts), energy production in stars by pop and carbon cycles (concepts). (2)

CONTENTS

UGC MODEL SYLLABUS (Course 7)

[Relativity, Quantum Mechanics, Atomic, Molecular and Nuclear Physics]

1. Relativity (15)

Reference systems, inertial frames, Galilean invariance and conservation laws, propagation of light, Michelson-Morley experiment; search for ether. (5)

Postulates for the special theory of relativity, Lorentz transformations, length contraction, time dilation, velocity addition theorem, variation of mass with velocity, mass-energy equivalence, particle with a zero rest mass, (10)

2. Quantum Mechanics (30)

Origin of the quantum theory: Failure of classical physics to explain the phenomena such as black-body spectrum, photoelectric effect, Ri tz combination principle in spectra, stability of an atom. Planck's radiation law, Einstein's explanation of photoelectric effect, Bonr's quantization of angular momentum and its applications to hydrogen atom, limitations of Bohr's theory (5)

Wave-particle duality and uncertainty principle: de Broglie's hypothesis for matter waves, the concept of wave and group velocities, evidence for diffraction and interference of particles', experimental demonstration of matter waves.

Consequene of de Broglie's concepts; quantisation in hydrogen atom; energies of a particle in a box, wave packets, Heisenberg's uncertainty relation for p and x, its extension to energy and time. (3)

Consequence of the uncertainty relation; gamma ray microscope, diffraction at a slit, particle in a box, position of electron in a Bohar orbit.

Quantum Mechanics: Schrödinger's equation. postulatory basis of quantum mechanics: operators, expectation values, transition probabilities, applications to particle in a one-and three-dimensional boxes, harmonic oscillator, reflection at a step potential, transmission across a potential barrier.

Hydrogen atom: natural occurrence of n, l and m quantum numbers, the related physical quantities, comparison with Bohr's theory.

3. Atomic Physics (15)

Spectra of hydrogen, deuteron and alkali atoms spectral terms, doublet fine structure, screening constants for alkali spectra for s, p, d, and f states, selection rules. (6)

Singlet and triplet fine structure in alkaline earth spectra, L-S and J-J couplings. (3)

Week spectra: continuous X-ray spectrum and its dependence on voltage, Duane and Hunt's, law. Characteristics X-rays. Moseley's law, doublet structure of X-ray spectra, X-ray absorption spectra.
(3)

4. Molecular Physics (15)

Discrete set of electronic energies of molecules, quantisation of vibrational and rotational energies, determination of internuclear distance, pure rotational and rotation vibration spectra. Dissociation limit for the ground and other electronic states, transition rules for pure vibration and electronic vibration spectra. (7)

Raman effect, Stokes and anti-Stokes lines, complimentary character of Raman and infrared spectra, experimental arrangements for Raman spectroscopy. (3)

Spectroscopic techniques: Sources of excitation, prism. and grating spectrographs for visible, UV and IR, absorption spectroscopy, double beam instruments, different recording systems. (5)

5. Nuclear Physics (15)

Interaction of charged particles and neutrons with matter, working of nuclear detectors, G-M counter, proportional counter and scintillation counter, cloud chambers, spark chamber, emulsions. (5)

Structure of nuclei, basic properties (I, l, Q and binding energy), deuteron binding energy. *p-p* and *n-p* scattering and general concepts of nuclear forces. Beta decay, range of alpha particle Geiger-Nuttal law. Gamow's explanation of beta decay, alpha decay and continuous and discrete spectra. (5)

Nuclear reactions, channels, compound nucleus, direct reaction (concepts) (3)

Shell model; liquid drop model, fission and fusion (concepts), energy production in stars by pop and carbon cycles (concepts). (2)

This Page has been intentionally left blank

1

SPECIAL THEORY OF RELATIVITY

1.1 Frame of Reference

Rest and motion are relative terms. To define motion, the observer must define a *frame of reference* relative to which the motion is considered.

A body in motion can be located with reference to some coordinate system called the frame of reference. If the co-ordinates of all the points of a body remain unchanged with time and with respect to the frame of reference, the body is said to be at rest. If, however, the co-ordinates of any point of the body change with time and with respect to the frame of reference, the body is said to be in motion.

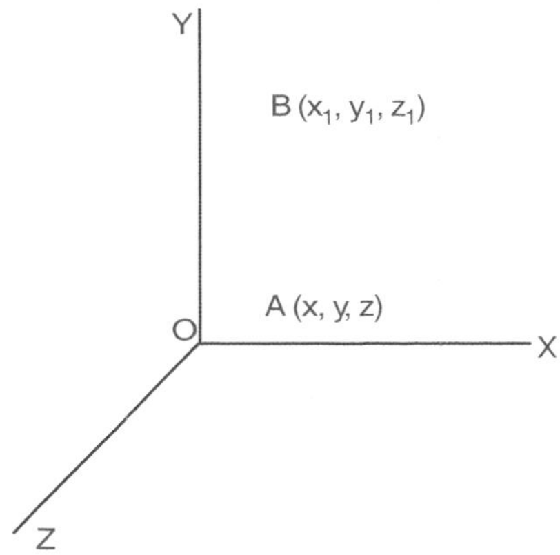

Fig. 1.1.

Suppose a body P is at the point A. Its co-ordinates are (x, y, z) with respect to the frame of reference (Fig. 1.1). If the body P always remains at A, it will be at rest with respect to the frame of reference. If another body Q is initially at A and after some time it is at $B(x_1, y_1, z_1)$, it is in motion with respect to the frame reference.

Now consider two frames of reference XYZ and $X' Y' Z'$ (Fig. 1.2).

The observer O considers the motion of P with respect to the frame of reference XYZ and the observer O' with respect to the frame of reference $X' Y' Z'$.

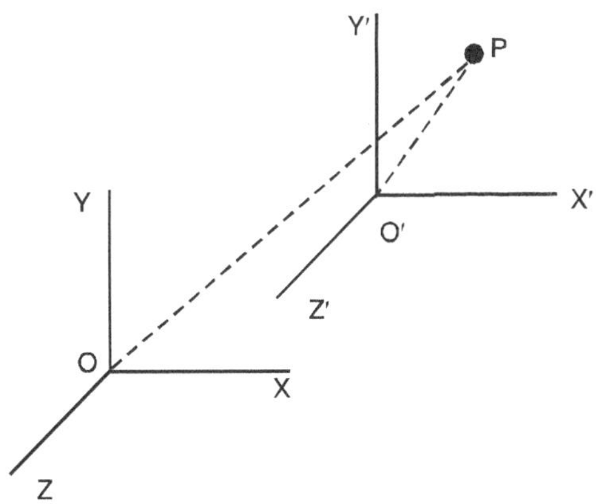

Fig. 1.2.

If O and O' are at rest with respect to each other, they will observe the same motion of P. If O and O' are in relative motion, their observation of motion will be different.

Examples

1. Consider two observers A and B. A is on the earth and B is on the sun. Both observe the motion of the moon. To the observer A, moon will appear to move along a circular path.

To the observer B, moon will appear to move in a wavy path.

2. Consider a car in motion. To an observer at the centre of wheel any point on the rim will appear to move along a circular path. But, to an observer on the ground the path of the point on the rim will appear to be a cycloid (Fig. 1.3).

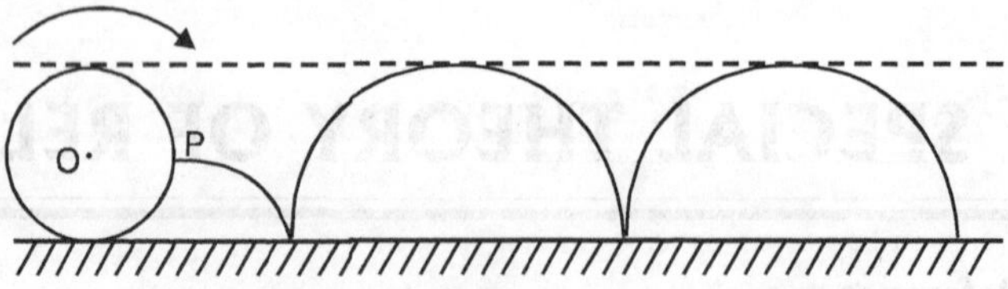

Fig. 1.3.

3. Consider that a train is moving with uniform velocity. A person sitting in the train drops a stone from the window. To this person the stone will appear to be falling vertically downwards. But, to a person standing near the track, the stone will appear to move along a parabolic path.

4. Consider a person A sitting in a train. All the windows of the train are closed and the person A cannot see anything outside. If the train is moving with uniform velocity, a stone thrown upward will return to the thrower. It means that this person cannot find the velocity of the train *i.e.*, the frame in which he is located. On the other hand, if the train is accelerated the stone thrown upward will not return to the thrower. This is the case of a non -inertial of reference.

1.2 Inertial Frames

We locate objects in space using a co-ordinate system. This co-ordinate system is referred to as **reference frame** or a **frame of reference.**

There are two types of reference frames :

(*i*) Inertial frame

(*ii*) Non-inertial or accelerated frame.

An inertial system is defined as a frame of reference in which the law of inertia holds *i.e.*, Newton's first law holds. Such a system is an unaccelerated system *i.e.*, it moves with a constant velocity (or is at rest).

Frames of reference which are accelerating with respect to some other frame are not inertial. In this frame the law of inertia is not valid.

It is convenient to take a fixed star (pole) as a standard inertial frame of reference. For practical purpose, the earth can be taken as an inertial frame of reference. Its rotation around its own axis can be taken to be negligibly small. In fact, it depends on the experiment to be performed whether earth can be taken as an inertial frame of reference or not.

1.3 Galilean Transformation Equations

Let an inertial frame S' move with a constant velocity $\vec{v}$ with respect to an inertial frame S. So that the relative motion is along the common X - X' axis. At time $t = O$, the two origins O and O' coincide. Consider an event to occur at some point P, whose space and time co-ordinates are measured in each inertial frame. The event is given by the co-ordinates x, y, z and time t to an observer in the S-frame and by x', y', z' and t to an observer attached to the S' frame.

To find the Galilean transformation equations *i.e.*, equations that relate the two co-ordinates x, y, z, t and x', y', z', t'. According to classical theory motion does not affect the lengths

As seen in Fig. 1.4

$$x' = x - vt$$
$$y' = y \qquad \text{...(1.1)}$$
$$z' = z$$

By using the universal nature of time we also have

$$t' = t$$

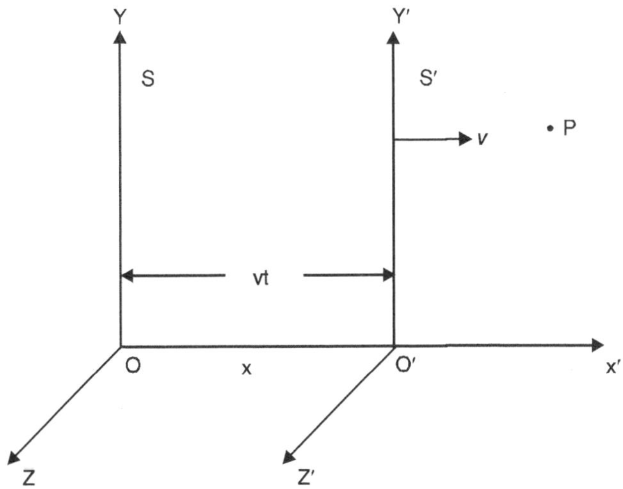

Fig. 1.4

i.e., Time measurements are independent of the motion of the observer.

The set of equations given in 1.1 are called the **Galilean transformations**.

1.4 Transformation Equations for the velocity

Now, the velocity is the time derivative of the position.

∴ The velocity transformation equations follows at once using equation $x' = x - vt$ and differentiating with respect to time gives

$$\frac{dx'}{dt'} = \frac{dx}{dt} - v \qquad \text{Since } t = t' \; \frac{d}{dt} \text{ is the same as } \frac{d}{dt'}$$

$$\therefore \quad \frac{dx'}{dt'} = \frac{dx'}{dt}$$

$$\therefore \quad \frac{dx'}{dt'} = \frac{dx}{dt} - v \text{ or}$$

$$u_x' = u_x - v \qquad \text{...(1.2)}$$

where, $u_x' = \dfrac{dx'}{dt'}$ and $u_x = \dfrac{dx}{dt}$.

Also, $$\frac{dy'}{dt'} = \frac{dy}{dt} \text{ and } \frac{dz'}{dt'} = \frac{dz}{dt}$$

Implies that $u_y' = u_y$ ...(1.3)

and $u_z' = u_z$ ...(1.4)

Combing (1.2), (1.3) and (1.4) we have

$$\vec{u'} = \vec{u} - \vec{v}$$

1.5 Transformation for acceleration and Force

To obtain the acceleration transformation equations, we differentiate the velocity with respect to time.

$\therefore$ From (1) $\therefore\ u_x' = u_x - v$

$$\therefore \quad \frac{d}{dt'}(u_x') = \frac{d}{dt}(u_x - v)$$

$$\therefore \quad \frac{d}{dt'}(u_x') = \frac{du_x}{dt} \qquad (v \text{ is a constant})$$

$$\therefore \quad a_x' = a_x \qquad ...(1.5)$$

Similarly $\dfrac{du'_y}{dt'} = \dfrac{du_y}{dt}$ implies $a'_y = a_y$ ...(1.6)

and $\dfrac{du'_z}{dt} = \dfrac{du_z}{dt}$ $\therefore\ a_z' = a_z$ ...(1.7)

Combing (1.5), (1.6) and (1.7) we have $\vec{a'} = \vec{a}$ ...(1.8)

Now since mass m is the same and is not affected by the motion of the reference frame.

We conclude that

$$\vec{F} = ma = \vec{F'} = ma'$$

$\therefore$ Newton's laws of motion and the equations of motion of a particle are exactly the same in all inertial frames.

1.6 Invariance of Length

Using the Galilean transformation equations one can find the length of an object in a moving co-ordinate system.

Any set of equations that retain the same form from one co-ordinate system to another is said to be **invariant** under such co-ordinate transformations.

If x_2 and x_1 are the lengths of an object in the S-frame and x_2' and x'_1 in the S' frame, then by Galilean transformation equations we have,

$$x_2' = x_2 - vt_2 \text{ and } x'_1 = x_1 - vt_1$$

$$\therefore \quad x_2' - x_1' = x_2 - x_1 - v(t_2 - t_1)$$

But $t_2 = t_1$ as the measurements of length have to be done **simultaneously** *i.e.*, at the same time. This is how the length of a moving object is defined.

$$\therefore \quad x'_2 - x'_1 = x_2 - x_1 \qquad ...(1.9)$$

Note that if the end points are not measured simultaneously but at different times; it would be like measuring the location of the rear end of the car at one time and its front end at another instant of time.

This would obviously give a wrong result for the length of the car.

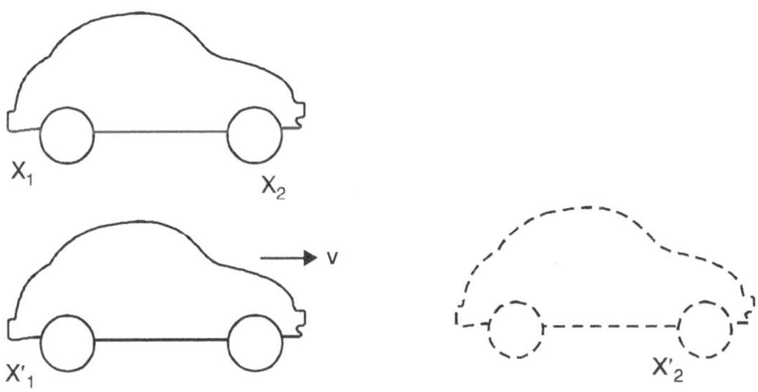

Fig. 1.5.

1.7 Conservation of Linear Momentum

Consider a head-on collision between two particles of mass m_1 and m_2 and moving along the x-axis of the S-frame with velocities u_1 *and* u_2 respectively. Let the velocity of the particles be v_1 and v_2 after the collision.

By conservation of linear momentum in S-frame we have,

Momentum before collision = Momentum after the collision.

$$\therefore \quad m_1 u_1 + m_2 u_2 = m_1 v_1 + m_2 v_2$$

Using Galilean transformation equations for the S' frame (which moves with a velocity v with respect to the S-frame) we have,

$$u_1 = u_1' + v \quad \text{and} \quad u_2 = u_2' + v$$

Similarly $$v_1 = v_1' + v \quad \text{and} \quad v_2' = v_2 + v$$

$\therefore$ In S' frame

$$m_1 (u_1' + v) + m_2 (u_2' + v) = m_1(v_1' + v) + m_2 (v_2' + v)$$

$$\therefore \quad m_1 u_1' + m_1 v + m_2 u_2' + m_2 v = m_1 v_1' + m_1 v + m_2 v_2' + m_2 v$$

$$\therefore \quad m_1 u_1' + m_2 u_2' = m_1 v_1' + m_2 v_2'$$

$\therefore$ Momentum is also conserved in S'-frame under Gallelian transformation. $\therefore$ Invariance of momentum since it retains the same form.

Note that even though momentum is conserved in both the frames the **values** of the momentum are different in the two frames.

1.8 Kinetic Energy Conservation

Now, let kinetic energy be conserved in S-frame in an elastic collision.

i.e., Kinetic energy before collision = K. E. after collision

$$i.e., \quad \tfrac{1}{2} m_1 u_1^2 + \tfrac{1}{2} m_2 u_2^2 = \tfrac{1}{2} m_1 v_1^2 + \tfrac{1}{2} m_2 v_2^2$$

Substituting $$u_1 = u_1' + v, \quad u_2 = u_2' + v$$

and $$v_1 = v_1' + v, \quad v_2 = v_2' + v$$

we get,

$$\tfrac{1}{2} m_1 (u_1' + v)^2 + \tfrac{1}{2} m_2 (u_2' + v)^2 = \tfrac{1}{2} m_1 (v_1' + v)^2 + \tfrac{1}{2} m_2 (v_2' + v)^2$$

Since conservation of momentum is also valid in an elastic collision we have,

$$\tfrac{1}{2} m_1 u_1'^2 + \tfrac{1}{2} m_2 u_2'^2 = \tfrac{1}{2} m_1 v_1'^2 + \tfrac{1}{2} m_2 v_2'^2$$

Kinetic energy is also conserved in S'-frame.

∴ It is also an elastic in this frame. Hence invariance under Galelian transformation.

Note again that through kinetic energy is conserved in both the frames, the values of the kinetic energies are different in the two frames.

1.9 Luminiferous Ether

Newton referred to the fixed stars as the frame of reference to explain his laws. A search for a new frame of reference which is more fixed than the stars was carried on for a number of years. According to Maxwell, light waves are electromagnetic waves. Light waves are transverse and sound waves are longitudinal waves. Polarization confirmes that light waves are transverse waves.

A material medium is a necessity for the propagation of a wave. For light waves a material medium called 'luminiferous ether' was supposed to fill the free space. It had zero density and perfect transparency.

Transverse waves require shearing forces and these can occur in solids only. It means that ether must be a rigid solid filling the wholespace. The velocity of wave propagation depends on the elasticity of the medium. As the velocity of light is very hight the elasticity of ether must be very high. Thus the whole free space must be filled up with this medium called ether which is difficult to conceive.

If light is propagated through the ether medium, then in optical experiments, one would expect change in velocity (drift) depending upon the direction in which the light is propagated with respect to the motion of apparatus. Considering earth to be moving in its orbit around the sun with a velocity 3×10^4 m/s and the velocity of light 3×10^8 m/s the ratio between the two is 10^{-4}. Earth can be considered to be moving through the stationary ether. This led to the experiments of Michelson and Morley in 1881 - 1887 to detect the luminiferous ether.

1.10 Michelson Morley Experiment

Light from a monochromatic source S is rendered parallel with the lens L. It is divided into two portions by the half silver plate A (Fig. 1.6).

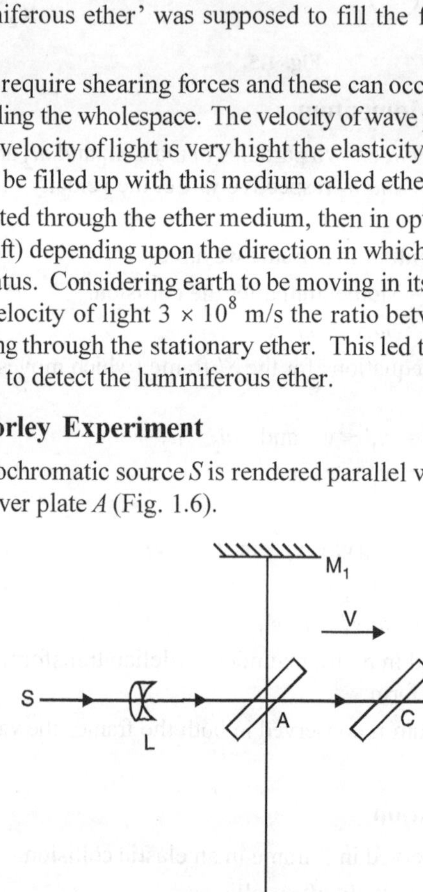

Fig. 1.6

One portion of the beam travels to the mirror M_1 and is reflected back. The other portion of the beam travels towards M_2 and is reflected back. The two reflected beams interfere and the interference fringes are viewed with the help of a telescope.

The apparatus is arranged to move along the direction of the earth's orbit round the sun. The speed of moment of the apparatus is equal to v, *i.e.*, speed of the earth in its orbit. The whole apparatus is floated on mercury and can adjusted such that it is always moving along the direction of the earth's orbit round the sun. Earth is assumed to be stationary.

According to the Galilean frame of reference, F is a fixed frame corresponding to the ether medium (Fig. 9.10). F' is the frame of reference moving with a velocity v in the direction of the movement of the apparatus *i.e.*, in the direction of the movement of the earth in its orbit round the sun. The velocity of light in the direction of h movement of the frame $F' = c - v$ and in the opposite direction it is $= c + v$. Let T_1 be the time taken by light to travel from A to M'_2 and T_2 the time taken from M'_2 to A' (Fig 1.7).

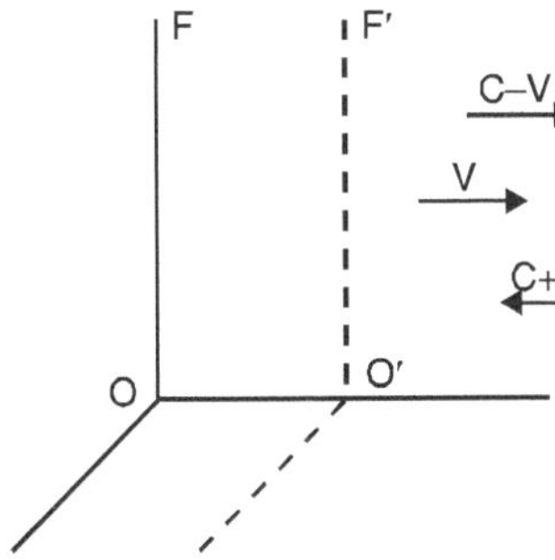

Fig. 1.7

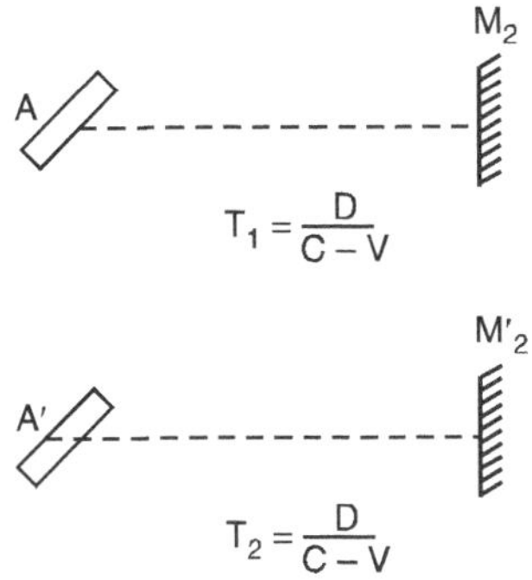

Fig. 1.8

Here $$D = AM_1 = AM_2$$

The total time taken by light (to and fro)

$$T = T_1 + T_2 = \frac{D}{c-v} + \frac{D}{c+v}$$

$$T = \frac{2Dc}{c^2 - v^2}$$

The total distance travelled by light

$$x_1 = T \times c = \frac{2Dc^2}{c^2 - v^2}$$

$$x_1 = 2D\left[1 + \frac{v^2}{c^2}\right] \quad \text{...(1.10) [neglecting the higher power]}$$

Let the time taken by light from A to M'_1 be T'. Here M_1 is shifted to M'_1 in the time A is shifted to P (Fig. 1.9).

$$AP = VT = vLc$$

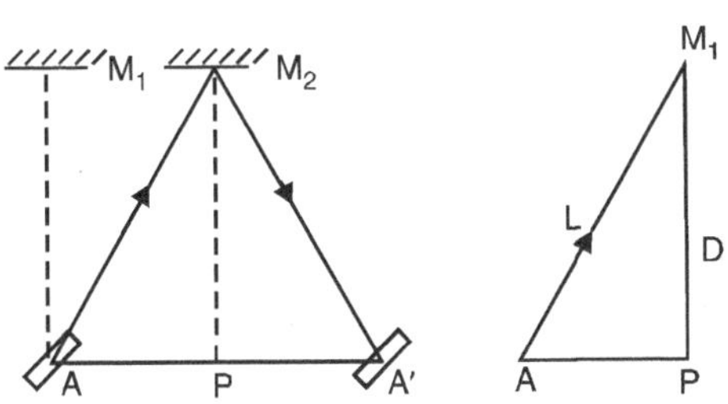

AP = VT = vL / c

Fig. 1.9

The distance $$AP = vt'$$

But $$T' = \frac{L}{c}$$

$$AP = \frac{vL}{c}$$

$$L^2 = D^2 + \left[\frac{vL}{c}\right]^2$$

$$L^2\left[1-\frac{v^2}{c^2}\right] = D^2$$

$$L^2 = \frac{D_2}{\left[1-\frac{v^2}{c^2}\right]}$$

$$L = D\left[1+\frac{v^2}{2c^2}\right]$$

∴ Total distance travelled by light in time T in going from A to M'_1 and back to A'

$$= 2L = 2D\left[1+\frac{v^2}{2c^2}\right]$$

$$x_2 = 2D\left[1+\frac{v^2}{2c^2}\right] \quad ...(1.11)$$

From equations (1.10) and (1.11), the path difference,

$$x_1 - x_2 = 2D\left[1+\frac{v^2}{c^2}\right] - 2D\left[1+\frac{v^2}{2c^2}\right]$$

$$= \frac{Dv^2}{c^2} + 1 \quad ...(1.12)$$

If the apparatus is turned through 90°, the path difference will be $= -\frac{Dv^2}{c^2}$. The displacement in the interference fringe is $\frac{Dv^2}{c^2}$.

The displacement expected is about 0.4 of a fringe width.

But, in this experiment no displacement of the fringes was observed. This shows that this experiment is a negative experiment. The experiment of Michelson and Morley was repeated with certain modifications by a number of workers but no drift was observed. Even with monochromatic light from a modern laser, the drift has not been detected.

This negative result suggests that the *velocity of light is invariable and remains constant in all directions.* Moreover, the effects of ether are undetectable. This has been verified by Doppler's principle *i.e.*, the speed of light is independent of the motion of the source or the observer.

Thus, all attempts to make ether as a fixed frame or reference failed.

1.11 Search for Ether

Bradley's Experiment on Stellar Aberration

To see a star directly overhead, one would have to point a telescope straight up. The rays of light coming from the star would go straight down through the telescope. Now if the observer and the earth had a velocity v, the light coming from the star will not be vertical. The telescope would have to be tilted as in [Fig. 1.10(*a*)] with the vertical.

Now, in the time the light proceeds straight down in the ether, the vertical distance travelled is $l = c\Delta t$, through the telescope. In this time Δt, the telescope moves a distance $v\Delta t$. The eyepiece at the time the light leaves the telescope is on the same vertical line as the objective when the ray entered. The angle of tilt of the telescope θ is given by

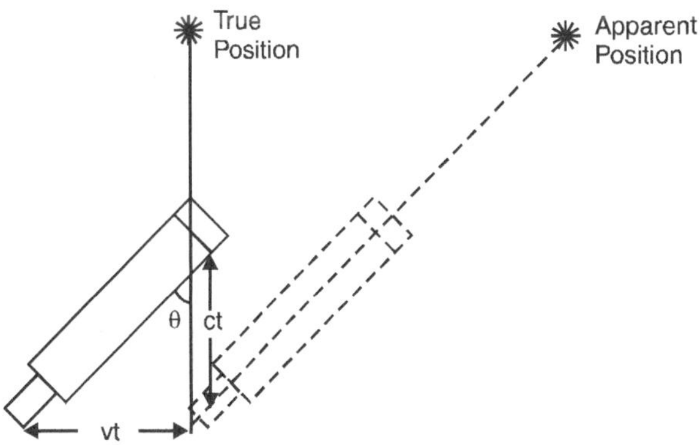

Fig. 1.10 (*a*)

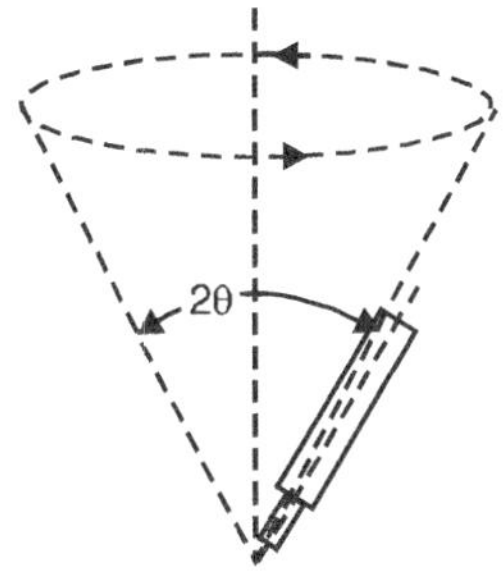

Fig. 1.10 (*b*)

$$\tan\theta = \frac{v\Delta t}{c\Delta t} = \frac{v}{c} \qquad \text{...(1.13)}$$

Since the earth goes round the sun the direction of observation reverses every six months. The telescope axis traces out a cone of aberration through the year. The angular diameter of the cone is $2\theta = 41$ seconds of arc in agreement with observations. Hence we see that if there is ether, the earth moves freely through it. There is no ether diag.

1.12 Einstein's Special Theory of Relativity

Einstein in 1905 propounded the special theory of relativity and in 1915 proposed the general theory of relativity. The special theory deals with the problems in which one frame of reference moves with a *constant linear velocity* relative to another frame of reference. The general theory of relativity deals with problems in which one frame of reference is associated with respect to another frame of reference. Einstein assumed that all the observers will notice that their motion in space will make no difference in the velocity of light with respect to them. He assumed that a fixed frame of reference cannot be located. All laws must be stated that they are applicable in any frame of reference.

According to Einstein's special theory of relativity :

(*i*) Observation on any particular frame of reference are not preferred to those of the other frames of reference. It means that the laws of physics are applicable equally well for an observer in the frame of reference A moving with a constant linear velocity with respect to an observer in another frame of reference B and *vice versa*, and

(*ii*) Velocity of light in free space is constant and is independent of the motion of the observer in any frame of reference.

Einstein proved the following facts based on his theory of relativity. Let v be the velocity of the spaceship with respect to a given frame of reference. The observations are made by an observer in that reference frame.

1.13 Lorentz Transformations

Consider two frames of reference A and B (Fig. 1.11). A is fixed and B is moving with a constant speed υ along the direction of the X-axis. Initially both the frames have the same origin of coordinates. After a time t the frame of reference B has moved a distance $OO' = \upsilon t$. For the point P in space, the coordinates are (x, y, z) with reference to the frame A and (x', y', z') with reference to the frame B.

According to Galilean frame of reference

$$x' = x - \upsilon t$$
$$y' = y$$
$$z' = z$$
$$t' = t \qquad ...(1.14)$$

Differentiating the first equation in (1.14)

$$\frac{dx'}{dt} = \frac{dx}{dt} - \upsilon \qquad ...(1.15)$$

$$c' = c - \upsilon \qquad ...(1.16)$$

It means that if a person is moving in a spaceship, the speed of the passing light will be $(c - \upsilon)$. But all attempts to show that the velocity of light changes with the motion of the observer have failed. The velocity of light remains constant in free space, according to the postulates of the special theory of relativity. The equation $x' = x - \upsilon t$ is in accordance with the ordinary laws of mechanics. The new transformation for the x-coordinate must be similar to this equation when the value of v is extremely small as compared to the velocity of light, c. The simplest possible form of this equation can be

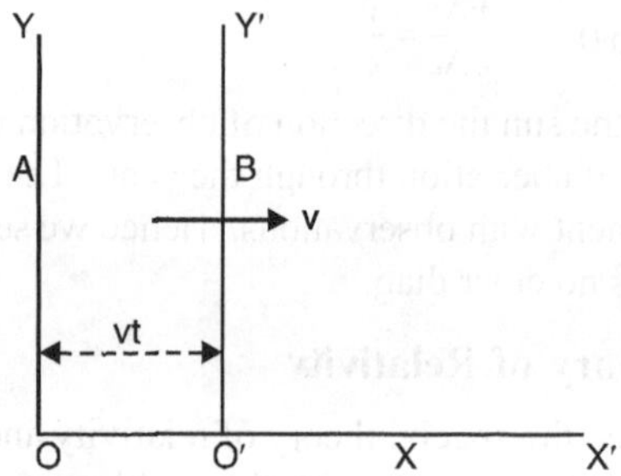

Fig. 1.11

$$x' = k(x - \upsilon t) \qquad ...(1.17)$$

Here k depends only on the value of υ and does not depend upon the values of x and t. Equation (1.17) is linear and x' has only one value for given value of x.

According to the first postulate of the special theory of relativity, observations made in the frame of reference B must be identical to those made in A, expect for a change in the sign of υ and having the same value, for the constant of proportionality k.

$$\therefore \quad x = k(x' + \upsilon t') \qquad ...(1.18)$$

As the relative motion of A and B is confined to only x

$$y' = y \qquad ...(1.19)$$

$$z' = z \qquad ...(1.20)$$

But this equality does not hold good for t and t'.

Equations (4) and (5) will not hold, if $t' = t$.

The value of x' from equation (4) is substituted in equation (1.18). This gives

$$x = k\,[k\,(x - \upsilon t) + \upsilon t']$$

$$x = k^2\,(x - \upsilon t) + k\upsilon t' \qquad ...(1.21)$$

$$k\upsilon t' = x - k^2\,(x - \upsilon t)$$

$$t' = \frac{x - k^2 x + k^2 \upsilon t}{k\upsilon} = \frac{k^2 \upsilon t}{k\upsilon} + \frac{x(1-k^2)}{k\upsilon}$$

$$t' = kt + \left(\frac{1-k^2}{kv}\right)x \quad \text{... (1.22)}$$

To find the value of k, consider two reference frames A and B. The spaceship is reference frame A measures the time t and the spaceship in reference frame B measures the time t', for a flash of light. The X-coordinates for both the ships will be

$$x = ct \quad \text{... (1.23)}$$

$$x' = ct' \quad \text{... (1.24)}$$

It is so because the value of c must remain constant in both the frames of reference according to the postulate of relativity.

Substituting the values of x' and t' in equation, (1.24)

$$k(x - \upsilon t) = c\left[kt + \left(\frac{1-k^2}{k\upsilon}\right)x\right]$$

$$kx - k\upsilon t = ckt + cx\left(\frac{1-k^2}{k\upsilon}\right)$$

$$kx - cx\left(\frac{1-k^2}{k\upsilon}\right) = ckt + k\upsilon t$$

$$x\left[k - c\left(\frac{1-k^2}{k\upsilon}\right)\right] = ctk\left[1 + \frac{\upsilon}{c}\right]$$

$$x = \frac{ckt\left[1+\frac{\upsilon}{c}\right]}{k\left[1 - c\left(\frac{1-k^2}{k^2\upsilon}\right)\right]}x = ct\left[\frac{1+\frac{\upsilon}{c}}{1-\left(\frac{c}{\upsilon}\right)\left(\frac{1}{k^2}-1\right)}\right]$$

This value of x must be equal to the value of x in equation (1.23).

$$\therefore \qquad ct = ct\left[\frac{1+\frac{\upsilon}{c}}{1-\left(\frac{c}{\upsilon}\right)\left(\frac{1}{k^2}-1\right)}\right]$$

$$1-\left(\frac{c}{\upsilon}\right)\left(\frac{1}{k^2}-1\right) = 1+\frac{\upsilon}{c}$$

$$-\left(\frac{c}{\upsilon}\right)\left(\frac{1}{k^2}-1\right) = \frac{\upsilon}{c}$$

$$1-\frac{1}{k^2} = \frac{\upsilon^2}{c^2}$$

$$\frac{1}{k^2} = 1-\frac{\upsilon^2}{c^2}$$

$$k^2 = \frac{1}{1-\frac{\upsilon^2}{c^2}}$$

$$k = \frac{1}{\sqrt{1-\frac{\upsilon^2}{c^2}}}$$

The value of k is in accordance with the experiment result. When the value of υ is extremely small in comparison to c, the value of k is practically equal to 1.

When the value of k is substituted in equations (1.17), (1.19), (1.20) and (1.22), we get

$$x' = k(x - \upsilon t)$$

$$x' = \frac{x - \upsilon t}{\sqrt{1-\frac{\upsilon^2}{c^2}}} \qquad \text{... (1.25)}$$

$$y' = y \qquad \text{... (1.26)}$$

$$z' = z \qquad \text{... (1.27)}$$

$$t' = kt + \left(\frac{1-k^2}{k\upsilon}\right)x = kt + \frac{x}{kv} - k\left(\frac{x}{\upsilon}\right)$$

$$t' = kt + \frac{x}{\upsilon}\left[\frac{1}{k} - k\right] = kt + k\frac{x}{\upsilon}\left(\frac{1}{k^2} - 1\right)$$

$$t' = \frac{t + \frac{x}{\upsilon}\left[1 - \frac{\upsilon^2}{c^2} - 1\right]}{\sqrt{1-\frac{\upsilon^2}{c^2}}}$$

$$t' = \frac{1 - \frac{\upsilon x}{c^2}}{\sqrt{1-\frac{\upsilon^2}{c^2}}} \qquad \text{... (1.28)}$$

Equations (1.25), (1.26), (1.27) and (1.28) are called Lorentz transformations.

These equations give the conversions for the measurements of time and space made in the stationary frame A to their counterparts in the moving frame B.

The inverse Lorentz transformation for measurements of time space made in the frame B to their counterparts in frame A can be written as follows :

$$x = \frac{x' + \upsilon t'}{\sqrt{1-\frac{\upsilon^2}{c^2}}}$$

$$y = y' \qquad \text{... (1.29)}$$

$$z = z' \qquad \text{... (1.30)}$$

and

$$t = \frac{t' + \frac{\upsilon x'}{c^2}}{\sqrt{1-\frac{\upsilon^2}{c^2}}} \qquad \text{...(1.31)}$$

1.14 Lorentz-Fitzerald Contraction

Measurement of space and time are not absolute but depend on the relative motion of the observer and the observed objects.

Consider a rod of length L_0 parallel to the X-axis and having the coordinates x_1 and x_2 in the frame A. An observer in the reference frame A measures the length of the rod as $L_0 = x_2 - x_1$.

Consider a second reference frame B moving with a velocity υ along the X-axis, with respect to the reference frame A. An observer in the reference frame B measures the end coordinates of the rod as x_1' and x_2'. The length L as observed by the observer in B is

$$L = x_2' - x_1'.$$

The relation between x_1 and x_1', and also between x_2 and x_2' according to inverse Lorentz transformation will be,

$$x_1 = \frac{x_1' + \upsilon t'}{\sqrt{1-\frac{\upsilon^2}{c^2}}} \text{ and } x_2 = \frac{x_2' + \upsilon t'}{\sqrt{1-\frac{\upsilon^2}{c^2}}}$$

$$L_0 = x_2 - x_1$$

$$L_0 = \frac{x_2' + \upsilon t'}{\sqrt{1-\frac{\upsilon^2}{c^2}}} - \frac{x_1' + \upsilon t'}{\sqrt{1-\frac{\upsilon^2}{c^2}}}$$

$$L_0 = \frac{x_2' - x_1'}{\sqrt{1-\frac{\upsilon^2}{c^2}}} = \frac{L}{\sqrt{1-\frac{\upsilon^2}{c^2}}}$$

$$L = L_0\sqrt{1-\frac{\upsilon^2}{c^2}} \qquad \text{...(1.32)}$$

This shows that the length of stationary object with respect to an observer in motion is shorter than length measured by the observer at rest. Similarly, when the object is in motion with respect to a stationary observer, again the object is shortened in length. This relativistic construction works both ways, *i.e.*, whether object is in motion or the observer is in motion. This is called Lorentz - Fitzerald contraction.

Lorentz-Fitzerald contraction is appreciable only when the velocity v is comparable to the velocity of light c.

Suppose $\upsilon = 0.6c$

$$L = L_0\sqrt{1-\frac{\upsilon^2}{c^2}}$$

$$L = L_0\sqrt{1-0.36}$$

$$L = 0.8\, L_0$$

The contraction in length

$$L_0 - L = 0.2\, L_0$$

i.e., contraction in length is 20%.

When the velocity of the body is negligibly small as compared to c, the contraction in length is negligible and $L = L_0$.

Suppose $\upsilon = 0.01c$

$$L = L_0\sqrt{1-(10^{-4})}$$

$$L = 0.9999 L_0$$

i.e., $L \approx L_0$

Due to this reason, the shortening in length cannot be visually observed for objects moving on the surface of earth. However, Lorentz - Fitzerald contraction has been verified by experimental measurement.

1.15 Time Dilation

A clock is measured to go at its fastest rate when it is at rest relative to the observer. When it moves with a velocity v relative to the observer, its rate is measured to have slowed down by a factor

$1/\sqrt{1-v^2/c^2}$ or $\dfrac{1}{\sqrt{1-\beta^2}}$ where $\beta = v/c$.

Let a clock be at rest at the position x' in the frame S′. In order to simplify matters, we consider the hand of the clock to go around once which we assume is unit time. Hence, the events we observe (*i.e.,* the two successive coincidences of the hand of the clock with a given marker on the face of the clock) span the time interval t' to $(t' + 1)$ in the primed co-ordinates S′. The S-frame observer records these events as occurring at times

$$t_1 = \frac{t' + vx'/c^2}{\sqrt{1-\beta^2}}$$

and

$$t_2 = \frac{(t' + 1) + vx'/c^2}{\sqrt{1-\beta^2}}$$

The clock in S′ frame is at a fixed position x', but time t_1 and t_2 are read by two different clocks in the S-frame, namely the stationary S-clock that happens to coincide with the moving clock at the beginning of the interval (t_1 or t'), and also, the stationary S-clock coincident with the moving clock at the end of the interval. These clocks are synchronised so that the time interval they record for the event would simply be

$$t_2 - t_1 = \frac{1}{\sqrt{1-\beta^2}} \qquad \text{...1.33}$$

Now, instead of unit time, this clock recorded a time interval $(t_2^1 - t_1^1)$, the S-clock would have recorded the corresponding interval as

$$t_2 - t_1 = \frac{t_2^1 - t_1^1}{\sqrt{1-\beta^2}} \qquad \text{...1.34}$$

Since $p > 1$

$$\therefore \qquad t_2' - t_1' > t_2 - t_1$$

Hence unit time measured on the S′ clock is recorded as a longer time on the S-clock. (Note $t_2 - t_1$ or $t_2' - t_1'$ is the time interval difference). From the point of view of observer S, the moving S′ clock appears slowed down *i.e.,* it runs at a rate which is slow by a factor $\sqrt{1-\beta^2}$. This result is true for all S′ clocks observed from S.

Suppose that in a certain inertial system we observe clocks which are moving relative to us in an arbitrary manner. At each instant of time that is different the motion can be regarded as uniform. Hence at each instant we introduce a co-ordinate system, rigidly linked to the moving clocks which constitutes an inertial reference frame.

During the course of an infinitesimal time interval dt (as read by the clock in the rest frame). The moving clocks go a distance $\sqrt{dx^2 + dy^2 + dz^2}$. Let dt' be the time interval indicated for this period by the moving clcoks. In a system of co-ordinates associated with the moving clocks, the latter are at rest *i.e.,* $dx' = dy' = dz' = 0$.

Due to the invariance of the interval.

$$ds^2 = c^2dt^2 - dx^2 - dy^2 - dz^2 = c^2dt'^2 \qquad \text{...1.35}$$

where $\qquad dt' = \dfrac{ds}{c} = \dfrac{1}{c}\sqrt{c^2\,dt^2 - dx^2 - dy^2 - dz^2}$

or $\qquad dt' = (1 - (dx^2 + dy^2 + dz^2) / c^2\,dt^2)^{1/2}\,dt.$

But $\dfrac{dx^2 + dy^2 + dz^2}{dt^2} = v^2$, where v = velocity of the moving clocks.

$$\therefore \qquad dt' = ds/c = dt\sqrt{1 - v^2/c^2} \qquad ...1.36$$

Integrating this equation we get the time interval indicated by the moving clocks for the corresponding elapsed time $t_2 - t_1$ for the stationary clocks.

$$\therefore \qquad t_2' - t_1' = \int_{t_1}^{t_2} dt\left(1 - v^2/c^2\right)^{1/2} \qquad ...1.37$$

The time read by a clock moving with a given frame is called the *profer time* for the frame. From the eqs. we see that, the profer time in a moving frame is always less than the corresponding interval in a rest frame.

1.16 Relative of Mass

According to relativity, the measurements of an object or an event will be different for different observers who are in relative motion. If the rest mass of the body is m_0 then mass of the body in motion is

$$m \frac{m_0}{\sqrt{1 - \frac{\upsilon^2}{c^2}}} \qquad .... (1.38)$$

Here υ is the velocity of the body in motion and c is the velocity of light.

It shows that $m > m_0$. Thus an object will appear to have more mass while in motion than when it is at rest.

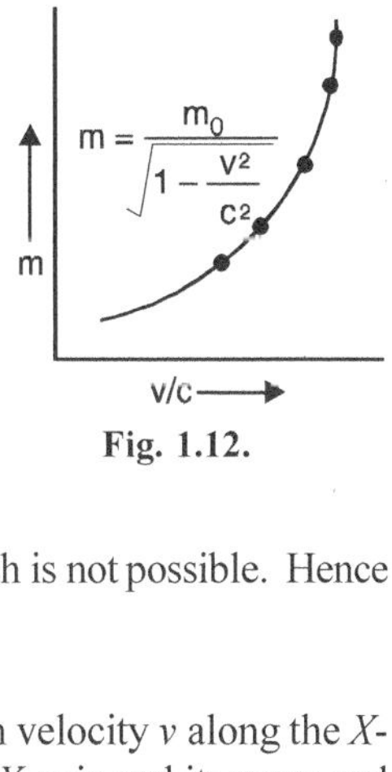

Fig. 1.12.

The relative mass of the electron at various speeds is shown in the graph (Fig. 1.12). The relative mass of the electron increases enormously when its velocity approaches the velocity of light.

If $\qquad \upsilon = c$

$$m = \frac{m_0}{\sqrt{1 - \frac{\upsilon^2}{c^2}}}$$

$$m = \frac{m_0}{0} = \infty$$

i.e., an electron moving with the speed of light will have an infinite mass which is not possible. Hence no material particle can have a velocity equal to the velocity of light.

1.17 Derivation for two frames

Consider two frames of reference A and B, and the frame B moving with velocity v along the X-axis with reference to frame A. Suppose that a particle is moving along the X-axis and its mass and velocity with respect to reference frame A are (*i*) m_1 and u_1 (*ii*) and with respect to reference frame B are m_2 and u_2.

Let $$k = \frac{1}{\sqrt{1 - \frac{\upsilon^2}{c^2}}}, \; k_1 = \frac{1}{\sqrt{1 - \frac{u_1^2}{c^2}}} \text{ and } k_2 = \frac{1}{\sqrt{1 - \frac{u_2^2}{c^2}}}$$

From relative velocity principle,

$$u_1 = \frac{u_2 + \upsilon}{1 + \left(\frac{\upsilon u_2}{c^2}\right)} \text{ and } u_2 = \frac{u_1 - \upsilon}{\left(1 - \frac{\upsilon u_1}{c^2}\right)}$$

$$k_2 u_2 = \frac{u_1 - \upsilon}{\left[1 - \left(\frac{u_2^2}{c^2}\right)\right]^{\frac{1}{2}}\left(1 - \frac{\upsilon u_1}{c^2}\right)} \qquad ...(1.39)$$

Substituting the value of u_2 on the right hand side

$$\therefore \quad k_2 u_2 = \frac{u_1 - \upsilon}{\left[1 - \dfrac{(u_1-\upsilon)^2}{c^2\left(1 - \dfrac{\upsilon u_1}{c^2}\right)^2}\right]^{\frac{1}{2}} \left(\dfrac{1 - \upsilon u_1}{c^2}\right)}$$

1.18 Velocity Addition Theorem

Consider a particle moving relative to both S and S'. An observer in S measures the velocity components as :

$$u_x = \frac{dx}{dt}, \quad u_y = \frac{dy}{dt} \quad \text{and} \quad u_z = \frac{dz}{dt}.$$

To an S' observe the components are :

$$u_x' = \frac{dx'}{dt'} \quad u_y' = \frac{dy'}{dt'} \text{ and } u_z' = \frac{dz'}{dt'}$$

Now

$$x' = \frac{x - \upsilon t}{\sqrt{1 - \dfrac{v^2}{c^2}}} \qquad \text{...(1.40)}$$

where v is the velocity of S' w.r.to. S'-frame.

Also

$$t' = \frac{t - \dfrac{v}{c^2}x}{\sqrt{1 - \dfrac{v^2}{c^2}}} \qquad \text{...(1.41)}$$

By taking differentials of both sides of Eqs. 1.40 and 1.41, we get

$$dx' = \frac{dx - vdt}{\sqrt{1 - \dfrac{v^2}{c^2}}}, \quad dy' = dy \text{ and } dz' = dz$$

and

$$dt' = \frac{dt - \dfrac{\upsilon}{c^2}dx}{\sqrt{1 - \dfrac{\upsilon^2}{c^2}}}$$

$$\therefore \quad u_x' = \frac{dx'}{dt'} = \frac{dx - vdt}{dt - \dfrac{vdx}{c^2}}, \text{ Divide Numerator and Denominator by } dt$$

$$u_x' = \frac{\dfrac{dx}{dt} - \upsilon}{1 - \dfrac{\upsilon}{c^2}\dfrac{dx}{dt}} = \frac{u_x - \upsilon}{1 - \dfrac{u_x \upsilon}{c^2}} \qquad \text{...(1.42)}$$

This equation is termed as the **Einstein's velocity Addition theorem.**

Similarly

$$u_y' = \frac{dy'}{dt'} = \frac{dy\sqrt{1 - \dfrac{\upsilon^2}{c^2}}}{dt - \dfrac{dxv}{c^2}}, \text{ Divide Numerator and Denominator by } dt$$

$$\therefore \quad \frac{\dfrac{dy}{dt}\sqrt{1 - \dfrac{\upsilon^2}{c^2}}}{1 - \dfrac{\upsilon}{c^2}\dfrac{dx}{dt}}$$

$$\therefore \qquad u_y' = \frac{u_y\sqrt{1-\frac{\upsilon^2}{c^2}}}{1-\frac{u_x\upsilon}{c^2}} \qquad \text{...(1.43)}$$

also it can be shown that

$$u_z' = \frac{u_z\sqrt{1-\frac{\upsilon^2}{c^2}}}{1-\frac{u_x\upsilon}{c^2}} \qquad \text{...(1.44)}$$

Eqs. (1.42), (1.43) and (1.44) constitute the realistic velocity transformation.

The inverse transformations are :

$$u_x = \frac{u_x'+\upsilon}{1+\frac{u_x'}{c^2}} \qquad u_y = \frac{u_y'\sqrt{1-\frac{\upsilon^2}{c^2}}}{1+\frac{u_x'\upsilon}{c^2}} \quad \text{and} \quad u_z = \frac{u_z'\sqrt{1-\frac{\upsilon^2}{c^2}}}{1+\frac{u_x'\upsilon}{c^2}}$$

1.19 Variation of mass with velocity

Consider an elastic collision between two identical balls A and B of masses m_A and m_B respectively. Let υ be the speed of S' the respect to S. Let A be moving along the $+y$-direction with a speed u in the S-frame and B in the –ve y-direction in S' frame with speed u'.

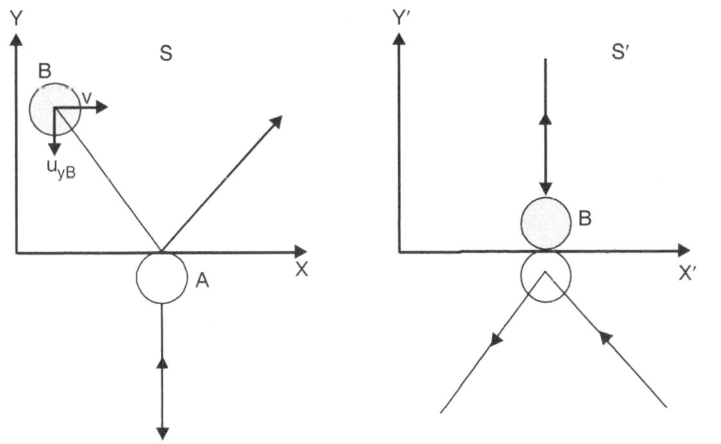

Fig. 1.13

We assume that $u' = u$ (small compared to υ)

After the elastic collision velocity of A is reversed in S and that of B in S' (See Fig.).

Now the y-component of the velocity of ball B is

$$u_{yB} = \frac{u'\sqrt{1-\frac{\upsilon^2}{c^2}}}{1+\frac{u_x'u}{c^2}} \qquad u_x' = 0$$

$$\therefore \qquad u_{yB} = u'\sqrt{1-\frac{\upsilon^2}{c^2}}$$

Since energy is conserved the y-components of the velocities reverse their directions while the x-components remain unchanged.

Now conservation of momentum requires that : y-component of momentum before = momentum after in the S-frame.

$$\therefore \quad m_A u - m_B u' \sqrt{1-\frac{\upsilon^2}{c^2}} = m_B u' \sqrt{1-\frac{\upsilon^2}{c^2}} - m_A u$$

$\therefore$ $u' = u$ by assumption in S-frame we find that m_B is not equal to m_A but equals $\dfrac{m_A}{\sqrt{1-\frac{\upsilon^2}{c^2}}}$.

Now replacing m_A by m_0 for one of the balls is at rest,

and if $m_B = m$ $\quad \therefore \; m = \dfrac{m_0}{\sqrt{1-\frac{\upsilon^2}{c^2}}}$ ...(1.45)

where m_0 = rest mass.

$\therefore$ The momentum $\vec{p}$ of a moving body is defined as $\vec{p} = \vec{mv} = \dfrac{m_0 \vec{\upsilon}}{\sqrt{1-\frac{\upsilon^2}{c^2}}}$.

1.20 Mass - Energy Equivalence

Now we know that mass is a function of $\vec{u}$

$$\therefore \quad \vec{F} = \frac{d}{dt}(m\vec{v}) = m\frac{d\vec{v}}{dt} + \upsilon\frac{dm}{dv}\cdot\frac{d\upsilon}{dt} = m\vec{a} + v\frac{dm}{dt}.$$

In relativity, work done by a force = $\int \vec{F} \cdot d\vec{s}$

$\therefore$ Relativistically kinetic Energy = K is given by

$$K = \int_{u=0}^{u=u} F.\,ds = \int_{u=0}^{u=u} \frac{d}{dt}(mu)\frac{ds}{dt}.\,dt \qquad \text{Since } \frac{ds}{dt} = u.$$

$$= \int u\frac{d}{dt}(mu)\,dt = \int u\,d(mu)$$

Now $\quad m = \dfrac{m_0}{1-\frac{u^2}{c^2}}$

i.e., $\quad m^2c^2 - m^2u^2 = m_0^2c^2$

Taking differentials on both sides, $2mc^2dm - m^2 2udu - u^2 2mdm = 0$

Dividing by $2m$, we get

$\therefore \quad ud\,(mu) = mudu + u^2dm = c^2dm$

$$\therefore \quad K = \int_{u=o}^{u=u} c^2 dm = c^2 \int_{m=m_0}^{m=m} dm = mc^2 - m_0c^2 \quad \text{Substituting for } m$$

$$K = m_0c^2\left[\frac{1}{\sqrt{1-\frac{u^2}{c^2}}} - 1\right] \qquad \text{...(1.46)}$$

This is the relativistic expression for kinetic Energy particles.

If we take $mc^2 = E$, where E is the total energy of the particle.

$\therefore \quad E = m_0c^2 + K$ (here m_0c^2 is the rest energy of the particle *i.e.*, $K = 0$)

Now if $u/c << 1$ then

$$K = m_0c^2\left[\frac{1}{\sqrt{1-\frac{u^2}{c^2}}}-1\right] = m_0c^2\left[\left(1-\frac{u^2}{c^2}\right)^{-\frac{1}{2}}-1\right]$$

By the binomial expansion of u/c gives

$$K = m_0c^2\left[1+\frac{1}{2}\left(\frac{u}{c}\right)^2+\frac{3}{8}\left(\frac{u}{c}\right)^4+.....-1\right]$$
$$= 1/2\ m_0u^2$$

which is the classical result as the 3rd term and others are negligibly small.

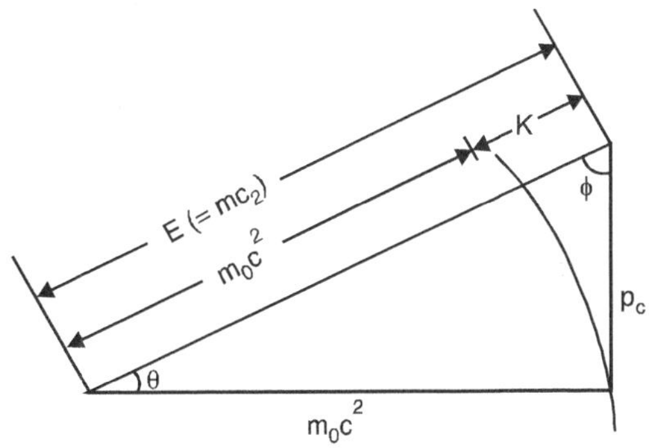

Fig. 1.14

1.21 Energy-Momentum Relation

If a body of mass m is with a velocity v, its momentum

$$p = mv \quad ...(1.47)$$
$$E = mc^2 \quad ...(1.48)$$

and
$$m = \frac{m_0}{\sqrt{1-\frac{\upsilon^2}{c^2}}} \quad ...(1.49)$$

$\therefore$
$$E^2 - c^2p^2 = m^2c^4 - c^2m^2\upsilon^2 \quad ...(1.50)$$

$$E^2 - c^2p^2 = \frac{m_0^2c^4}{\left[1-\frac{\upsilon^2}{c^2}\right]} - \frac{c^2m_0^2\upsilon^2}{\left[1-\frac{\upsilon^2}{c^2}\right]}$$

$$E^2 - c^2p^2 = \frac{m_0^2c^4\left[1-\frac{\upsilon^2}{c^2}\right]}{\left[1-\frac{\upsilon^2}{c^2}\right]}$$

$$E^2 - c^2p^2 = m_0^2c^4$$

or
$$E^2 = m_0^2c^4 + c^2p^2$$
$$E = \sqrt{m_0^2c^4+c^2p^2}$$
$$\boxed{E = c\sqrt{m_0^2c^2+p^2}} \quad ...(1.51)$$

1.22 Massless Particles

Can we have a massless particle ? In other words, can a particle with no rest mass exist at the same time having energy and momentum ?

In relativistic mechanics; since $p = \dfrac{m\upsilon}{\sqrt{1-\dfrac{\upsilon^2}{c^2}}}$ and $E = \dfrac{mc^2}{\sqrt{1-\dfrac{\upsilon^2}{c^2}}}$,

We see that when $m = 0$, p and E are zero. If v is **also equal** to c then $E = 0/0$ and $p = 0/0$ which become indeterminate quantities. Hence the two relations for p and E point to the fact that massless particles can exist **provided** they travel with the speed of light.

Hence for a **massless particle** $E = pc$. A **photon** is a massless particle travelling at the speed of light.

On the other hand for a particle with rest mass like the electron the relation for $E = \sqrt{m^2c^4 + p^2c^2}$.

Example 1.1. *One Cosmic ray particle approaches the earth along its axis with a velocity of 0.87C towards the north pole and another with a velocity of 0.63 C towards the south pole. What is the relative speed of approach of one partcle with respect to the other ?*

Solution. Here

$$u = 0.87C$$

$$\upsilon = -0.63C$$

$$V = \frac{u-\upsilon}{1-\dfrac{u\upsilon}{c^2}} = \frac{0.87c+0.63c}{1+\dfrac{0.87c \times 0.63c}{c^2}}$$

$$= \mathbf{0.9689\ C.}$$

Example 1.2. *Two spaceships A and B are moving in opposite directions, each with a speed of 0.90c. Find the relative speed of B with respect to A.*

Solution. Here

$$u = 0.9\ C$$

$$\upsilon = -0.9\ C$$

$$V_x = \frac{u-\upsilon}{1-\dfrac{u\upsilon}{c^2}} = \frac{0.9\ c+0.9\ c}{1+\dfrac{0.9\ c \times 0.9\ c}{c^2}}$$

$$= \mathbf{0.994C}$$

Example 1.3. *What is the length of a metre stick moving parallel to its length when its mass 3/2 of its rest mass ?* [*Delhi Engineering, 1985*]

Solution. Here

$$m = \frac{m_0}{\sqrt{1-\dfrac{\upsilon^2}{c^2}}} \quad ...(i)$$

and

$$\frac{m}{m_b} = \frac{1}{\sqrt{1-\dfrac{\upsilon^2}{c^2}}}$$

and

$$L = L_0\sqrt{1-\frac{\upsilon^2}{c^2}} \quad ...(ii)$$

Dividing (*ii*) by (*i*)

$$\frac{L}{m} = \left(\frac{L_0}{m_0}\right)\left(1-\frac{\upsilon^2}{c^2}\right)$$

$$L = \left(\frac{m}{m_0}\right) L_0 \left(1 - \frac{\upsilon^2}{c^2}\right)$$

$$\therefore \quad L = \left(\frac{m}{m_0}\right) L_0 \left(\frac{m_0}{m}\right) = \left(\frac{m_0}{m}\right) L_0$$

Here $\frac{m_0}{m} = \frac{2}{3}$

and $L_0 = 1.$

$$\therefore \quad L = \frac{2}{3} \times 1$$

$$= \mathbf{0.667\ m.}$$

Example 1.4. *Show that* $\left(\frac{E^2}{c^2}\right) - p^2$ *is invariant.*

Solution. Consider two frames of reference A and B.

The reference frame B is moving with a velocity υ *along the X-axis with respect to frame A.*

Let p be the mometum and E the Energy of a particle in reference frame A and P' and E' in reference frame B.

$$p^2 = p_x^2 + p_y^2 + p_z^2 \quad \text{...}(i)$$

and

$$p'^2 = p_x'^2 + p_y'^2 + p_z'^2 \quad \text{...}(ii)$$

$$k = \frac{1}{\sqrt{1 - \frac{\upsilon^2}{c^2}}} \quad \text{...}(iii)$$

and

$$p_x' = k\left(p_x - \frac{\upsilon E}{c^2}\right)$$

$$p_y' = p_y, p_z' = p_z$$

and

$$E' = k\,(E - Vp_x)$$

$$\therefore \quad \frac{E'^2}{c^2} - p'^2 = \frac{k^2}{c^2}(E - \upsilon p_x)^2 - \left[p_x'^2 + p_y'^2 + p_z'^2\right]$$

$$= \frac{k^2}{c^2}\left[(E - \upsilon p_x)^2 - c^2\left(p_x - \frac{\upsilon E}{c^2}\right)^2\right] - p_y^2 - p_z^2$$

$$= \frac{k^2}{c^2}\left[(E - Vp_x)^2 - c^2\left(p_x - \frac{VE}{c^2}\right)^2\right] - p_y^2 - p_z^2$$

$$= \frac{k^2}{c^2}\left[E^2 + \upsilon^2 p_x^2 - 2\upsilon EP_x - p_x^2 c^2 - \frac{\upsilon^2 E^2}{c^2} + 2\upsilon Ep_x - p_y^2 - p_z^2\right]$$

$$= \frac{k^2}{c^2}\left[\left(E^2 - \frac{\upsilon^2 E^2}{c^2}\right) + \left(-p_x^2 c^2 + \upsilon^2 p_x^2\right)\right] - p_y^2 - p_z^2$$

$$= \frac{k^2}{c^2}\left[E^2\left(1 - \frac{\upsilon^2}{c^2}\right) - p_x^2 c^2\left(1 - \frac{\upsilon^2}{c^2}\right)\right] - p_y^2 - p_z^2$$

But $$\left(1-\frac{\upsilon^2}{c^2}\right) = \frac{1}{k^2}$$

$$\therefore \quad \frac{E'^2}{c^2} - p'^2 = \frac{k^2}{c^2}\left[\frac{E^2}{k^2} - \frac{p_x^2 c^2}{k^2}\right] - p_y^2 - p_z^2$$

$$= \frac{E^2}{c^2} - p_x^2 - p_y^2 - p_z^2 = \frac{E^2}{c^2} - \left(p_x^2 + p_y^2 + p_z^2\right)$$

$$\frac{E'^2}{c^2} - p'^2 = \frac{E^2}{c^2} - p^2 \qquad \text{...(iii)}$$

Hence $\frac{E^2}{c^2} - p^2$ is an invariant.

Example 1.5. *The total electrical energy generated in a country in a particular year was 7.5 × 10^{11} kWh. How much mass was converted into energy ?* *[Delhi 1985]*

$$U = mc^2$$

Solution. Here $U = 7.5 \times 10^{11}$ kWH

$$U = 7.5 \times 10^{11} \times 1000 \times 3600 \text{ J}$$

$$c = 3 \times 10^8 \text{ m/s}$$

$$\therefore \quad m = \frac{U}{c^2} = \frac{7.5\times10^{11}\times1000\times3600}{\left(3\times10^8\right)^2}$$

$$m = \mathbf{30kg}$$

Example 1.6. *If 1 kg of a substance is fully converted into energy, how much energy is produced ?*

Solution. Energy,

$$U = mc^2$$

Here $m = 1$ kg

$$c = 3 \times 10^8 \text{ m/s}$$

$$U = 1 \times (3 \times 10^8)^2$$

$$U = 9 \times 10^{16} \text{ J}$$

Example 1.7. *A particle of mass 10^{-24} kg is moving with a speed of 1.8 × 10^8 m/s. Calculate its mass when it is in motion.*

Solution. $$m = \frac{m_0}{\sqrt{1-\frac{\upsilon^2}{c^2}}}$$

Here $m_0 = 10^{-24}$ kg

$$\upsilon = 1.8 \times 10^8 \text{ m/s}$$

$$c = 3 \times 10^8 \text{ m/s}$$

$$\frac{\upsilon}{c} = 0.6$$

$$m = \frac{10^{-24}}{\sqrt{1-(0.6)^2}}$$

$$= \mathbf{1.25 \times 10^{-24}} \text{ kg.}$$

Example 1.8. *10^6 kg of water is heated from 273 K to 373 K. Calculate the increase in the mass of water.*

Specific heat of water $= 10^3$ *calories/ kg-K.*

Solution. Heat required to raise the temperature of 10^6 kg water from 273 to 373 K.

$$= 10^6 \times 10^3 \times 100$$

$$= 10^{11} \text{ calories}$$

But $\quad 1 \text{ calorie} = 4.18 \text{ J}$

$\therefore \quad \text{Increase in energy} = 10^{11} \times 4.18 \text{ J}$

$$U = 4.18 \times 10^{11} \text{ J}$$

Also $\quad U = mc^2$

Increase of mass, $\quad m = \dfrac{U}{c^2}$

$$\therefore \quad m = \frac{4.18 \times 10^{11}}{\left(3 \times 10^8\right)^2}$$

$$= \mathbf{4.64 \times 10^{-6} \ kg.}$$

Example 1.9. *A particle is moving with a speed of 0.5 c. Calculate the ratio of the rest mass and the mass while in motion.*

Solution. $\quad m = \dfrac{m_0}{\sqrt{1 - \dfrac{\upsilon^2}{c^2}}}$

or $\quad \dfrac{m_0}{m} = \sqrt{1 - \dfrac{\upsilon^2}{c^2}}$

Here $\quad \upsilon = 0.5\, c$

or $\quad \dfrac{\upsilon}{c} = 0.5$

$$\frac{m_0}{m} = \sqrt{1 - (0.5)^2}$$

$$\frac{m_0}{m} = \mathbf{0.8666}$$

Example 1.10. *4.18×10^{-3} kg of a substance is fully converted to heat energy. Calculate the amount of heat produced.*

Solution. $\quad U = mc^2$

Here $\quad m = 4.18 \times 10^{-3}$ kg

$$c = 3 \times 10^8 \text{ m/s}$$

$$U = 4.18 \times 10^{-3} \times (3 \times 10^8)^2 \text{ J}$$

$$\text{Heat produced} = \frac{U}{4.18} \text{ calories} = \frac{4.18 \times 10^{-3} \times (3 \times 10^8)^2}{4.18}$$

$$= \mathbf{9 \times 10^{13} \ calories.}$$

Example 1.11. *The length of a spaceship is measured to be exactly half its actual length. Calculate (i) the speed of the spaceship and (ii) the time dilation corresponding to one second on the spaceship.*

Solution. (*i*)

$$L = L_0\sqrt{1-\frac{\upsilon^2}{c^2}}$$

or

$$\frac{L}{L_0} = \sqrt{1-\frac{\upsilon^2}{c^2}}$$

Here

$$\frac{L}{L_0} = 0.5, \upsilon = ?$$

$$c = 3 \times 10^8 \text{ m/s}$$

$$0.5 = \sqrt{1-\frac{\upsilon^2}{c^2}}$$

$$\frac{\upsilon^2}{c^2} = 0.75$$

$$\upsilon = 0.866\ c = 0.866 \times 3 \times 10^8$$

$$\upsilon = \mathbf{2.598 \times 10^8 \text{ m/s}}$$

(*ii*)

$$t_0 = t\sqrt{1-\frac{\upsilon^2}{c^2}}$$

$$t = \frac{t_0}{\sqrt{1-\frac{\upsilon^2}{c^2}}}$$

Here

$$t_0 = 1 \text{ s},$$

$$\frac{\upsilon^2}{c^2} = 0.75$$

$$t = \frac{1}{\sqrt{1-0.75}}$$

$$t = \mathbf{2 \text{ s}}$$

Example 1.12. *The mean life of π meson is 2×10^{-8} s. Calculate mean life of a meson moving with a velocity of 0.8 c.*

Solution.

$$t = t_0\sqrt{1-\frac{\upsilon^2}{c^2}}$$

$$t_0 = 2 \times 10^{-8} \text{ S}$$

$$\upsilon = 0.8 \text{ c}$$

$$t = 2 \times 10^{-8}\sqrt{1-(0.8)^2}$$

$$t = \frac{2\times 10^{-8}}{0.6}$$

$$t = \mathbf{3.33 \times 10^{-8} \text{s.}}$$

Example 1.13. *Calculate the velocity that one atomic mass unit have if it has a kinetic energy equal to twice the rest mass energy.*

Solution. Here

$$2m_0c^2 = mc^2 - m_0c^2$$

$$3m_0c^2 = mc^2$$

$$m = 3m_0$$

But
$$m = \frac{m_0}{\sqrt{1-\frac{\upsilon^2}{c^2}}}.$$

$\therefore$
$$3m_0 = \frac{m_0}{\sqrt{1-\frac{\upsilon^2}{c^2}}}$$

or
$$1-\frac{\upsilon^2}{c^2} = \frac{1}{9}$$

$$\upsilon^2 = \left(\frac{8}{9}\right)c^2$$

$$\upsilon = \mathbf{0.941\ c.}$$

Example 1.14. *The rest mass of a proton is 2000 times the rest mass of an electron. Calculate the speed at which the electron should move so that its mass will be equal to the rest mass of the proton.* [*Lucknow, 1984*]

Solution.
$$c = 3 \times 10^8 \text{ m/s}$$

$$m = \frac{1}{\sqrt{1-\frac{\upsilon^2}{c^2}}}$$

$$\frac{m}{m_0} = \frac{1}{\sqrt{1-\frac{\upsilon^2}{c^2}}}$$

Here
$$\frac{m}{m_0} = 2000$$

$$2000 = \frac{1}{\sqrt{1-\left(\frac{\upsilon^2}{c^2}\right)}}$$

$$1-\frac{v^2}{c^2} = \left(\frac{1}{2000}\right)^2 = \frac{1}{4\times 10^6}$$

$$\frac{\upsilon^2}{c^2} = \left(1-\frac{1}{4\times 10^6}\right)$$

$$\upsilon = [1-1.25\times 10^{-7}]\text{c}$$

$$\upsilon = [1-1.25\times 10^{-7}]\times 3\times 10^8$$

$$\upsilon = \mathbf{(3\times 10^7 - 37.5)\ m/s}$$

Example 1.15. *The total energy of a particle is exactly twice its energy. Calculate its speed.* [*Lucknow, 1985*]

Solution. Here
$$mc^2 = 2m_0c^2$$

$$\frac{m}{m_0} = 2$$

Also

$$m = \frac{m_0}{\sqrt{1-\left(\frac{\upsilon^2}{c^2}\right)}}$$

$$\frac{m}{m_0} = \frac{1}{\sqrt{1-\left(\frac{\upsilon^2}{c^2}\right)}}$$

$$2 = \frac{1}{\sqrt{1-\left(\frac{\upsilon^2}{c^2}\right)}}$$

$$1-\left(\frac{\upsilon^2}{c^2}\right) = 0.25$$

$$\frac{\upsilon^2}{c^2} = 0.75$$

$$\upsilon = \mathbf{0.866c.}$$

Example 1.16. *Two particles each of rest mass 3 × 10^{-25} approach each other in head on collision. If each particle has an initial velocity of 2 ×10^8 m/s, calculate the velocity of one particle as seen the other. What is the relativistic mass of one as seen by the other.* [*Bombay 1984*]

Solution. Here

$$u = 2 \times 10^8 \text{ m/s}$$

$$\upsilon = 2 \times 10^8 \text{ m/s}$$

$$c = 3 \times 10^8 \text{ m/s}$$

$$u' = \frac{u-\upsilon}{1-\frac{u-\upsilon}{c^2}}$$

$$u' = \frac{4\times10^8}{1+\left(\frac{4\times10^{16}}{9\times10^{16}}\right)} = \left(\frac{36}{13}\right)\times10^6$$

$$\mathbf{u' = 2.769 \times 10^8 \text{ m/s}}$$

$$\frac{u'}{c} = \frac{36\times10^8}{13\times3\times10^8} = \frac{12}{13}$$

$$m_0 = 3\times10^{-25} \text{ kg}$$

$$m = \frac{m_0}{\sqrt{1-\left(\frac{u'}{c}\right)^2}} = \frac{3\times10^{-25}}{\sqrt{1-\left(\frac{12}{13}\right)^2}}$$

$$m = \mathbf{7.8 \times 10^{-25} \text{ kg.}}$$

Example 1.17. *The density of gold is 19.3 × 10 kg/m^3 in a frame S which is at rest. Calculate its density that an observer in a frame S' would determine of the frame S' is moving along the x-axis with a speed of 0.9c.*

(*i*)

$$L = L_0\sqrt{1-\frac{\upsilon^2}{c^2}}$$

Solution.

Here $\frac{\upsilon}{c} = 0.9$

$$L = L_0\sqrt{1-(0.9)^2}$$

$$L = 0.4349\ L_0$$

Volume $V_0 = L_0 \times y \times z$

$$V = L \times y \times z = 0.4359\ L_0 \times y \times z \qquad ...(i)$$

$$V = 0.4359\ V_0$$

Mass $m = \frac{m_0}{\sqrt{1-\left(\frac{\upsilon^2}{c^2}\right)}}$

$$m = \frac{m_0}{\sqrt{1-(0.9)^2}} = \frac{m_0}{0.4539}$$

Original density, $\rho_0 = \frac{m_0}{V_0}$

Final density, $\rho = \frac{m}{V}$

$$\frac{\rho}{\rho_0} = \left(\frac{m}{m_0}\right)\left(\frac{V_0}{v}\right) = \left(\frac{1}{0.4359}\right) \times \left(\frac{1}{0.4359}\right)$$

$$\rho = 19.3 \times 10^3 \left(\frac{1}{0.4359}\right)^2$$

$$= \mathbf{101.6 \times 10^3\ kg/m^3}$$

Example 1.18. *Electrons are accelerated upto a kinetic energy of $10^9 eV$. Find (i) the ratio of their mass to the rest mass (ii) the ratio of their velocity to the velocity of light and (iii) the ratio of their energy to the rest mass energy.* [*Bombay, 1983*]

Solution. $U = 10^9 eV$

$$= 10^9 \times 1.6 \times 10^{-19}\ \text{J}$$

Also $U = mc^2$

$$m = \frac{U}{c^2} = \frac{10^9 \times 1.6 \times 10^{-19}}{\left(3 \times 10^8\right)^2}$$

$$= \mathbf{1.77 \times 10^{-27}\ kg.}$$

Rest mass of electron, $m_0 = 9 \times 10^{-31}$ kg

(*i*) $\frac{m}{m_0} = \frac{1.77 \times 10^{-27}}{9 \times 10^{-31}}$

$$\frac{m}{m_0} = \mathbf{1.95 \times 10^3}$$

(*ii*) $m = \frac{m_0}{\sqrt{1-\left(\frac{\upsilon^2}{c^2}\right)}}$

$$1.95 \times 10^3 = \frac{1_0}{\sqrt{1-\left(\frac{\upsilon^2}{c^2}\right)}}$$

$$1-\left(\frac{\upsilon^2}{c^2}\right) = \left(\frac{1}{1.95\times10^3}\right)^2$$

$$\frac{\upsilon^2}{c^2} = 1-\left(\frac{1}{1.95\times10^3}\right)^2$$

$$\frac{\upsilon}{c} = \left[\left(1-\frac{1}{1.95\times10^3}\right)^2\right]^{\frac{1}{2}}$$

$$\frac{\upsilon}{c} = \left(1-\frac{1}{2\times(1.95)^2\times10^6}\right)$$

$$\frac{\upsilon}{c} = [1-1.315\times10^{-7}]$$

(*iii*) Rest mass energy

$$U_0 = m_0 c^2$$

$$U_0 = 9\times10^{-31}\times(3\times10^8)^2$$

$$U_0 = 8.1\times10^{-14}\text{ J}$$

$$\frac{U}{U_0} = \frac{10^9\times1.6\times10^{-19}}{8.1\times10^{-14}}$$

$$= \mathbf{1.975\times10^3}$$

Example 1.19. *Calculate the speed of an electron so that it has double its rest mass.*

Solution.

$$m = \frac{m_0}{\sqrt{1-\frac{\upsilon^2}{c^2}}}$$

$$\sqrt{1-\frac{\upsilon^2}{c^2}} = \frac{m_0}{m} = \frac{1}{2}$$

$$\upsilon = c\left[1-\left(\frac{m_0}{m}\right)^2\right]^{\frac{1}{2}} = 3\times10^8\left[1-\left(\frac{1}{2}\right)^2\right]^{\frac{1}{2}}$$

$$= 3\times10^8\,[0.75]^{\frac{1}{2}}$$

$$= \mathbf{2.598\times10^8\,m/s}$$

Example 1.20. *At what speed will an electron move in order to double its relative mass to the rest mass ?* [*IAS, 1984*]

Solution.

$$m = 2m_0$$

$$m = \frac{m_0}{\left[1-\frac{\upsilon^2}{c^2}\right]^{\frac{1}{2}}}$$

$$\left(1-\frac{\upsilon^2}{c^2}\right)^{\frac{1}{2}} = \frac{m_0}{m} = \frac{1}{2}$$

$$1-\frac{\upsilon^2}{c^2} = \frac{1}{4}$$

$$\upsilon^2 = 0.75\, c^2$$

$$\upsilon = 0.866\, c$$

$$\upsilon = 0.866 \;\; 3 \times 10^8 \text{m/s}$$

$$= \mathbf{2.598 \times 10^8\ m/s}$$

Example 1.21. *What fraction of the speed of light is the speed of electron whose kinetic energy is 0.5 Mev ?* [*IAS, 1979*]

Solution. Here

$$E = (m - m_0)c^2$$

$$E = 0.5 \text{ MeV}, E = 0.5 \times 10^6 \times 1.6 \times 10^{-19}$$

$$E = 0.8 \times 10^{-13} \text{ J}$$

$$m_0 = 9.1 \times 10^{-31} \text{kg.}$$

$$c = 3 \times 10^8 \text{ m/s}$$

$$\therefore \quad 0.8 \times 10^{-13} = (m - 9.1 \times 10^{-31})\,(3 \times 10^8)^2$$

$$m = 8.9 \times 10^{-31} + 9.1 \times 10^{-31}$$

$$m = 18 \times 10^{-31} \text{kg}$$

$$\frac{m}{m_0} = \frac{1}{\left[1-\left(\frac{\upsilon^2}{c^2}\right)\right]^{\frac{1}{2}}}$$

$$1-\left(\frac{\upsilon^2}{c^2}\right) = (0.5055)^2$$

$$\left(\frac{\upsilon^2}{c^2}\right) = 0.7445$$

$$\frac{\upsilon}{c} = \mathbf{0.863}$$

Example 1.22. *An electron has a total energy of 2MeV. Calculate the effective mass of the electron in kg and also its speed. Assume rest mass of the electron to be 0.511 MeV.* [*IAS*]

Solution. Here

$$U = 2\text{MeV}$$

$$U = 2 \times 10^6 \times 1.6 \times 10^{-19} = 3.2 \times 10^{-13} \text{ J}$$

$$U = mc^2$$

(*i*)

$$m = \frac{U}{c^2} = \frac{3.2 \times 10^{-13}}{\left(3 \times 10^8\right)^2}$$

$$= 35.6 \times 10^{-31} \text{ kg.}$$

(*ii*) Rest mass

$$m_0 = \frac{U}{c^2} = \frac{0.511 \times 10^6 \times 1.6 \times 10^{-19}}{\left(3 \times 10^8\right)^2}$$

$$= \mathbf{9.1 \times 10^{-31}\ kg.}$$

(*iii*)
$$\frac{m}{m_0} = \frac{1}{\sqrt{1-\left(\frac{\upsilon^2}{c^2}\right)}}$$

$$\frac{1}{\sqrt{1-\frac{\upsilon^2}{c^2}}} = \frac{m}{m_0} = \frac{9.1\times10^{-31}}{35.6\times10^{-31}} = 0.2556$$

$$\upsilon^2 = 0.93467\ c^2 \quad \text{or} \quad \upsilon = \mathbf{0.96\ c}$$

Example 1.23. *Calculate in MeV the energy released in the fission of $_{92}U^{238}$ into two equal fragments assuming it to arise from Coulomb repulsion when the fragments are just separated by distance equal to the sum of their radii. The nuclear radius is given by the relation*

$$R = r_0 A^{1/3}$$

when A is the mass number and

$$r_0 = 1.2 \text{ fermi.}$$

[*IAS, 1977*]

Solution. Here charge on each fragment = 46 *e*

$$q_1 = 46 \times 1.6 \times 10^{-19}\ c = 73.6 \times 10^{-19}\ c$$

Also
$$q_2 = 73.6 \times 10^{-19}\ c$$

$$R = r_0 A^{1/3}, \quad \text{or} \quad 1.2 \times 10^{-15}\ \text{m}$$

$$R = (1.2 \times 10^{-15})\,(119)^{1/3}$$

$$R \approx \mathbf{5.9 \times 10^{-15}\ m}$$

$$\text{Distance} = 2R = 11.8 \times 10^{-15}\ \text{m}$$

Energy released
$$U = \frac{q_1 q_2}{4\pi\varepsilon_0\,(2R)}$$

$$= \frac{(73.6\times10^{-19})\times(73.6\times10^{-19})\times9\times10^9}{11.8\times10^{-15}} = 4.13 \times 10^{-11}\ \text{J}$$

$$U = \frac{4.13\times10^{-11}}{1.6\times10^{-19}\times10^6}\ \text{MeV} = \mathbf{258\ MeV.}$$

Example 1.24. *What must be the speed of particle of rest mass $\frac{1}{2}$ MeV/c^2 be so that its relativistic mass is $\frac{1}{\sqrt{2}}$ MeV/c^2 ?* [*IAS*]

Solution. Here
$$m = \frac{m_0}{\left[1-\left(\frac{\upsilon^2}{c^2}\right)\right]^{\frac{1}{2}}}$$

Here
$$m_0 = \frac{1}{2}\ MeV/c^2$$

$$m = \frac{1}{\sqrt{2}}\ \text{MeV/c}^2$$

$$\left(\frac{m_0}{m}\right)^2 = 1-\left(\frac{\upsilon^2}{c^2}\right)$$

$$\upsilon^2 = c^2\left[1-\left(\frac{m_0}{m}\right)^2\right]$$

$$\upsilon = c\left[1-\left(\frac{m_0}{m}\right)^2\right]^{\frac{1}{2}}$$

Here $$\frac{m_0}{m} = \frac{1}{2} / \frac{1}{\sqrt{2}} = \frac{\sqrt{2}}{2} = \frac{1.414}{2} = 0.707$$

$$\upsilon = 3 \times 10^8 [1-(0.707)^2]^{1/2}$$

$$\upsilon = \mathbf{2.12 \times 10^8\ m/s.}$$

Example 1.25. *A muon (μ-meson) is formed high up in the atmosphere and travels towards the earth with a speed of 0.992 c. It decays after travelling a distance of 60 km. In what time does the muon decays as measured (i) by us and (ii)* in its own frame reference ? What is the distance covered by the muon in its own frame of reference ? [*I.A.S, 1987*]

Solution. $S = 60 \text{ km} = 6 \times 10^4 \text{ m}$

$$\upsilon = 0.992 \text{ c} = 0.992 \times 3 \times 10^8 = 2.976 \times 10^8 \text{ m/s}$$

(*i*) Time measured by us

Here $$t' = \frac{6 \times 10^4}{2.976 \times 10^8}$$

$$t' = 2.016 \times 10^{-4}\ S$$

(*ii*) Time measured in its own frame $= t$

$$t' = \frac{t}{\left[1-\frac{\upsilon^2}{c^2}\right]^{\frac{1}{2}}}$$

$$t = t'\left[1-\frac{\upsilon^2}{c^2}\right]^{\frac{1}{2}}$$

$$t = 2.016 \times 10^{-4} [1-(0.992)^2]^{1/2}$$

$$t = \mathbf{2.55 \times 10^{-5}\ s}$$

(*iii*) Distance measured, $s = v \times t$

$$= 2.976 \times 10^8 \times 2.55 \times 10^{-5}$$

$$= 7.59 \times 10^3 \text{ m}$$

$$= \mathbf{7.59\ km.}$$

Example 1.26. *The time of an unstable particle at rest is 10^{-6} s. If at the instant of creation it moves with a speed of 0.9 c, what is the distance it will traverse before decaying ?* [*Osmania, 1991*]

Solution. Here $$t = \frac{t_0}{\left[1-\left(\frac{\upsilon^2}{c^2}\right)\right]^{\frac{1}{2}}}$$

$$\frac{\upsilon}{c} = 0.9$$

∴ $$t_0 = 10^{-6} s$$

∴ $$t = \frac{10^{-6}}{\left[1-(0.9)^2\right]^{\frac{1}{2}}}$$

$$t = \mathbf{2.294 \times 10^{-6}}\ \text{s}$$

$$\text{Distance travelled} = v \times t$$
$$= 0.9 \times 3 \times 10^8 \times 2.294 \times 10^{-6}$$
$$= \mathbf{6.19 \times 10^2} \text{ m}$$

EXERCISE I

1. Explain the term frame or reference. Give examples.

2. What is inertial frame of reference ?

3. Discuss Galilean transformation for (*i*) position (*ii*) velocity (*iii*) acceleration.

4. Define and discuss Galilean transformations. Prove that Newton's law of motion are invariate under these transformations. [*Delhi, 1980*]

5. What are fictitious forces ? Give examples.

6. Discuss uniformly rotating frame of reference. Explain how velocity and acceleration are transformed in such a frame of reference.

7. Discuss the effect of rotation of the earth on the value of *g*.

8. Describe Foucalt's pendulum and explain when the time period of the plane of oscillation becomes infinite.

9. What is a Foucalt's pendulum ? How does it demonstrate the rotation of the earth about its own axis ? What will be period of rotation of the plane of oscillatiuon of Foucalt's pendulum : (*i*) in latitude 30º, (*ii*) at the equation and (*ii*) at the north pole ? [*Delhi (Hons.) 1980*]

10. Describe Michelson-Morley experiment. Discuss the results obtained.

11. State the postulates of the special theory of relativity.

12. Describe briefly Lorentz transformation for (*i*) length and (*ii*) time. [*Delhi 1984*]

13. Show that the length of an object appars to contract when moving with a velocity comparable to velocity of light.

14. Explain the term time dilation.

15. Show that an interval of time observed in a moving frame of reference will be less than the same onterval of time observed in a stationary frame of reference.

16. What do you understand by the relativity of mass ? Show that the mass of a body becomes infinite when it moves with the velocity light. What do you understand by this result ?

17. Derive Einstein's mass-energy relation. [*Delhi 1978, 1983*]

18. Derive Energy-momentum relation.

19. Show that $\left(\dfrac{E^2}{c^2}\right) - p^2$ is invariant.

20. Describe Michelson-Morely experiment. With what aim in view was this experiment performed ? How did Einstein explain negative result of this experiment ?
[*Delhi, 1976; Delhi (Hons.) 1979, 1982*]

21. Derive an equation for the reativistic kinetic energy of a body. How does this equation lead to the equivalence of mass and energy ? [*Delhi (Hons.) 1978*]

22. Calculate the velocity that one atomic mass unit will have if it had a kinetic energy three times its rest mass energy. [*Delhi (Hons.) 1978*]

23. Define an inertial frame of reference. Show that the laws of physics remain unchanged whether a system is referred to an inertial frame *S* or to another frame *S′* moving with uniform velocity *v* was respect to the frame *S*. [*Delhi (Hons.) 1979*]

24. What are the fictitious forces introduced in mechanics ? Derive an expression for the Coriolis force acting on a particle in a rotating frame of reference. [*Delhi (Hons.) 1977*]

25. What is an inertial frame ? Explain how a moving rod appears shorter to an observer at rest ? [*Delhi 1978, Delhi (Hons.) 1978*]

26. What is Galilean invariance ? Prove that the Galilean invariance leads to the law of conservation of linear momentum. [*Delhi (Hons.), 1978*]

27. Derive the equations of Lorentz transformations and obtain the formula for the addition of velocities. [*Delhi (Hons.) 1983*]

28. Deduce the expression for the variation of mass of a particle with velocity. [*Delhi 1982*]

29. State the postulates of special theory of relativity and deduce Lorentz transformations. [*Delhi Engineering, 1980*]

30. A solid cube has a mass M_0 and length L_0 when at rest. What will be its volume as viewed from a reference frame moving with a constant velocity v in direction parallel to one of its edges ? What is its density change in this frme ? [*Delhi Engineering, 1983*]

31. Two bodies A and B each of mass 1 kg are moving along the X-axis with speeds 12 m/s and – 5 m/s in a frame of reference S. Assuming that the collision satisfies the principles of conservation of energy and momentum, calculate the speeds of two bodies after collision. [**Ans.** (*i*) — 5 m/s, (*ii*) 12 m/s]

32. Two bodies P and Q of equal mass move along the X-axis with velocities 16 m/s and –12 m/s in a given reference frame. Using the principles of converstion of energy and momentum, calculate their speed after collision. [**Ans.** (*i*) — 12 m/s, (*ii*) 16 m/s]

33. A particle is moving with a speed of 2.4×10^8 m/s. Velocity of light = 3×10^8 m/s. Calculate the ratio of the mass of the body in motion and the rest mass. [**Ans.** 1.67]

34. A particle is moving with a speed of 0.6 c. Calculate the ratio of the rest mass and the mass while in motion. [**Ans.** 0.8]

35. A particle of rest mass 10^{-27} kg is moving with a speed of 2.4×10^8 m/s. Calculate the mass of the particle while in motion. [**Ans.** 1.67×10^{-27} kg]

36. If 10^{-3} kg of substance is fully converted into energy, how much energy is liberated ? [**Ans.** 9×10^{13}]

37. 8.36×10^{-6} kg of a substance is fully converted to heat energy. Calculate the amount of heat produced. [**Ans.** 1.8×10^{11}cal]

38. 10^{12} kg of water is heated from 0° C to 100°C. Calculate the increase in the mass of water. Specific heat of water = 10^3 calories/kg. – °C. [**Ans.** 4. 64 kg.]

39. What speed must have a body have for its mass to be doubled ? [*Delhi (Hons.) 1983*] [**Ans.** 0.866 *c*]

40. Derive Lorentz space-time transformation for two inertial frames. Hence explain 'time dilation' and 'length contraction' [*Gauhati, 1991*]

41. A certain quantity of ice at 0°C melts into water at 0°C in so doing gains 1 kg of mass. What was its initial mass ? [**Ans.** 2.7×10^{11} kg]

42. Derive the relativistic acceleration transformation

$$a'_x = \frac{a_x\left(1-\frac{\upsilon^2}{c^2}\right)^{\frac{3}{2}}}{\left(1-\frac{u_x\upsilon}{c^2}\right)^3}, \text{ where } a_x = \frac{du_x}{dt}, \frac{du'_x}{dt'}$$

Hint.
$$\frac{du_x}{dt'} = \left(\frac{du_x}{dt'}\right)\left(\frac{dt}{dt'}\right)$$

2

QUANTUM MECHANICS

2.1 Origin of Quantum Theory

Relativity and Quantum Theory clarified a number of unresolved problems that could not be explained by the classical theory.

The distribution of energy among the various wavelengths of a black body was one major problem.

Atomic spectroscopy became one of the most active fields of experimental physics. Balmer's empirical formula was recognized only after Bohr's theory.

The solution to these problems came from Max Planck's effort to deduce a theoretical formula for the spectral distribution of black body radiation. Hence we look at the radiation laws of a black body and their short comings.

2.2 Black Body

A perfectly black body is one which absorbs all the heat radiations (corresponding to all wavelengths) incident on it. When such a body is placed inside an isothermal enclosure, it will emit the full radiation of the enclosure after it is in equilibrium with the enclosure. These radiations are independent of the nature of the substance. Such heat radiations in a uniform temperature enclosure are known as **black body** radiations. Also the black body completely absorbs heat radiations of all wavelengths. Thus the black body also emits completely the radiation of all wavelengths at that temperature.

In practice, a perfectly black body is not available. A body showing close approximation to a perfectly black body can be constructed.

A hollow copper sphere is taken and is coated with lamp black on its inner surface (Fig. 2.1). A fine hole is made and a pointed projection is made just in front of the hole. When the radiations enter the hole, they

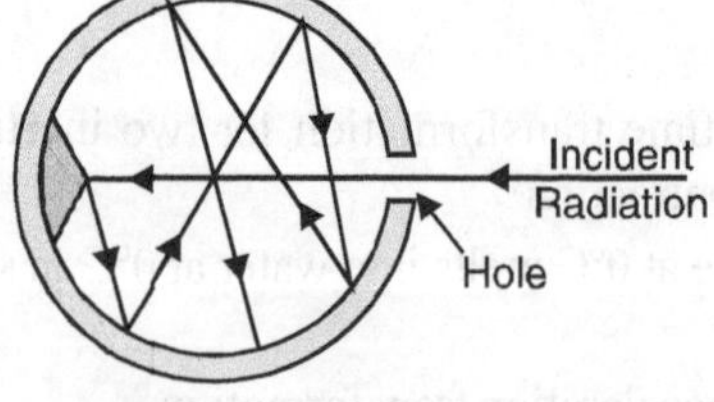

Fig. 2.1.

suffer multiple reflections and are completely absorbed. This body acts as a black body absorber. When this body is placed in a bath at a fixed temperature, the heat radiations come out of the hole (Fig. 2.2). The hole acts as a black body radiator. It should be remembered that only the hole and not the walls of the body, acts as the black body radiator.

Wien also constructed a black body in the form of a cylinder. This black body is commonly used now-a-days.

It consists of a hollow metallic cylinder and fitted with heating coils wound around it. The inner surface of the cylinder is coated with lamp black. The cylinder is placed in concentric porcelain tubes (Fig. 2.2).

The temperature is measured with the help of a thermocouple arrangement. Heat radiations emerge out of the hole. The radiations from the inner chamber can be limited with the help of diaphragms provided on the inner side. This hole will act as a black body radiator.

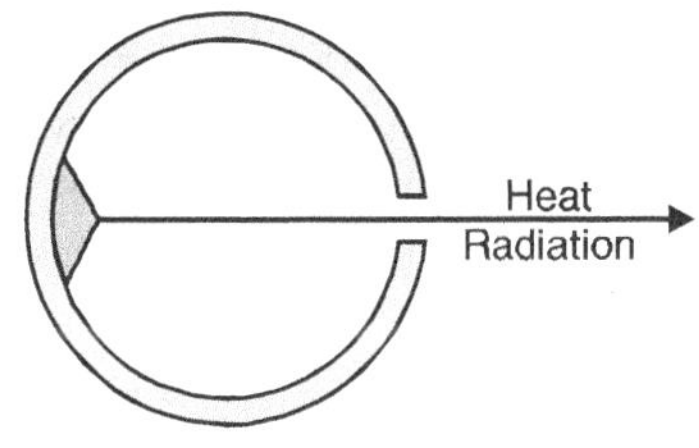

Fig. 2.2.

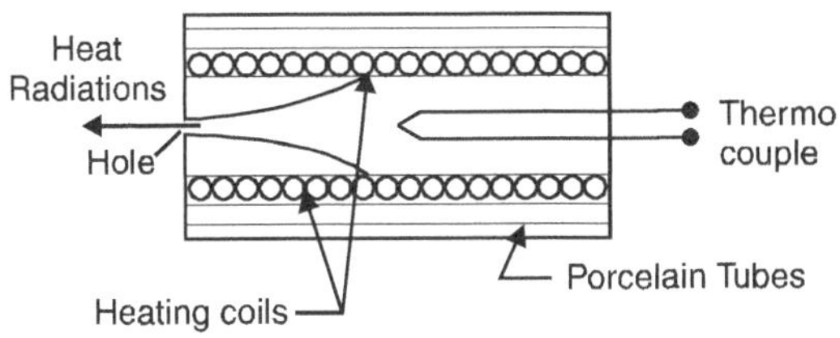

Fig. 2.3.

2.3 Distribution of Energy in the Spectrum of a Black body

Lummer and Pringsheim investigated the distribution of energy amongst the different wavelengths of a thermal spectrum of a black body radiation (full emitter).

The experiment arrangement consists of a carbon tube heated electrically (Fig. 2.4). The radiations from this tube are allowed to be incident on a reflector M_1 through a slit S_1. The parallel reflected beam is dispersed by a fluorspar prism. The dispersed beam is focused on a line bolometer with the help of the mirror M_2. The intensity of the radiations corresponding to different wavelengths is measured with the help of bolometer. The deflection in the galvanometer (of the balometer) is proportional to the energy of the heat radiations.

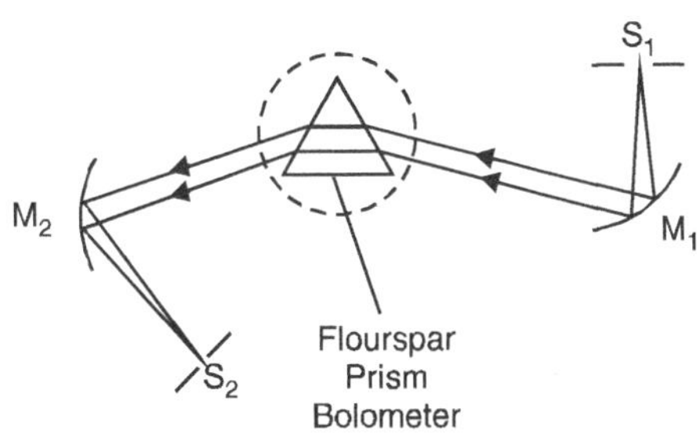

Fig. 2.4.

The find the energy distribution for different wavelengths the fluorspar prism is gradually rotated.

The results obtained by Lummer and Pringsheim are shown in Fig. 2.5. Each curve represents the intensity of the radiation (E_λ) with wavelength for a given temperature of the source.

Results : (*i*) At a given temperature, the energy is not uniformly distributed in the radiation spectrum of a hot body.

(*ii*) At a given temperature the intensity of radiations increases with increase in wavelength and at a particular wavelength its value is maximum. With further increase in wavelength the intensity of heat radiations decreases.

(*iii*) With increase in temperature, λ_m decreases. λ_m is the wavelength at which maximum emission of energy takes place. The points on the dotted line represent λ_m at various temperatures.

(*iv*) For all wavelengths, an increase in temperature causes an increase in the energy emission.

(*v*) The area under each curve represents the total energy emitted for the complete spectrum at a particular temperature. This area increases with increase in temperature of the body. It is found that the area is directly proportional to the fourth power of the temperature of the body,

i.e., $$E \propto T^4 \qquad ...(2.1)$$

This represents Stafan Boltzmann's law.

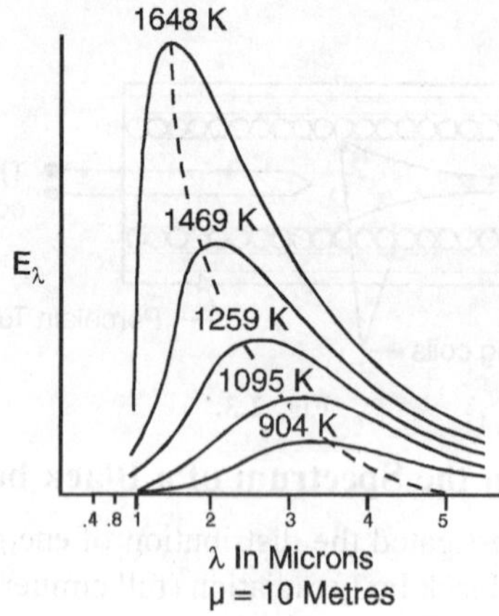

Fig. 2.5.

2.4 Wien's Displacement Law

From thermodynamical considerations, Wien has shown that the product of the wavelength corresponding to maximum energy and absolute temperature is constant.

$$\lambda_m \times T = \text{constant} = 0.2892 \text{ cm-K} \qquad ...(2.2)$$

This is called Wien's displacement law. It also shows that with increase in temperature, λm decreases. Wien has also shown that the energy $E_{\max}$ is directly proportional to the fifth power of the absolute temperature

i.e.
$$E_m \propto T^5$$
$$E_m = \text{constant} \times T^5 \qquad ...(2.3)$$

Wien deduced the radiation law for the energy emitted at a given wavelength λ and a temperature T.

$$E_\lambda = \lambda^{-5} f(\lambda T) \qquad ...(2.4)$$

With certain assumptions Wien was able to deduce the relation

$$f(\lambda T) = C_1 e^{-\frac{C_2}{\lambda T}}$$

$$\therefore \quad E_\lambda = C_2 \lambda^{-5} e^{-\frac{C_2}{\lambda T}} \qquad ...(2.5)$$

Here C_1 and C_2 are constants.

Equation (2.5) represents the Wien's law of distribution of energy.

2.5 Rayleigh-Jean's Law

The energy distribution in the thermal spectrum according to Rayleigh, is given by the formula

$$E_\lambda = \frac{8\pi kT}{\lambda^2} \qquad ...(2.6)$$

Here K is the Boltzmann's constant.

The experimental results of Lummer and Pringsheim in the infrared region, however, show that the Wien's law holds good only in the region of shorter wavelengths. It does not hold good at longer wavelengths. The Rayleigh-Jean's law holds good in the region of longer wavelengths but not for shorter wavelengths. This was shown by Rubens and Kurlbaum. Thus Wein's law and Rayleigh-Jean's law do not precisely agree with the experimental results.

The difficulty recognised by Rayleigh and Jean's is that according to the equation (2.6) the energy increases without limit as the wavelength decreases.

$\therefore$ E_λ diverges to ∞ as $\lambda \rightarrow 0$.

This "unphysical" consequence of equation came to be known as the **ultraviolet catastrophe**.

Hence we see that a satisfactory distribution law should approach the Wien distribution at low wavelengths and Rayleigh-Jeans law at higher wavelengths.

2.6 Photo-electric Effect

Hertz in 1887 discovered that a metallic surface is capable of emitting electrons when light of very short wavelength falls on it. He noted that the air in the spark gap became a better conductor when it was illuminated by ultra violet light from an arc lamp. Hallwachs in 1888 found that :

1. when ultra-violet light was incident on a neutral zinc plate, the plate became positively charged.
2. when ultra-violet light was incident on a negatively charged zinc plate, it lost its charge rapidly, and
3. when ultra-violet light was incident on a positively charged zinc plate, it became more positively charged.

He came to the conclusion that only negatively charged particles can be emitted by the surface under the action of ultra-violet light. Thomson in 1898 showed that the $\frac{e}{m}$ value of emitted particles was the same as that for cathode rays. Einstein in 1916 studied the effect of visible light of a range of frequencies on sodium, potassium, cesium, rubidium and lithium and found that the electrons were ejected out of these metals.

The electrons ejected out of the metal under the action of light are known as *photo-electrons* and this phenomenon is known as '*photo-electric effect*'.

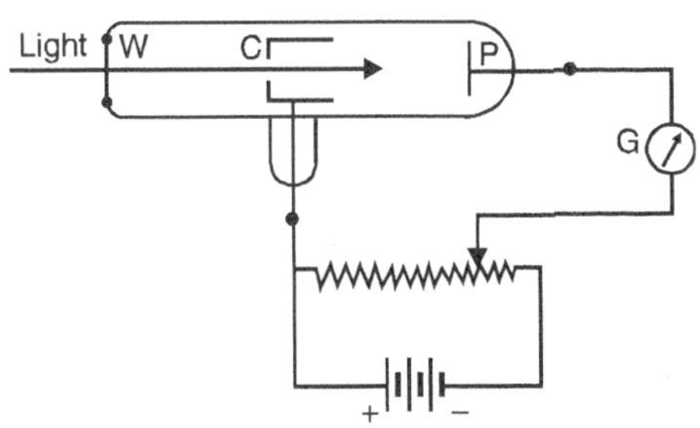

Fig. 2.6

Expt. Photo-electric effect is studied with the help of the following experiment. An evacuated glass tube has a photo-electrically sensitive plate P. It may be a highly polished zinc plate. A hollow cylinder C has a small hole that permits the incident light to fall on the plate P. A quartz window W allows the light to enter the tube. P is connected to the negative and C is connected to the positive terminal of the battery. A sensitive galvanometer is used in the circuit as shown in (Fig. 2.6).

When light from some source such as mercury arc lamp falls on the plate P, photo-electrons are ejected out of the plate. These photo-electrons are attracted by the positively charged cylinder *C*. The galvanometer shows deflection and thus photo-electric current is produced in the circuit. The effect of light of a range of frequencies was studied. Based on the results obtained, the following laws of photo-electric emission were formulated.

2.7 Laws of Photo-electric Emission

(1) For a given metallic surface, there is a minimum frequency for which the incident light can eject the photo-electrons out of the metal. Light of frequency smaller than this particular value cannot eject electrons, no matter how long it falls on the surface or how high is its intensity.

(2) The number of photo-electrons ejected depends upon the intensity of the incident light. Thus, the photo-electric current depends upon the intensity of the incident light.

(3) Light of frequency higher than the *critical frequency or threshold frequency* ejects electrons of different velocities. The maximum velocity with which an electron is ejected, depends upon the frequency of the incident light and not on the intensity of the incident light. The *threshold frequency* is the minimum frequency which can eject an electron out of the metal. Light of frequency less than the threshold frequency cannot eject an electron out of the metal, howsoever intense the light may be.

(4) The maximum kinetic energy of the ejected electrons increases linearly with the frequency of the incident light. It does not depend on the intensity of the incident light.

The above experimental facts could not be explained on the basis of the wave theory of light. Einstein, in 1905, explained the photo-electric effect based on Planck's idea of quantum theory of light. According to quantum theory, light consists of quanta or corpuscles of energy $h\nu$ which move through space with the velocity of light. These particles are known as photons and each photon has an energy = $h\nu$, where h is the Planck's constant and ν is the frequency of light. When a photon collides with an electron in the metal, it may transfer energy to the electron. This transfer is an "all or none" process, *i.e.*, either the photon gives the whole of its energy $h\nu$ to the electron or no energy to the electrons.

2.8 Ritz Combination Principle

Using Rydberg's formula for the spectral lines, Ritz came up with a principle to extend it to the spectral lines. Series other than Balmer for hydrogen was predicated.

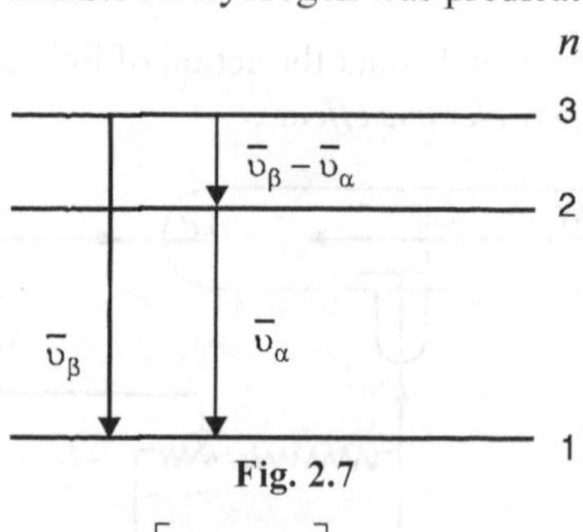

Fig. 2.7

Now $$\bar{\upsilon}_\alpha = R\left[\frac{1}{2^2} - \frac{1}{3^2}\right]$$

and $$\bar{\upsilon}_\beta = R\left[\frac{1}{2^2} - \frac{1}{4^2}\right]$$

$$\therefore \quad \bar{\upsilon}_\alpha - \bar{\upsilon}_\beta = R\left[\left(\frac{1}{2^2} - \frac{1}{4^2}\right) - \left(\frac{1}{2^2} - \frac{1}{3^2}\right)\right]$$

$$= R\left[\frac{1}{3^2} - \frac{1}{4^2}\right] \qquad \text{...(2.7)}$$

This gave a new line in the spectra in the infra-red region discovered latter by Paschen. Ritz made this discovery empirically much before Bohr's theory was known. Similarly other lines like Brackett were obtained using this principle. Such lines are known as **intercombination lines or Series.**

This principle was useful for X-rays as well. It also gave Bohr a clue to interpret the spectra on the basis of the Quantum theory.

2.9 Planck's Radiation Law

Max-Planck's in 1900 deduced an empirical relation to explain the distribution of energy in the spectrum of a black body observed experimentally. The law is derived with the following basic assumptions :

(1) A black body radiator contains simple harmonic oscillators capable of vibration with all possible frequencies. The oscillators are of molecular dimensions.

(2) The oscillators do not radiate or absorb energy continuously.

(3) The oscillators can exchange energy (emission or absorption) within surrounding in discrete values viz.

$$E = 0, h\nu, 2\,h\nu, 3\,h\nu \text{ etc.}$$

These are the only available energy states.

Here h is Planck's constant.

and ν is the frequency.

The average energy per oscillator is given by where $\varepsilon = hr$

$$\frac{\sum_{n=0}^{n=\infty} n \in e^{-n\in/kT}}{\sum_{n=0}^{n=\infty} e^{-n\in/kT}}$$

Taking $x = e^{-\in/KT}$, we get

$$\text{average energy} = \frac{\sum_{n=0}^{n=\infty} n \in x^n}{\sum_{n=0}^{n=\infty} x^n} = \frac{\in \sum_{n=0}^{n=\infty} n\,x^n}{\sum_{n=0}^{n=\infty} x^n}$$

But

$$\sum_{n=0}^{n=\infty} nx^n = x + 2x^2 + 3x^3 + 4x^4 + \ldots$$

$$= x\,[1 + 2x + 3x^2 + 4x^3 + \ldots]$$

$$= \frac{x}{(1-x)^2}$$

and

$$\sum_{n=0}^{n=\infty} x^n = 1 + x + x^2 + x^3 + \ldots$$

$$= \left(\frac{1}{1-x}\right)$$

Substituting these values in the main equation,

$$\text{Average energy} = \frac{\in x/(1-x)^2}{\left(\frac{1}{1-x}\right)}$$

$$= \frac{\in x}{1-x} = \frac{\in e^{-\in/KT}}{(1-e^{-\in/KT})}$$

$$= \frac{\in}{(e^{\in/KT}-1)} \qquad ...(2.8)$$

where $\in$ is small.

$$e^{\in/KT}-1 = 1+\frac{\in}{kT}-1 = \frac{\in}{kT}$$

$$\therefore \quad \text{Average energy} = \frac{\in}{\in/kT} = kT \qquad ...(2.9)$$

This value is in agreement with Maxwell's law of equipartition of energy. Therefore, in general, average energy of an oscillator is not kT but it is equal to $\frac{\in}{(e^{\in/kT}-1)}$.

But $\in = hv$

Therefore average energy of an oscillator

$$= \frac{hv}{(e^{hv/kT}-1)}$$

The number or resonators per unit volume in the frequency υ and $\upsilon + d\upsilon$ is given by

$$N = \frac{8\pi v^2}{c^2}dv$$

Therefore energy density

$$E_\upsilon d_\upsilon = \frac{8\pi v^2}{c^2}dv\left[\frac{hv}{(e^{h\upsilon/KT}-1)}\right] \qquad ...(2.10)$$

$$E_\upsilon = \frac{8\pi v^3}{c^2}\left[\frac{1}{(e^{h\upsilon/KT}-1)}\right] \qquad ...(2.11)$$

This expression represents Planck's radiation law, and this law explains all experimental results.

In terms of wavelength, $c = v\lambda$ and $\upsilon = \frac{c}{\lambda}$

$$v = c\,\lambda^{-1}$$

$$dv = -c\lambda^{-2}d\,\lambda$$

Numerically $dv = c\lambda^{-2}\,d\lambda$

Substituting the values in equation (2.10) we get

$$E_\lambda\, d\lambda = \frac{8\pi\, hc}{\lambda^2}d\lambda\left[\frac{1}{(e^{hc/\lambda KT}-1)}\right]$$

or

$$E_\lambda = \frac{8\pi hc}{\lambda^2}\left[\frac{1}{(e^{hc/\lambda kT}-1)}\right] \qquad ...(2.12)$$

(*i*) For shorter wavelengths

$$e^{h\upsilon/KT} \geq 1$$

From equation (2.12)

$$\therefore \quad E_\lambda = 8\pi hc\lambda^{-5}\, e^{-\frac{h\upsilon}{KT}}$$

$$E_\lambda = C_1\lambda^{-5}\, e^{-\frac{C_2}{KT}} \qquad \text{...(2.13)}$$

where, $C_1 = 8\pi\, hc$ and $C_2 = \dfrac{hc}{\lambda}$.

Equation (2.13) represents Wien's radiation law.

(*ii*) For longer wavelengths, $\dfrac{h\upsilon}{kT'}$ is small.

From equation (2.12), expanding $e^{h\upsilon/KT}$ and neglecting higher powers,

$$E_\lambda = \frac{8\pi\, hc}{\lambda^2\left[1+\dfrac{h\upsilon}{kT}-1\right]}$$

$$= \frac{8\pi\, hc\,\lambda\, kT}{\lambda^5 hc}$$

$$E_\lambda = \frac{8\pi\, kT}{\lambda^4} \qquad \text{...(2.14)}$$

Equation (2.14) represents Rayleigh-Jeans's law.

Thus Planck's formula for the energy distribution in a thermal spectrum is applicable for all wavelengths.

2.10 Rigid Rotator

A rigid rotator is considered to be dumbell shaped and consisting of two small masses m_1 and m_2 connected together by a thin rod (Fig. 2.8).

Fig. 2.8.

The system is capable of rotation about an axis passing through the centre of mass of the system and perpendicular to line joining two masses. In such a system the energy is only the kinetic energy of rotation. In such a system the moment of interia I is given by

$$I = m_1 r_1^2 + m_2 r_2^2 \qquad \text{...(2.15)}$$

The kinetic energy of the system

$$E = \frac{1}{2} I\,\omega^2 = \frac{1}{2} I\left(\frac{d\theta}{dt}\right)^2 \qquad \text{...(2.16)}$$

The phase integral for quantum-mechanical energy levels is given by

$$J = \oint P_\theta d\,\theta = 2\pi\, P_\theta$$

According to Wilson-Sommerfeld rule

$$J = nh$$

$$\therefore \quad 2\pi\, P_\theta = nh$$

$$P_\theta = \frac{nh}{2\pi} = n\hbar$$

The energy of the system

$$E_n = \frac{1}{2} I\omega^2 = \frac{P^2 \omega^2}{2I}$$

$$E_n = \frac{(P_n)^2}{2I}$$

$$E_n = \frac{n^2 h^2}{2I}$$

$$E_n = \frac{n^2 h^2}{8\pi^2 I} \qquad ...(2.17)$$

The energy values are discrete

$$E_n \propto n^2$$

For $n = 1$

$$E_1 = \frac{h^2}{8\pi^2 I}$$

For $n = 2$

$$E_2 = \frac{4h^2}{8\pi^2 I}$$

For $n = 3$

$$E_3 = \frac{9h^2}{8\pi^2 I}$$

2.11 Einstein's Photo Electric Equation

According to Einstein's, when light of frequency υ is incident on a metallic surface, photons of energy hv strike the metal. The energy of a quantum or a photon is found to be directly proportional to the frequency of the incident radiation. The entire energy of the photon is transferred to a single electron in the metal. If the energy $h\upsilon$ of the striking photon is less than the energy required to eject an electron out of the metal, no electron is ejected. The minimum energy required = $h\upsilon_0$, where υ_0 is the threshold frequency and it depends on the nature of the metal. If the frequency of the incident light is just equal to the threshold frequency *i.e.*, $\upsilon = \upsilon_0$ then the electron is just ejected out of the metal but its velocity is zero. If υ is greater than v_0 then the difference in energy $(h\upsilon - h\upsilon_0)$ is used in giving kinetic energy to the electron. If the mass of an electron is m and velocity is υ, then,

$$hv - \mathrm{hv}_0 = \frac{1}{2} m\upsilon^2 \qquad ...(2.18)$$

where $hv_0 = W$ = Work function of the metal.

This is the photo-electric equation due to Einstein.

This equation was verified experimentally by Millikan in 1916.

2.12 Millikan's Experiment for Verification of Einstein's Photo-electric Equation

The apparatus consists of an evacuated chamber C (Fig. 2.9). At the centre of the chamber, a drum D is kept which can rotate freely about a vertical axis. Four cylindrical blocks of different materials (sodium, potassium, lithium etc.) are fixed on the surface of the drum. K is a sharp knife edge which can scrap the surface of the metal. W is a quartz window through which monochromatic light of known frequency is allowed to pass, F is a Faraday cylinder connected to a quadrant electrometer. With the help of a sensitive potential divider arrangement, suitable positive or negative potential can be applied to the drum with respect to the cylinder F.

Initially the surface of any one of the metal blocks is scraped with the knife edge *K* and then by rotating the drum *D*, the block is turned towards the window *W*.

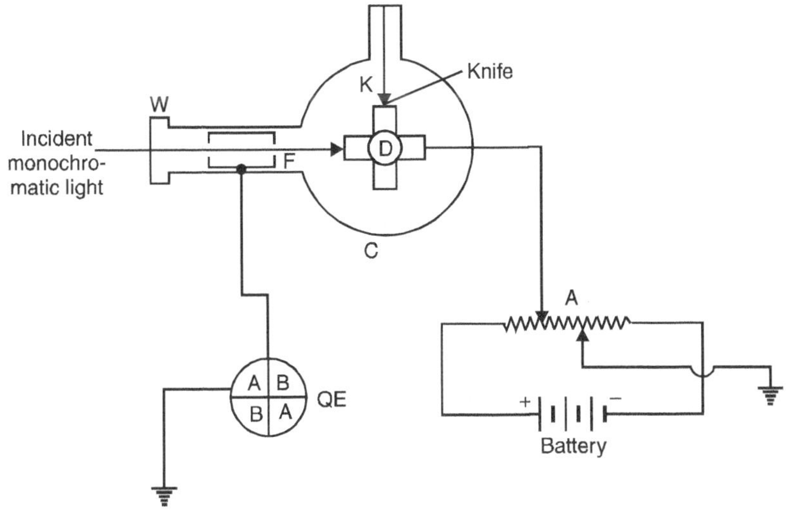

Fig. 2.9

Allow monochromatic radiation of known frequency υ to be incident on the surface of the metal block. When the drum is kept at a negative potential with respect to the cylinder *F*, the photo-electrons ejected from the surface are accelerated towards the cylinder and the quadrant electrometer records the deflection. When a small positive potential is applied to the drum, only the fast moving electrons reach the cylinder *F*, the deflection in the electrometer decreases. Now, the positive potential of the drum is gradually increased till the deflection in the quadrant electrometer is zero. At this stage no photo-electron reaches the cylinder. This particular potential V_0 is called the stopping potential. Even when the intensity of light is increased for the same frequency, the deflection remains zero for the same stopping potential V_0. **It means that the stopping potential V_0 is the same** and **is independent of the intensity of incident light.** The experiment is repeated for different frequencies for the same metal. It is found that the stopping potential V_0 increases with increase in frequency for a given metal. According to Einstein's photo-electric equation

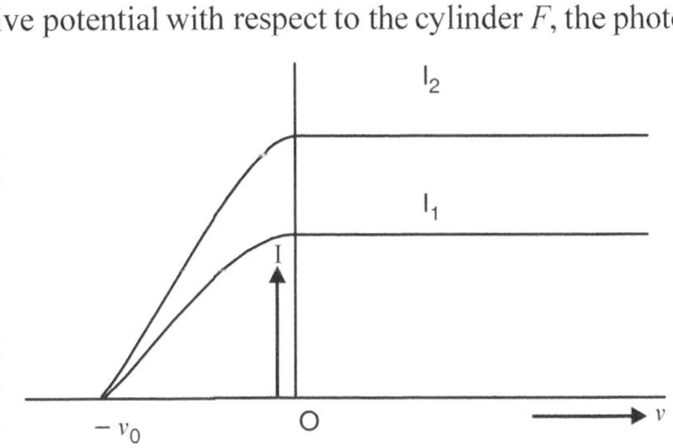

Fig. 2.10

$$hv - \phi = \frac{1}{2}m\upsilon^2$$

Here $$\phi = hv_0$$

∴ $$hv = hv_0 = \frac{1}{2}m\upsilon^2$$

Also $$ev_0 = \frac{1}{2}m\upsilon^2$$

$$hv - hv_0 = eV_0$$

$$V_0 = \frac{h}{e}[v - v_0]$$

The experiment is repeated for different metals and graphs are drawn between frequency υ along the X-axis and the stopping potential V_0 along Y-axis. The graph is a straight line (Fig. 2.11). A straight line graph is obtained in all cases and this is in agreement with Einstein's theory.

Determination of Planck's Constant

As discussed earlier, the stopping potentials of a particular metal are determined for different frequencies. A graph is plotted between frequency υ along the X-axis and stopping potential υ_0 along the X-axis. Take a point *A* on the graph and measure υ_0. Draw *AB* parallel to Y-axis. The point *B* gives the value of υ. The graph meets the X-axis at the point *C* corresponding to the threshold frequency υ_0. The slope of the graph

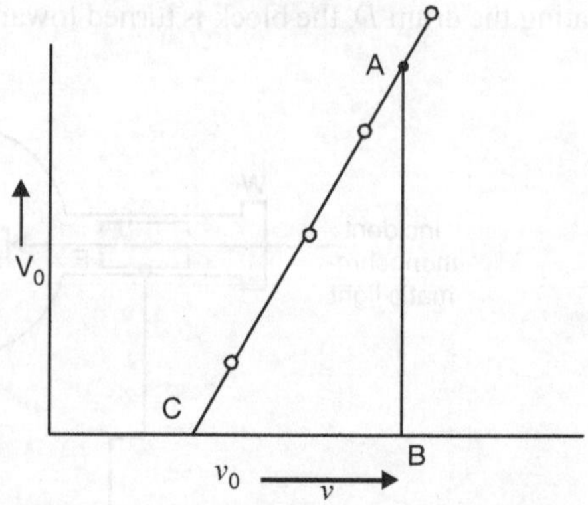

Fig 2.11.

$$= \frac{AB}{BC} = \left(\frac{V_0}{\upsilon - \upsilon_0}\right)$$

But $$V_0 = \frac{h}{e}[\upsilon - \upsilon_0]$$

$$\therefore \quad h = e\left[\frac{V_0}{\upsilon - \upsilon_0}\right]$$

$$h = e\,[\text{slope of the graph}]$$

Here *e* is the charge of the electron and it is equal to 1.6×10^{-19} C. Hence the value of Planck's constant *h* can be determined. The value of *h* was found to be $[6.624 \pm 0.01] \times 10^{-34}$ J–s. This value is in agreement with the value determined by other methods.

Example 2.1. *What is the threshold wavelength for a tungsten surface whose work function is 4.5 eV.* [*Delhi (Hons.) 1980*]

Solution. Here $$\phi = 4.5 \text{ eV}$$
$$= 4.5 \times 1.6 \times 10^{-19} \text{ Joule}$$

But $$\phi = h\nu_0$$

or $$\phi = \frac{hc}{\lambda_0}$$

$$\lambda_0 = \frac{hc}{\phi}$$

Here h = 6.6×10^{-34} Joule-second

$$c = 3 \times 10^8 \text{ m/s}$$

$$\therefore \quad \lambda_0 = \frac{6.6 \times 10^{-34} \times 3 \times 10^8}{4.5 \times 1.6 \times 10^{-19}}$$

$$\lambda_0 = 9.640 \times 10^{-7} \text{ m}$$

$$\lambda_0 = \mathbf{9640\ Å.}$$

Example 2.2. *A photoelectric surface has a work function of 4 eV. What is the maximum velocity of the photoelectrons emitted by light of frequency 10^{15} hertz incident on the surface ?*

$$h = 6.6 \times 10^{-34} \text{ Joule-second}$$

$$e = 1.6 \times 10^{-19} \text{ Coulomb}$$

$$m = 9 \times 10^{-31} \text{ kg}$$

Solution. Here

$$\phi = 4 \text{ eV}$$
$$= 4 \times 1.6 \times 10^{-19} \text{ Joule}$$
$$= 6.4 \times 10^{-19} \text{ Joule}$$

$$\frac{1}{2}m\upsilon^2 = h\upsilon - \phi = 6.6 \times 10^{-34} \times 10^{-15} - 6.4 \times 10^{-19}$$

$$\frac{1}{2}m\upsilon^2 = 0.2 \times 10^{-19}$$

$$\upsilon = \sqrt{\frac{0.4 \times 10^{-19}}{m}} = \sqrt{\frac{0.4 \times 10^{-19}}{9 \times 10^{-31}}}$$

$$\upsilon = \mathbf{2.107 \times 10^5 \text{ m/s.}}$$

Example 2.3. *Calculate the energy in electron-volts of the photoelectrons from the surface of a tungsten emitter when it is irradiated with light of wavelength 1800 Å, given that the threshold wavelength for photoelectric emission in this case is 2300 Å.* [*Delhi (Hons.) 1981*]

Solution.

$$E = h(\upsilon - \upsilon_0)$$

$$E = hc\left[\frac{1}{\lambda} - \frac{1}{\lambda_0}\right] = hc\left[\frac{\lambda_0 - \lambda}{\lambda\,\lambda_0}\right]$$

Here

$$h = 6.6 \times 10^{-34} \text{ J-s}$$
$$c = 3 \times 10^8 \text{ m/s}$$
$$\lambda = 1800 \text{ Å} = 18 \times 10^{-18} \text{ m}$$
$$\lambda = 2300 \text{ Å} = 23 \times 10^{-8} \text{ m}$$

$$E = 6.6 \times 10^{-34} \times 3 \times 10^8 \left[\frac{5 \times 10^{-8}}{18 \times 10^{-8} \times 23 \times 10^{-8}}\right]$$

$$E = \frac{6.6 \times 10^{-34} \times 3 \times 10^8 \times 5}{18 \times 23 \times 10^{-18}}$$

$$E = 2.4 \times 10^{-19} \text{ Joule}$$

But

$$1 \text{ eV} = 1.6 \times 10^{-19} \text{ Joule}$$

or

$$1 \text{ Joule} = \frac{1}{1.6 \times 10^{-19}} \text{ eV}$$

$\therefore$

$$E = \frac{2.4 \times 10^{-19}}{1.6 \times 10^{-19}} = \mathbf{1.5 \text{ eV}}$$

Hence the energy of each photoelectron emitted from the surface is 1.5 eV.

Example 2.4. *Calculate the longest wavelength of the incident radiation which will eject electrons from a metal work function is 6 electron volts. Planck's constant $h = 6.624 \times 10^{-34}$ joule-second.* [*Delhi 1982*]

Solution. Here, work function $\phi = h\upsilon_0 = 6$ electron volts

$$= 6 \times 1.6 \times 10^{-19} \text{ joule}$$

$$\upsilon_0 = \frac{6 \times 1.6 \times 10^{-19}}{h} = \frac{6 \times 1.6 \times 10^{-19}}{6.624 \times 10^{-34}} \text{ hertz}$$

$$\lambda_0 = \frac{c}{\upsilon_0} = \frac{3 \times 10^8 \times 6.624 \times 10^{-34}}{6 \times 1.6 \times 10^{-19}}$$

$$\lambda = 2.07 \times 10^{-7} \text{ m} = \mathbf{2070 \text{ Å}.}$$

Example 2.5. *Calculate the threshold frequency and the corresponding wavelength of radiation incident on a certain metal whose work function is 3.31 × 10^{-19} J. Given, Planck's constant = 6.62 × 10^{-34} J-s.* [*Delhi (Sub.) 1985*]

Solution. Here, work function

$$\phi = h\upsilon_0 = 3.31 \times 10^{-19} \text{ J}$$

$$h = 6.62 \times 10^{-34} \text{ J-s}$$

$$\upsilon_0 = \frac{\phi}{h} = \frac{3.31 \times 10^{-19}}{6.62 \times 10^{-34}}$$

$$= 5 \times 10^{14} \text{ hertz}$$

$$\lambda_0 = \frac{c}{\nu_0} = \frac{3 \times 10^8}{5 \times 10^{14}} = 6 \times 10^{-7} \text{ m}$$

$$= \mathbf{6000\ Å}$$

Example 2.6. *Light of wavelength 4300 Å is incident on (a) nickel surface of work function 5 electron volts and (b) a potassium surface of work function 2.3 electron volts. Find out, if electrons will be emitted, and if so, the maximum velocity of the emitted electrons in each case.* [*Delhi 1985*]

Solution. (*a*) For the nickel surface

$$\phi = h\upsilon_0 \text{ electron volts}$$

$$= 5 \times 1.6 \times 10^{-19} \text{ Joule}$$

$$\upsilon_0 = \frac{\phi}{h} = \frac{5 \times 1.6 \times 10^{-19}}{6.624 \times 10^{-34}} \text{ hertz}$$

$$\lambda_0 = \frac{c}{\nu_0} = \frac{3 \times 10^8 \times 6.624 \times 10^{-34}}{8 \times 10^{-19}}$$

$$= 2484 \times 10^{-10} \text{ m}$$

$$= \mathbf{2484\ Å.}$$

As λ_0 is less than the wavelength of the incident radiation, (λ = 4300 Å), electrons **will not** be emitted.

(*b*) For the potassium surface,

$$\phi = 2.3 \text{ electrons volts}$$

$$= 2.3 \times 1.6 \times 10^{-19} \text{ joule}$$

$$\therefore \quad \lambda_0 = \frac{ch}{\phi}$$

$$= \frac{3 \times 10^8 \times 6.624 \times 10^{-34}}{2.3 \times 1.6 \times 10^{-19}}$$

$$= 4389 \times 10^{-10} \text{ m}$$

$$= \mathbf{4389\ Å.}$$

As λ_0 is greater than λ electrons will be emitted. Let the maximum velocity of the ejected electrons be υ_0

$$\therefore \quad \frac{1}{2} m\upsilon^2 = h\upsilon - h\upsilon_0$$

$$= hc\left[\frac{1}{\lambda} - \frac{1}{\lambda_0}\right]$$

$$\frac{1}{2}m\upsilon^2 = \frac{hc(\lambda_0 - \lambda)}{\lambda\lambda_0}$$

$$\upsilon = \left[\frac{2hc(\lambda_0 - \lambda)}{m\lambda\lambda_0}\right]^{\frac{1}{2}}$$

$$\upsilon = \left[\frac{2 \times 6.624 \times 10^{-34} \times 3 \times 10^8 \times 89 \times 10^{-10}}{9.1 \times 10^{-31} \times 4300 \times 10^{-10} \times 4389 \times 10^{-10}}\right]$$

$$\upsilon = \mathbf{1.423 \times 10^5\ m/s.}$$

Example 2.7. *The wavelength of the photoelectric threshold for silver is 3250 Å. Determine the maximum energy if the ejected electrons from a silver surface by light of wavelength 2537 Å.* [*I.A.S. 1983*]

Solution.

$$U = h(\upsilon - \upsilon_0)$$

$$U = hc\left[\frac{1}{\lambda} - \frac{1}{\lambda_0}\right] = \left[\frac{\lambda_0 - \lambda}{\lambda \times \lambda_0}\right]$$

Here

$$h = 6.6 \times 10^{-34} \text{ J–s and } c = 3 \times 10^{-8} \text{ m/s}$$

$$\lambda = 2537 \text{ Å} = 2537 \times 10^{-10} \text{ m}$$

$$\lambda_0 = 3250 \text{ Å} = 3250 \times 10^{-10} \text{ m}$$

$$U = 6.6 \times 10^{-34} \times 3 \times 10^8 \left[\frac{3250 \times 10^{-10} - 2537 \times 10^{-10}}{3250 \times 10^{-10} \times 2537 \times 10^{-10}}\right]$$

$$U = \mathbf{1.625 \times 10^{-19}\ J}$$

Example 2.8. *A surface having work function 1.51* eV *is illuminated by light of wavelength 4000 Å. Calculate (i) the maximum kinetic energy of the ejected electrons and (ii) the stopping potential.* [*I.A.S.*]

Solution. Here

$$\phi_0 = 1.51 \text{ eV}$$

$$= 1.51 \times 1.6 \times 10^{-19} \text{ J}$$

$$= 2.416 \times 10^{-19} \text{ J}$$

$$\lambda = 4000 \text{ Å} = 4 \times 10^{-7} \text{ m}$$

$$\nu = \frac{c}{\lambda} = \frac{3 \times 10^8}{4 \times 10^{-7}} = 7.5 \times 10^{14} \text{Hz}$$

(*i*)

$$h\upsilon - \phi_0 = \frac{1}{2}m\upsilon^2 = U$$

$$U = 6.624 \times 10^{-34} \times 7.5 \times 10^{14} - 2.416 \times 10^{-19}$$

$$U = 4.968 \times 10^{-19} - 2.416 \times 10^{-19}$$

$$U = \mathbf{2.552 \times 10^{-19}\ J}$$

$$U = \frac{2.552 \times 10^{-19}}{1.6 \times 10^{-19}} \text{ eV}$$

$$U = 1.595 \text{ eV}$$

(*ii*) Also $h\upsilon - \phi_0 = e\upsilon = U$

stopping potential

$$V = \frac{U}{e}$$

$$V = \frac{2.552 \times 10^{-19}}{1.6 \times 10^{-19}}$$

$$V = \mathbf{1.595\ volts.}$$

Example 2.9. *The stopping potential for the electrons emitted from a metal due to photoelectric effect is found to be 1 V for light of 2500 Å. Calculate the work function of the metal in* eV.

Solution. Here

$$V = 1 \text{ Volt}$$

$$\lambda = 25000 \text{ Å} = 2.5 \times 10^{-7} \text{ m}$$

$$\upsilon = \frac{c}{\lambda} = \frac{3 \times 10^8}{2.5 \times 10^{-7}}$$

or

$$\upsilon = 1.2 \times 10^{15} \text{ Hz}$$

$$\phi = ?$$

$$e = 1.6 \times 10^{-19} \text{ C}$$

$$h\upsilon = \phi = eV$$

$$\phi = h\upsilon - eV$$

$$= 6.624 \times 10^{-34} \times 1.2 \times 10^{15} - 1.6 \times 10^{-19} \times 1$$

$$= 7.9488 \times 10^{-19} - 1.6 \times 10^{-19}$$

$$= 6.3488 \times 10^{-19} \text{ J}$$

$$= \frac{6.3488 \times 10^{-19}}{1.6 \times 10^{-19}} \text{eV}$$

$$= \mathbf{3.968\ eV.}$$

Example 2.10. *The photoelectric threshold wavelength for Tungsten is 2300 Å. Find the energy in* eV *of the emitted electrons from the surface by ultraviolet light of wavelength 1800 Å.*

[*Delhi, 1990*]

Solution.

$$E = h(\upsilon - \upsilon_0)$$

$$E = hc\left[\frac{1}{\lambda} - \frac{1}{\lambda_0}\right]$$

$$E = hc\left[\frac{\lambda_0 - \lambda}{\lambda\lambda_0}\right]$$

Here

$$h = 6.6 \times 10^{-34} \text{ J-s}$$

$$c = 3 \times 10^8 \text{ m/s}$$

$$\lambda = 1800 \text{ Å} = 1800 \times 10^{-10} \text{ m} = 18 \times 10^{-8} \text{ m}$$

$$\lambda_0 = 2300 \text{ Å} = 2300 \times 10^{-10} \text{ m} = 23 \times 10^{-8} \text{ m}$$

$\therefore$

$$E = 6.6 \times 10^{-34} \times 3 \times 10^8 \left[\frac{23 \times 10^{-8} - 18 \times 10^{-8}}{18 \times 10^{-8} \times 23 \times 10^{-8}}\right]$$

$$E = \frac{6.6 \times 10^{-34} \times 3 \times 10^8 \times 5}{18 \times 23 \times 10^{-8}}$$

$$E = 2.4 \times 10^{-19} \text{ J}$$

But

$$1 \text{ eV} = 1.6 \times 10^{-19} \text{ J}$$

or

$$1 \text{ J} = \left(\frac{1}{1.6 \times 10^{-19}}\right) \text{eV}$$

$$\therefore \qquad E = \frac{2.4 \times 10^{-19}}{1.6 \times 10^{-19}}$$

$$E = \mathbf{1.5\ eV}$$

Hence, the energy of each photoelectron emitted from the surface is **1.5 eV.**

Example 2.11. *Calculate the energy of a photon having the same momentum as that of 10 MeV proton.* [*Osmania, 1992*]

Solution. Mass of proton, $m = 1.67 \times 10^{-27}$ kg

Energy of proton, $U = \frac{1}{2}m\upsilon^2 = 10^7 \times 1.6 \times 10^{-19}$ J

$$U = 1.6 \times 10^{-12} \text{ J}$$

$$\upsilon = \left[\frac{2U}{m}\right]^{\frac{1}{2}}$$

Momentum, $p = m\upsilon = m\left[\frac{2U}{m}\right]^{\frac{1}{2}}$

$$p = [2mU]^{\frac{1}{2}}$$

Energy of photon, $E = pc = [2mU]^{\frac{1}{2}}$

$$E = \left[2 \times 1.67 \times 10^{-27} \times 1.6 \times 10^{-12}\right]^{\frac{1}{2}} \times 3 \times 10^8 \text{ J}$$

$$E = \mathbf{21.9 \times 10^{-12}\ J}$$

$$E = \frac{21.9 \times 10^{-12}}{1.6 \times 10^{-19}} = \mathbf{13.69 \times 10^7\ eV.}$$

Example 2.12. *The wavelength of light falling on the surface of a metal of work function 2.3* eV *is 4300* Å. *With what velocity will the electron be emitted.* [*Delhi (Hons.) 1992*]

$$m = 9.1 \times 10^{-31} \text{ kg}$$

Solution. Here $\phi = 2.3$ eV

$$= 2.3 \times 1.6 \times 10^{-19} \text{ J}$$

$$= \mathbf{3.68 \times 10^{-19}\ J}$$

$$\lambda = 4300 \text{ Å} = 4.3 \times 10^{-17} \text{ m}$$

$$\frac{1}{2}m\upsilon^2 = h\upsilon - \phi = \left(\frac{hc}{\lambda}\right) - \phi$$

$$= \left(\frac{6.624 \times 10^{-34} \times 3 \times 10^8}{4.3 \times 10^{-7}}\right) - 3.68 \times 10^{-19}$$

$$= 4.62 \times 10^{-19} - 3.68 \times 10^{-19}$$

$$= 0.94 \times 10^{-19} \text{ J}$$

$$\upsilon = \left[\frac{2 \times 0.94 \times 10^{-19}}{m}\right]^{\frac{1}{2}} = \left[\frac{2 \times 0.94 \times 10^{-19}}{9.1 \times 10^{-31}}\right]^{\frac{1}{2}}$$

$$\upsilon = \mathbf{4.55 \times 10^5\ m/s.}$$

Example 2.13. *Calculate, in electron volt, the energy of a quantum of light of wavelength 5.3×10^{-7} m.* *[Delhi, 1992]*

Solution. Here $\lambda = 5.3 \times 10^{-7}$ m

$$\text{Energy, } E = h\nu = \frac{hc}{\lambda} = \frac{6.624 \times 10^{-34} \times 3 \times 10^{8}}{5.3 \times 10^{-7}} \text{ J}$$

$$= 3.75 \times 10^{-19} \text{ J} = \frac{3.75 \times 10^{-19}}{1.6 \times 10^{-19}} \text{ eV}$$

$$= \mathbf{2.34\ eV.}$$

2.13 Bohr's Model of the Atom

In 1913, Bohr suggested a model of the atom for which he was awarded Nobel Prize for Physics in 1922. He suggested that the electrons are negatively charged particles and move round the nucleus in various orbits. *The orbit of an electron is fixed.* He gave the name *Energy Levels* to these orbits. The electron cannot emit any energy when it moves in its own fixed orbit known as stationary energy level. In Fig. 2.12 the dotted line orbit is ruled out in Bohr's model and only thick line orbits are present. It means that the orbits of definite radii are present and orbits of all radii are ruled out. The electron gives out energy in the form of electromagnetic radiations of definite frequency when it jumps from higher energy level to lower energy level (Fig. 2.12). The Bohr model opened the way to new concepts in the twentieth century.

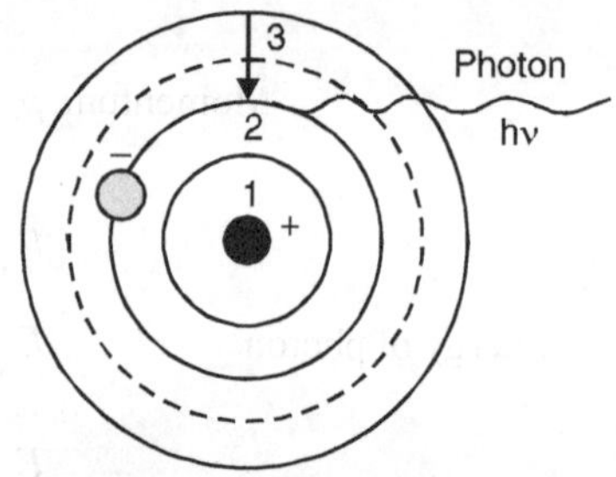

Fig 2.12. Bohr's model of the atom.

2.14 Bohr's Theory of the Hydrogen Spectrum

Rutherford's experiment on the scattering of α-particles led Bohr to the conclusion that the atom consists of a positively charged nucleus at its centre. Moreover, Bohr applied the quantum theory of radiation as developed by Planck and Einstein to the Rutherford's model. His theory is mainly based on the following postulates.

(1) An Atom consists of a positively charged nucleus at the centre.

(2) The negatively charged particles known as electrons move round the nucleus in various orbits known as stationary energy levels. The electrons cannot emit radiations when moving in their stationary levels.

(3) The Coulobian and Newtonian forces are applicable in the domain of the atom.

(4) The electrons revolve round the nucleus in various circular orbits and the angular momentum $m\upsilon r = \frac{nh}{2\pi}$

where n (= 1, 2, 3, 4 etc.) is called the quantum number, and h is the Planck's constant.

(5) When an electron jumps from a higher energy level to a lower energy level, it give out electromagnetic radiations of a particular frequency.

where W_{n_2} is the energy of the n_2 energy level, W_{n_1} is the energy of the n_1 energy level and υ is the frequency of the radiation.

Bohr's postulates are the combinations of some of the ideas of classical physics and quantum physics. While in some ways they agree with classical physics but in some aspects they are contradictory to it. *Here the energy levels are quantized.* It means that the electrons can move only in some particular orbits of *definite radii* as suggested by Rutherford's model.

Consider an atom whose nucleus has a positive charge E and let an electron of negative charge e move round the nucleus is an orbit of radius r:

$$E = Ze, \text{ where } Z = 1 \text{ for hydrogen.}$$

The force of attraction between the nucleus and the electron

$$= \frac{Ee}{4\pi \in_0 r^2} = \frac{Ze^2}{4\pi \in_0 r^2}$$

As the electrons move along the circular orbit, the centripetal force $\left(= \frac{m\upsilon^2}{r}\right)$ is provided by the attractive force on the electron.

$$\therefore \quad \frac{m\upsilon^2}{r} = \frac{Ze^2}{4\pi \in_0 r^2}$$

$$\therefore \quad m\upsilon^2 = \frac{Ze^2}{4\pi \in_0 r} \qquad ...(2.19)$$

$$\text{The angular momentum, } m\upsilon r = \frac{nh}{2\pi}$$

$$\therefore \quad \upsilon = \frac{nh}{2\pi\, mr} \qquad ...(2.20)$$

Substituting the value of v in equation (2.19),

$$m\left(\frac{nh}{2\pi\, mr}\right)^2 = \frac{Ze^2}{4\pi \in_0 r}$$

$$\therefore \quad r = \frac{\in_0 n^2 h^2}{\pi\, mZe^2} \qquad ...(2.21)$$

Substituting this value of r in equation (2.20)

$$\upsilon = \frac{Ze^2}{2 \in_0 nh} \qquad ...(2.22)$$

From equation (2.21), we find that $r \propto n^2$. The radii of the orbits are in the ratio of 1 : 4 : 9 : 16 : 25 etc. The radius of the second orbit is four times the radius of the first orbit. For hydrogen atom, the radius of the first orbits

$$r = \frac{\in_0 n^2 h^2}{\pi\, mZe^2}, \quad \text{Here } n = 1 \text{ and } Z = 1.$$

Equation (2.22) shows that the velocity of the electron decreases with the increase in the order of the orbit. The velocity of the electron in the first orbit $\upsilon = \frac{Ze^2}{2 \in_0 h}$.

Substituting the values of n, h, Z and e

$$\upsilon = 2.2 \times 10^6 \text{ m/s}$$

Let us consider the energy of an electron in the orbit. The energy is partly potential and partly kinetic. From definition, the potential energy of an electron $= \int_{\infty}^{r} \frac{Ze^2}{4\pi \in_0 r^2}\, dx = \frac{-Ze^2}{4\pi \in_0 r}$

$$\therefore \quad \text{Potential energy of the electron} = \frac{-Ze^2}{4\pi \in_0 r}$$

$$\text{Kinetic energy} = \frac{1}{2} m\upsilon^2$$

From equation (*ii*), $$mv^2 = \frac{-Ze^2}{4\pi \epsilon_0 r}$$

$$\text{Kinetic energy} = \frac{Ze^2}{8\pi \epsilon_0 r}$$

The potential energy is negative because energy must be given to the electron to bring it far away from the nucleus to zero energy level.

$$\text{Total energy} = \text{potential energy} + \text{kinetic energy}$$

$$= \frac{-Ze^2}{4\pi \epsilon_0 r} + \frac{Ze^2}{8\pi \epsilon_0 r} = -\frac{Ze^2}{8\pi \epsilon_0 r} \quad ...(2.23)$$

Substituting the value of r, we get

$$W_n = \frac{-me^4 Z^2}{8 \epsilon_0^2 n^2 h^2} \text{ Joule} \quad ...(2.24)$$

Here $m = 9.1 \times 10^{-31}$ kg, $e = 1.6 \times 10^{-19}$ coulomb

$$\epsilon_0 = 8.85 \times 10^{-12} \frac{\text{Coulomb}^2}{\text{Newton-m}^2}$$

$$h = 6.624 \times 10^{-34} \text{ Joule-second.}$$

For an electron to be removed from its first orbit in a hydrogen atom, energy must be supplied. Therefore ionization potential for hydrogen,

$$\phi = \frac{me^4 Z^2}{8 \epsilon_0^2 n^2 h^2}.$$

For hydrogen, $Z = 1, n = 1$

$$\therefore \quad \phi = \frac{me^4}{8 \epsilon_0^2 h^2} \text{ Joules} \quad ...[2.25\,(a)]$$

$$\phi = \frac{me^4}{8 \epsilon_0^2 h^2 \times 1.6 \times 10^{-19}} \text{ electrons volts} \quad ...[2.25\,(b)]$$

When an electron jumps from energy level n_2 to energy level n_1

$$W_{n_2} - W_{n_1} = \frac{-me^4 Z^2}{8 \epsilon_0^2 n_2^2 h^2} + \frac{me^4 Z^2}{8 \epsilon_0^2 n_1^2 h^2}$$

$$W_{n_2} - W_{n_1} = \frac{-me^4 Z^2}{8 \epsilon_0^2 h^2} \left(\frac{1}{n_1^2} - \frac{1}{n_2^2} \right)$$

But, $$W_{n_2} - W_{n_1} = h\nu$$

$$h\nu = \frac{me^4 Z^2}{8 \epsilon_0^2 h^2} \left(\frac{1}{n_1^2} - \frac{1}{n_2^2} \right)$$

and $$\nu = \frac{me^4 Z^2}{8 \epsilon_0^2 h^3} \left(\frac{1}{n_1^2} - \frac{1}{n_2^2} \right)$$

In the case of hydrogen atom $Z = 1$

$$\therefore \quad \nu = \frac{me^4}{8 \epsilon_0^2 h^3} \left(\frac{1}{n_1^2} - \frac{1}{n_2^2} \right)$$

But, $c = \nu\lambda, \therefore \nu = \frac{c}{\lambda}$

$$\frac{c}{\lambda} = \frac{me^4}{8\in_0^2 h^3}\left(\frac{1}{n_1^2} - \frac{1}{n_2^2}\right)$$

$$\frac{1}{\lambda} = \frac{me^4}{8\in_0^2 ch^3}\left(\frac{1}{n_1^2} - \frac{1}{n_2^2}\right)$$

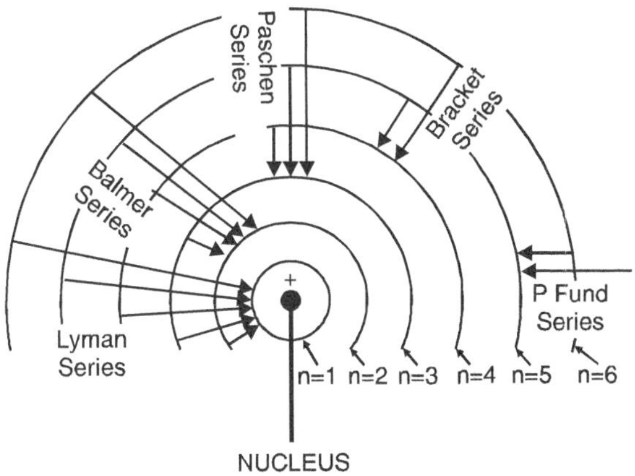

Fig. 2.13. Spectral Series of Hydrogen

$$\frac{1}{\lambda} = \bar{\nu} \text{ called the wave number}$$

$$\bar{\nu} = \frac{me^4}{8\in_0^2 ch^3}\left(\frac{1}{n_1^2} - \frac{1}{n_2^2}\right)$$

$$\frac{me^4}{8\in_0^2 ch^3} = R \text{ and is known as Rydberg constant.}$$

Substituting the value, $R = 109678 \text{ cm}^{-1}$

$$= 1.096 \times 10^7 \text{ per metre}$$

$$\bar{\nu} = R\left(\frac{1}{n_1^2} - \frac{1}{n_2^2}\right) \qquad ...(2.26)$$

2.15 Spectral Series of Hydrogen Atom

(1) Lyman Series. When an electron jumps from the outer orbits to the first orbit, the spectral lines are in the ultra-violet region (Fig. 2.13).

Here $n_1 = 1, n_2 = 2, 3, 4, 5.....$

$$\bar{\nu}_1 = R\left(\frac{1}{1} - \frac{1}{2^2}\right) = \frac{3}{4}R$$

$$\bar{\nu}_2 = R\left(\frac{1}{1} - \frac{1}{3^2}\right) = \frac{8}{9}R$$

$$\bar{\nu}_3 = R\left(\frac{1}{1} - \frac{1}{4^2}\right) = \frac{15}{16}R$$

(2) Balmer Series. When an electron jumps from outer orbits to the second orbit,

$$n_1 = 2, n_2 = 3, 4, 5..... \text{ etc.}$$

$$\therefore \quad \bar{\nu}_1 = R\left(\frac{1}{2^2} - \frac{1}{3^2}\right) = \frac{5}{36}R$$

$$\bar{\nu}_2 = R\left(\frac{1}{2^2} - \frac{1}{4^2}\right) = \frac{3}{16}R$$

This series lies in the visible region of the spectrum.

(3) Paschen Series. When an electron jumps from the outer orbits to the third orbit,

$$n_1 = 3, n_2 = 4, 5, 6, \text{ etc.}$$

$$\therefore \quad \bar{\nu}_1 = R\left(\frac{1}{3^2} - \frac{1}{4^2}\right) = \frac{9}{144}R$$

$$\bar{\nu}_2 = R\left(\frac{1}{3^2} - \frac{1}{5^2}\right) = \frac{16}{225}R$$

(4) Bracket Series. When an electron jumps from outer orbits to the fourth orbit.

$$n_1 = 4, n_2 = 5, 6, 7..... \text{ etc.}$$

$$\bar{\nu}_1 = R\left(\frac{1}{4^2} - \frac{1}{5^2}\right) = \frac{9}{400}R$$

$$\bar{\nu}_2 = R\left(\frac{1}{4^2} - \frac{1}{6^2}\right) = \frac{5}{144}R$$

(5) Pfund Series. When an electron jumps from outer orbits to the fifth orbit.

$$n_1 = 5, n_2 = 6, 7, 8..... \text{ etc.}$$

$$\bar{\nu}_1 = R\left(\frac{1}{5^2} - \frac{1}{6^2}\right) = \frac{11}{900}R$$

$$\bar{\nu}_2 = R\left(\frac{1}{5^2} - \frac{1}{7^2}\right) = \frac{24}{1225}R$$

The last three series are in the infra-red region. When the Bohr's theory was suggested, only the Balmer and Paschen series for hydrogen atom were known. The other series as suggested theoretically according to equation (2.26) were verified experimentally also. The Lyman series (1916), the Bracket series (1922), and Pfund series (1929) studied experimentally were found to show the same wave numbers as suggested theoretically. When an electron absorbs energy, it moves from lower energy level to higher level as in the absorption spectrum. When an electron jumps from higher energy level to the lower energy level, it emits radiation as in the emission spectrum.

Example 2.14. *The Rydberg constant for hydrogen is 1.09678 ×10^7/m and and for ionized helium 1.09722 × 10^7/m. Calculate the ratio of the mass of the proton to that of the electron, assuming the helium nucleus to be four times the mass of proton.* *[I.A.S. 1976]*

Solution. The nucleus of the atom has motion though it is too heavy as compared to the electron. Due to this reason the reduced mass of the electron is

$$m = \left[\frac{M}{M+m}\right] = \frac{1}{1+\frac{m}{M}}.$$

Here m is the mass of the electron and M is the mass of the nucleus. For hydrogen, Rydberg constant,

$$R_H = \frac{R}{\left[1+\frac{m}{M_H}\right]} \quad ...(2.27)$$

For ionized helium, Rydberg constant

$$R_{He} = \frac{R}{\left[1+\frac{m}{M_{\text{He}}}\right]} \qquad \text{...(2.28)}$$

Here $$R = \frac{me^4}{8\in_0^2 ch^3} \text{ per metre}$$

Dividing (*ii*) by (*i*)

$$\frac{R_{He}}{R_H} = \frac{\left[1+\frac{m}{M_H}\right]}{\left[1+\frac{m}{M_{He}}\right]}$$

Taking $M_{\text{He}} = 4M_H$ and simplifying,

$$\frac{M_H}{m} = \frac{R_H - \frac{1}{4}R_{He}}{R_{He} - R_H}$$

Here $R_H = 1.09678 \times 10^7$/m and $R_{He} = 1.09722 \times 10^7$/m

$$\frac{M_H}{m} = \frac{10^7}{10^7}\left[\frac{1.09678 - \frac{1}{4}\times 1.09722}{1.09722 - 1.09722 - 1.09678}\right] = 1869$$

2.16 Energy Levels of Hydrogen Atom

In the case of Hydrogen atom, the work function for the electron in the first orbit is given by

$$\phi = \frac{me^4}{8\in_0^2 h^3}$$

This value of $\phi = 13.60$ eV.

The energy of electron in the first orbit $U_1 = -13.60$ eV.

Similarly for the second, third etc. orbits, the energy of the electron will be

$$U_2 = -\frac{13.6}{2^2} = -3.4 \text{ eV}$$

$$U_3 = -\frac{13.6}{3^2} = -1.511 \text{ eV}$$

$$U_4 = -\frac{13.6}{4^2} = -0.85 \text{ eV}$$

$$U_5 = -\frac{13.6}{5^2} = -0.544 \text{ eV}$$

... ...

... ...

The energy level diagram is shown in Fig. 2.13.

The energy is inversely proportional to the square of the number of the orbit and is negative. The lowest energy state ($n = 1$) is called the ground state. With increase in the order of the orbit, the energy of the orbit increases. If $n = \infty$, the energy is zero. In all other orbits, the energy is negative. Due to this reason the electron is bound to the nucleus.

The different spectral series are shown in the Fig. 2.13.

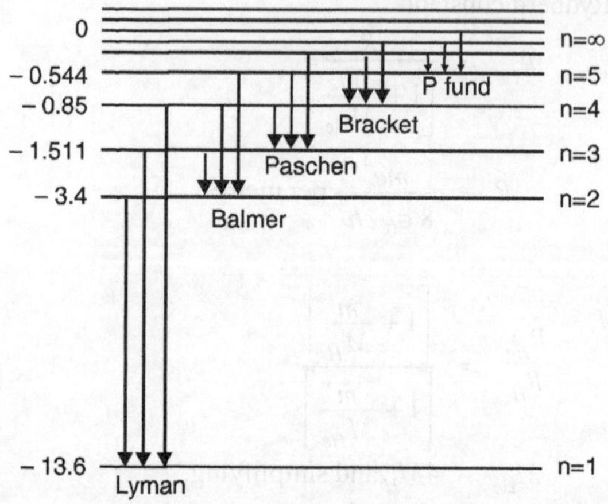

Fig. 2.14.

Example 2.15. *Calculate the radius of the hydrogen atom. Show that the velocity of the electron in the first Bohr orbit in hydrogen atom is (1/37) c, where c is velocity of light.*

[Delhi (Hons) 1922], [Delhi 1992]

Solution. (*i*) Radius, $r = \dfrac{\epsilon_0 n^2 h^2}{\pi mZe^2}$

Here

$$n = 1$$
$$h = 6.624 \times 10^{-34} \text{ J-s}$$
$$\epsilon_0 = 8.85 \times 10^{-12} \text{ C}^2/\text{N-m}^2$$
$$m = 9.1\times 10^{-31} \text{ kg}$$
$$e = 1.6 \times 10^{-19} \text{ C}$$

$$r = \frac{8.85 \times (6.624 \times 10^{-34})^2}{\pi \times 9.1 \times 10^{-31} \times (1.6 \times 10^{-19})^2}$$

$$r = \mathbf{5.29 \times 10^{-11} \text{ m}}$$
$$= \mathbf{0.529 \text{ Å}}$$

Velocity, $\upsilon = \dfrac{Ze^2}{2\epsilon_0 hn}$

Here $Z = 1, n = 1$

$$\therefore \quad \upsilon = \frac{(1.6 \times 10^{-19})^2}{2 \times 8.85 \times 10^{-12} \times 6.624 \times 10^{-34}}$$

$$\frac{\upsilon}{c} = \frac{(1.6 \times 10^{-19})^2}{2 \times 8.85 \times 10^{-12} \times 6.624 \times 10^{-34} \times 3 \times 10^8}$$

$$\frac{\upsilon}{c} = \frac{1}{137}$$

$$\therefore \quad \upsilon = \left(\frac{1}{137}\right)c$$

2.17 Resonance, excitation and Ionization Potentials

Resonance Potential. The minimum potential required to provide energy to the electron in the ground state to the first excited state *i.e.*, from $n = 1$ to $n = 2$ is called resonance potential.

The energy of electron in the ground state of hydrogen atom is –13.6 eV and first excited state ($n = 2$) is –3.4 eV. Therefore, the energy required to move the electron from the ground state to the first excited state is (–3.4) eV – (–13.6) eV = 10.2 eV.

Therefore the resonance potential for hydrogen is 10.2 V.

Excitation Potential. The state n > 1 are called excited states. The energy required to move the electron to the first, second, third excited states is given by

$$E_1 = 10.2 \text{ eV}$$

$$E_2 = -\left(\frac{13.6}{3^2}\right) - (-13.6)$$

$$= -1.51 + 13.60 = 12.09 \text{ eV}$$

$$E_3 = -\left[\frac{13.6}{3^2}\right] - (-13.6)$$

$$= -0.85 + 13.60 = 12.75 \text{ eV}$$

Therefore the successive excitation potentials are 10.2 V, 12.09 V, 12.75 V and so on.

The excitation potential is the potential required to provide energy to raise the electron from the ground state to the state $n > 1$ *i.e.*, $n = 2, 3, 4...$

Ionization Potential. It is the minimum potential required that provides energy to bring the electron from the ground state out of the atom.

For hydrogen atom, ionization potential = 13.6 V

The energy to the electron in the atom can be provided by electron emitted from a hot filament and accelerated though a creation potential V.

The electron will move with a velocity υ where $\upsilon = \left[\frac{2eV}{m}\right]^{\frac{1}{2}}$.

(1) If the energy of the striking electron is just equal to energy required by the electron in atom to move from the ground state to the first state, the electron will move to first excited state and striking electron gives whole of its energy to the electron in the atom.

(2) If the energy of the electron is just equal to or more than the energy required by the electron to come out of the atom, the electron in the atom absorbs energy and comes out of the atom.

Example 2.16. *Wavelength of Balmer H_α line is 6563 Å. Calculate the wavelength of H_β line.* [*Delhi (Hons.) 1982*]

Solution. For H_α line of Balmer series *i.e.*, the first number,

$$\bar{\nu}_1 = R\left[\frac{1}{2^2} - \frac{1}{3^2}\right]$$

$$\frac{1}{\lambda_1} = \frac{5}{36}R \qquad ...(i)$$

For H_β line of Balmer series *i.e.*, the second number,

$$\bar{\nu}_2 = R\left[\frac{1}{2^2} - \frac{1}{4^2}\right]$$

$$\frac{1}{\lambda_2} = \frac{3}{16}R \qquad ...(ii)$$

Dividing (*i*) by (*ii*)

$$\frac{\lambda_2}{\lambda_1} = \frac{20}{27}$$

$$\lambda_2 = \left(\frac{20}{27}\right)\lambda_1$$

But $\lambda_1 = 6563 \text{ Å}$

$$\lambda_2 = \left(\frac{20}{27}\right)6563$$

$\therefore$ $\lambda_2 = \mathbf{4861\ Å}$

Example 2.17. *Calculate the energy required to excite the hydrogen atom from the ground state (n = 1) to the first excited state (n = 2).* [*Delhi (Hons.)*]

Solution. Energy required $U = \frac{me^4}{8\epsilon_0^2 h^2}\left[\frac{1}{n_1^2} - \frac{1}{n_2^2}\right]$

$$U = \frac{me^4}{8\epsilon_0^2 h^2}\left[\frac{1}{1^2} - \frac{1}{2^2}\right]$$

$$U = \frac{3me^4}{32\epsilon_0^2 h^2}$$

Here $m = 9.1 \times 10^{-31}$ Kg, $e = 1.6 \times 10^{-19}$ C

$\epsilon_0 = 8.85 \times 10^{-12}$ C^2/N-m^2

$h = 6.624 \times 10^{-34}$ J-s

$$U = \frac{3 \times 9.1 \times 10^{-31} \times (1.6 \times 10^{-19})^4}{32 \times (8.85 \times 10^{-12})^2 \times (6.624 \times 10^{-34})^2} \text{ Joules}$$

$$U = \frac{3 \times 9.1 \times 10^{-31} \times (1.6 \times 10^{-19})^4}{32 \times (8.85 \times 10^{-12})^2 \times (6.624 \times 10^{-34})^2 \times 1.6 \times 10^{-15}} \text{ eV}$$

$U = \mathbf{10.13\ eV.}$

Example 2.18. *The wavelength of sodium D_1 line is 590 nm. Calculate the difference in energy levels involved in the emission or absorption of this line.*

Solution. Here $E_{(n_2-n_1)} = h\upsilon = \frac{hc}{\lambda}$

Here, $h = 6.624 \times 10^{-34}$ J-s

$c = 3 \times 10^8$ m/s

$h = 590 \text{ nm} = 590 \times 10^{-9}$ m

$$\therefore \quad E_{(n_2-n_1)} = \frac{(6.624 \times 10^{-34}) \times 3 \times 10^8}{590 \times 10^{-9}}$$

$= \mathbf{3.37 \times 10^{-19}\ J}$

Example 2.19. *The wavelength of the second line of the Balmer series in the hydrogen spectrum is 4861 Å. Calculate the wavelength of the first line.* [*Rajasthan*]

Solution. For the second line,

$$\bar{\nu} = R\left[\frac{1}{2^2} - \frac{1}{4^2}\right]$$

$$\bar{\nu}_2 = \frac{3}{16}R$$

$$\frac{1}{\lambda_2} = \frac{3}{16}R \qquad ...(i)$$

For the first line,

$$\bar{\nu} = R\left[\frac{1}{2^2} - \frac{1}{3^2}\right]$$

$$\bar{\nu}_1 = \frac{5}{36}R$$

or $$\frac{1}{\lambda_1} = \frac{5}{36}R \qquad ...(ii)$$

Dividing (*i*) by (*ii*)

$$\frac{\lambda_1}{\lambda_2} = \frac{27}{20}$$

$\therefore$ $$\lambda_1 = \frac{27\,\lambda_2}{20}$$

But $$\lambda_2 = 4861 \text{ Å}$$

$$\lambda_1 = \frac{27 \times 4861}{10}$$

or $$\lambda_1 = \mathbf{6563 \text{ Å}.}$$

Example 2.20. *A beam of electrons is used to bombard gaseous hydrogen. What is the minimum energy in* ***electron-volts*** *the electrons must have if the first number of the Balmer series corresponding to a transition from $n_2 = 3$ state to $n_1 = 2$ state is to be emitted ?*

$$h = 6.6 \times 10^{-34} \text{ Joule second.}$$

[*Mumbai, 1981, Kolkata, 1992*]

Solution. Energy required $= \dfrac{me^4}{8\,\epsilon_0^2\,h^2}\left[\dfrac{1}{n_1^2} - \dfrac{1}{n_2^2}\right]$

Here $n_1 = 2, n_2 = 3$

$\therefore$ Energy required $= \left(\dfrac{me^4}{8\,\epsilon_0^2\,h^2}\right) \times \left(\dfrac{5}{36}\right)$ Joule

But $1 \text{ eV} = 1.6 \times 10^{-19}$ Joule

$\therefore$ Energy required $= \dfrac{5me^4}{8 \times 36 \times \epsilon_0^2\,h^2 \times (1.6 \times 10^{-19})}$ electron-volts

$$= \frac{5 \times 9 \times 10^{-31} \times (1.6 \times 10^{-19})^4}{8 \times 36(8.85 \times 10^{-12})^2 \times (6.6 \times 10^{-34})^2 \times (1.6 \times 10^{-19})}$$

$$= \mathbf{1.88 \text{ eV}.}$$

Example 2.21. *Calculate the ionization potential in electron volts for hydrogen atom. Given that*

$$e = 1.6 \times 10^{-19} \text{ coulomb}$$
$$m = 9 \times 10^{-31} \text{ kg}$$
$$h = 6.6 \times 10^{-34} \text{ Joule-second}$$
$$\epsilon_0 = 8.85 - 10^{-19} \text{ coulomb}^2/\text{Newton-m}^2$$

Solution. Work function,

$$\phi = \frac{me^4}{8\,\epsilon_0^2\,h^2} \text{ Joules}$$

$$\phi = \frac{9\times10^{-31}\times(1.6\times10^{-19})^4}{8\times(8.85\times10^{-12})^2\times(6.6\times10^{-34})^2}\text{ Joule}$$

But $\quad 1\text{eV} = 1.6\times10^{-10}$ Joule

$$\therefore \quad \phi = \frac{9\times10^{-31}\times(1.6\times10^{-19})^4}{8\times(8.85\times10^{-19})^2\times(6.6\times10^{-34})^2\times1.6\times10^{-19}}$$

$$\phi = \mathbf{13.51\ eV.}$$

Example 2.22. *The wavelength of the first number of Balmer series of hydrogen is 6563×10^{-10} m.*

Calculate the wavelength of its second number. [*Delhi, 1982*]

Solution.

For the first number $\quad \bar{\nu}_1 = R\left[\frac{1}{2^2}-\frac{1}{3^2}\right]$

$$\bar{\nu}_1 = \frac{5}{36}R$$

or $\quad \frac{1}{\lambda_1} = \frac{5}{36}R \quad ...(i)$

For the second number,

$$\bar{\nu}_2 = R\left[\frac{1}{2^2}-\frac{1}{4^2}\right]$$

or $\quad \frac{1}{\lambda_2} = \frac{3}{16}R \quad ...(ii)$

Dividing (*i*) by (*ii*)

$$\frac{\lambda_2}{\lambda_1} = \frac{20}{27}$$

$$\lambda_2 = \frac{20}{27}\lambda_1$$

But $\quad \lambda_1 = 6563\times10^{-19}$ m

$$\therefore \quad \lambda_2 = \frac{20\times6563\times10^{-10}}{27}$$

$$= \mathbf{4861\times10^{-10}\ m}$$

Example 2.23. *The ionization potential of atomic hydrogen is 13.6* V. *Calculate the wavelength of light emitted in a transition starting at the first excited state of hydrogen atom.*

[*I.A.S. 1985*]

Solution. Ionization potential = 13.6 V

Energy of electron in the first orbit

$$U_1 = -13.6\text{ eV}$$

Energy of electron in the second orbit

$$U_2 = \frac{-13.6}{n^2} = \frac{-13.6}{4} = -3.4\text{ eV}$$

$$U_2 - U_1 = -3.4 + 13.6$$

$$h\nu = 10.2\text{ eV}$$

$$\frac{hc}{\lambda} = 10.2\times1.6\times10^{-19}\text{ J} = 1.632\times10^{-18}\text{ J}$$

$$\lambda = \frac{hc}{1.632 \times 10^{-18}}$$

$$= \frac{6.624 \times 10^{-34} \times 3 \times 10^{8}}{1.632 \times 10^{-18}}$$

$$= 1.217 \times 10^{-7} \text{ m}$$

$$= \mathbf{1217\ Å.}$$

Example 2.24. *Find the radius and speed of the electron in the first Bohr orbit of the hydrogen atom. How will the radius and speed of electron change with the increase in atomic number of the atom ?* [*I.A.S.*]

Solution. (*i*) The radius of the Bohr orbit is given by

$$r = \frac{\epsilon_0 n^2 h^2}{\pi m Z e^2}.$$

For hydrogen,

$$Z = 1, n = 1$$

$$r_H = \frac{\epsilon_0 h^2}{\pi m e^2}$$

Here

$$\epsilon_0 = 8.85 \times 10^{-12} \text{ C}^2/\text{N-m}^2$$

$$m = 9.1 \times 10^{-31} \text{ kg}$$

$$e = 1.6 \times 10^{-19} \text{ C}$$

$$h = 6.624 \times 10^{-34} \text{ J-s}$$

$$\therefore \quad r_H = \frac{8.85 \times 10^{-12} \times (6.624 \times 10^{-34})^2}{\pi \times 9.1 \times 10^{-31} \times (1.6 \times 10^{-19})^2}$$

$$r_H = \mathbf{5.30 \times 10^{-11}\ m}$$

Also

$$r = \frac{r_H}{Z}$$

Therefore radius of the orbit will decrease with increase in atomic number Z.

(*ii*) Velocity of the electron

$$\upsilon = \frac{Ze^2}{2 \epsilon_0 nh}$$

In hydrogen

$$Z = 1, n = 1$$

$$\upsilon_H = \frac{Ze^2}{2 \epsilon_0 h} = \frac{(1.6 \times 10^{-19})^2}{2 \times 8.85 \times 10^{-12} \times 6.624 \times 10^{-34}}$$

$$= \mathbf{2.2 \times 10^{6}\ m/s}$$

Also

$$\upsilon = Z\upsilon_H$$

Therefore velocity of the electron in the orbit will increase with increase in atomic number Z.

Example 2.25. *Calculate the radius of the first Bohr orbit for (i) H and (ii) He atoms and also the velocity of the electron in these orbits as compared to the velocity of light.* [*I.A.S.*]

Solution. For hydrogen atom, the radius of the orbit is

$$r_H = \frac{\epsilon_0 n^2 h^2}{\pi m Z e^2}$$

Here

$$n = 1, Z = 1$$
$$\epsilon_0 = 8.85 \times 10^{-12}\ C^2/N\text{-}m^2$$
$$m = 9.1 \times 10^{-31}\ kg$$
$$e = 1.6 \times 10^{-19}\ C$$
$$h = 6.624 \times 10^{-34}\ J\text{-}s$$

$$\therefore \quad r_H = \frac{8.85 \times 10^{-12} \times 1 \times (6.624 \times 10^{-34})^2}{3.142 \times 9.1 \times 10^{-31} \times (1.6 \times 10^{-19})^2}$$

$$r_H = \mathbf{5.3 \times 10^{-11}\ m}$$

For helium, atom, $n = 1, Z = 2$

$$r_H = \frac{r_H}{Z} = \frac{5.30 \times 10^{-11}}{2}$$
$$= \mathbf{2.65 \times 10^{-11}\ m}$$

(*ii*) Velocity of the electron,

$$\upsilon = \frac{Ze^2}{2\epsilon_0\, nh}$$

For hydrogen $\quad Z = 1$ and $n = 1$

$$\upsilon_H = \frac{e^2}{2\epsilon_0\, h}$$
$$= \frac{(1.6 \times 10^{-19})^2}{2 \times 8.85 \times 10^{-12} \times 6.624 \times 10^{-34}}$$
$$= \mathbf{2.2 \times 10^6\ m/s}$$

Also, $\quad \dfrac{\upsilon_H}{c} = \dfrac{2.2 \times 10^6}{3 \times 10^8} = \mathbf{7.33 \times 10^{-3}}$

For helium $\quad \upsilon_{He} = \dfrac{Ze^2}{2\epsilon_0\, nh}$

Here $n = 1, Z = 2$

$$\upsilon_{He} = \frac{Ze^2}{2\epsilon_0\, h} = 2\upsilon_H$$
$$\upsilon_{He} = 2 \times 2.2 \times 10^6$$
$$= \mathbf{4.4 \times 10^6\ m/s}$$
$$\frac{\upsilon_{He}}{c} = \frac{4.4 \times 10^6}{3 \times 10^8}$$
$$= \mathbf{1.466 \times 10^{-2}}$$

Example 2.26. *Calculate the difference in wavelength in the spectra of hydrogen and heavy hydrogen corresponding to the first line on the long wave side of Balmer series.*

$$\left[\begin{array}{l} R_H = 1.097 \times 10^7 / m \\ m = 0.000549\, M_H \end{array}\right]$$ [*I.A.S.*]

Solution. $$R_H = \frac{R}{1 + \dfrac{m}{M_H}} \quad ...(i)$$

$$R_D = \frac{R}{1+\frac{m}{M_D}}$$

$$\frac{m}{M_H} = 0.000549$$

$$\frac{m}{M_D} = \frac{m}{2\,M_H} = 0.000274$$

$$\frac{R_D}{R_H} = \frac{1+\frac{M}{M_H}}{1+\frac{m}{M_D}} = \frac{1+0.000549}{1+0.000274}$$

$$\frac{R_D}{R_H} = 1.00027$$

$$R_D = (R_H)\,1.00027$$

$$R_D = 1.097 \times 10^7 \times 1.00027$$

$$R_D = 1.0973 \times 10^7/\text{m}$$

In the case of Balmer series, for the first member

$$n_1 = 2, n_2 = 3$$

$$\frac{1}{\lambda_H} = R_H\left[\frac{1}{n_1^2} - \frac{1}{n_2^2}\right] = \left(\frac{5}{36}\right) R_H$$

$$\lambda_H = \frac{36}{5R_H} = \frac{1}{5 \times 1.097 \times 10^7}$$

$$\lambda_H = 6.5633 \times 10^{-7}\ \text{m}$$

Similarly $$\frac{1}{\lambda_D} = R_D\left[\frac{1}{n_1^2} - \frac{1}{n_2^2}\right] = \left(\frac{5}{36}\right) R_D$$

$$\lambda_D = \frac{36}{5R_D}$$

$$\lambda_D = \frac{36}{5 \times 1.0973 \times 10^7}$$

$$\lambda_D = 6.5615 \times 10^{-7}\ \text{m}$$

$\therefore$ $$\lambda_D = \lambda_H - \lambda_D$$

$$= 6.5633 \times 10^{-7} - 6.5615 \times 10^{-7}$$

$$= 0.0018 \times 10^{-7}\ \text{m}$$

$$= \mathbf{1.8\ Å}$$

Example 2.27. *In hydrogen atom the electron is replaced by a muon whose mass is 200 times that of an electron and charge is same as that of electron, calculate the ionization on the basis of Bohr's theory.* *[I.A.S. 1989]*

Solution. In the case of hydrogen atom, having an electron, the ionization potential

$$\phi = \frac{me^4}{8\,\epsilon_0^2\,h^2} \quad ...(i)$$

when electron is replaced by muon.

$$m_1 = 200\ \text{m}$$

$$\phi_1 = \frac{m_1 e^4}{8 \epsilon_0^2 h^2} = \frac{(200\ m)e^4}{8 \epsilon_0^2 h^2} \quad ...(ii)$$

Dividing (*ii*) by (*i*)

$$\frac{\phi_1}{\phi} = 200$$

But $\phi = 13.6$ eV

$\therefore$ $\phi_1 = 200 \times 13.6$

$= 2.72 \times 10^3$ eV

Example 2.28. *What is the energy, momentum and wavelength of the photon emitted by a hydrogen atom when an electron makes a transition from n = 2 to n = 1 state ? Given ionization potential = 13.6 eV.*

Solution. Energy of electron in the first orbit of hydrogen atom,

$$E_1 = -13.6 \text{ eV}$$

Energy of electron in second orbit

$$E_2 = \frac{E_1}{n^2} = \frac{E_1}{4} = \frac{-13.6}{4}$$

$$E_2 = -3.4 \text{ eV}$$

(1) Energy of photon emitted = $E_2 - E_1$

$$E = -3.4 + (13.6)$$

$$= \mathbf{10.2\ eV}$$

$$= 10.2 \times 1.6 \times 10^{-19} \text{ J}$$

$$= 16.32 \times 10^{-19} \text{ J}$$

(2) Momentum, $P = \frac{E}{c} = \frac{16.32 \times 10^{-19}}{3 \times 10^8}$

$$P = 5.44 \times 10^{-27} \text{ kg-m/s}$$

(3) Wavelength, $\lambda = \frac{h}{p} = \frac{6.624 \times 10^{-34}}{5.44 \times 10^{-27}}$

$$= 1.218 \times 10^{-7} \text{ m}$$

$$= \mathbf{1218\ Å.}$$

Example 2.29. *Calculate the energy required to create a vacancy in*

(i) K shell of copper atom

(ii) L shell of copper atom

Given, Ionization potential of hydrogen atom = 13.6 eV

For copper Z = 29.

Solution. (*i*) The energy required to create a vacancy in the first orbit (n = 1) *i.e.*, K shell of copper atom.

$$E_1 = \frac{(Z^2) \times 13.6}{n^2}$$

$$= \frac{(29)^2 \times 13.6}{1}$$

$$= \mathbf{11.44 \times 10^3\ eV}$$

(*ii*) Energy required to create a vacancy in the second orbit ($n = 2$) *i.e.*, L shell of copper atom

$$E_2 = \frac{(29)^2 \times 13.6}{2^2}$$

$$E_2 = \mathbf{2.86 \times 10^3\ eV}$$

Example 2.30. *The radiations given off by the Hg atoms returning to their normal states were studied by Frank and Hertz and a line was observed at 2537 Å. Calculate excitation potential for Hg.* [*Mumbai, 1991*]

Solution. Here $\lambda = 2537\text{ Å} = 2537 \times 10^{-10}$ m

Let excitation potential be V.

$$\therefore \quad eV = h\nu = \frac{hc}{\lambda}$$

$$\therefore \quad V = \frac{hc}{e\lambda}$$

Here

$$h = 6.624 \times 10^{-34}\text{ J-s}$$

$$c = 3 \times 10^8\text{ m/s}$$

$$e = 1.6 \times 10^{-19}\text{ C}$$

$$\lambda = 2537 \times 10^{-19}\text{ m}$$

$$\therefore \quad V = \frac{6.624 \times 10^{-34} \times 3 \times 10^8}{1.6 \times 10^{-19} \times 2537 \times 10^{-10}}$$

$$V = \mathbf{4.9\ volts.}$$

Example 2.31. *An X-ray line of wave length 0.53832 Å is found to be emitted from an X-ray tube with Zinc ($Z = 30$) target in addition to the characteristic K_α line of Zinc of wavelength 1.43603 Å. If the unknown line is due to an impurity in the target, obtain the atomic number of impurity.* [*Kolkata, 1992*]

Solution. Here

$$\lambda_1 = 1.43603\text{ Å}$$

$$Z_1 = 30$$

$$\lambda_2 = 0.53832\text{ Å}$$

$$Z_2 = ?$$

$$\frac{1}{\lambda_1} \propto Z_1^2$$

and

$$\frac{1}{\lambda_1} \propto Z_2^2$$

$$\therefore \quad \left(\frac{Z_2}{Z_1}\right)^2 = \left(\frac{\lambda_1}{\lambda_2}\right)$$

$$Z_2 = Z_1\left[\frac{\lambda_1}{\lambda_2}\right]^{\frac{1}{2}}$$

$$Z_2 = 30\left[\frac{1.43603}{0.53832}\right]^{\frac{1}{2}}$$

$$Z_2 = \mathbf{49.}$$

Example 2.32. *In a μ-mesonic atom, a muon with a charge -e and mass 200 m_e moves in a circular orbit around the nucleus of charge + 3e. Assuming that Bohr's model of atom is applicable to this system.*

(i) Derive the expression for the radius of n th Bohr orbit.

(ii) Find the value of n for which the radius of the orbit is approximately the same as that of Bohr's first orbit of hydrogen atom. Bohr radius = 0.53 × 10^{-10} m.

(iii) Calculate the first excitation potential of the atom. [Kolkata, 1991]

Solution. (*i*) Here

$$Z = 3$$

$$m = 200\ me$$

Here

$$\frac{m\upsilon^2}{r} = \frac{Ze^2}{4\pi \in_0 r^2}$$

$$m\upsilon^2 = \frac{Ze^2}{4\pi \in_0 r} \qquad ...(i)$$

Also

$$m\upsilon r = \left(\frac{nh}{2\pi}\right) \qquad ...(ii)$$

∴

$$\upsilon = \left(\frac{nh}{2\pi\, mr}\right)$$

∴

$$m\left(\frac{nh}{2\pi\, mr}\right)^2 = \frac{Ze^2}{4\pi \in_0 r}$$

$$r = \frac{\in_0 n^2 h^2}{\pi\, m\, Ze^2} \qquad ...(iii)$$

(*ii*)

$$r = \frac{\in_0 n^2 h^2}{\pi\, m\, Ze^2}$$

Here

$$m = 3$$

$$r = \frac{\in_0 n^2 h^2}{\pi \times 200\, m_e \times 3\, e^2}$$

$$r = \frac{n^2}{600}\left[\frac{\in_0 h^2}{\pi m_e e^2}\right]$$

For hydrogen atom, for the first Bohr orbit

$$r_H = \left(\frac{\in_0 h^2}{\pi\, m_e\, e^4}\right) = 0.53 \times 10^{-10}\ \text{m}$$

$$r = \left(\frac{n^2}{600}\right) r_{\text{H}}$$

But

$$r = r_{\text{H}}\ \text{(given)}$$

∴

$$\frac{n^2}{600} = 1,\ n_2 = 600$$

$$n = 24$$

(*iii*)

$$h\nu = \frac{me^4 Z^2}{8 \in_0^2 h^2}\left(\frac{1}{n_1^2} - \frac{1}{n_2^2}\right)$$

Here

$$n_1 = 1,\ n_2 = 2$$

$$U = \frac{200 m_e\, e^4 (3^2)}{8 \in_0^2 h^2}\left(\frac{1}{1} - \frac{1}{4}\right)$$

$$U = \left(\frac{1800 \times 3}{4}\right)\left(\frac{m_e e^4}{8 \epsilon_0^2 h^2}\right)$$

Ex. 2.33. Show that the velocity of the electron in the first orbit of hydrogen atom is $(1/137)c$ where c is the velocity of light.

Solution.

$$m_e v r_n = n \hbar \text{ where } \hbar = h/2\pi$$

and
$$r_n = \frac{n^2\hbar^2}{Km_e e^2} \quad \therefore \quad v = \frac{n\hbar}{m_e r_n}$$

or
$$v_n = \frac{n\hbar}{m_e} \times \frac{Km_e e^2}{n^2\hbar^2} = \frac{ke^2}{n^2\hbar^2}$$

$$= \frac{2\pi e^2 \times K}{nh} = \frac{2 \times 3.142\,(1.6 \times 10^{-19})^2 \times 9 \times 10^9}{6.6 \times 10^{-34}}$$

$\therefore$ $n = 1$

$\therefore$ $v_1 = 2.19 \times 10^6 \text{m/s}.$

$\therefore$
$$\frac{v_1}{c} = \frac{2.19 \times 10^6}{3 \times 10^8} = \frac{0.73}{100} = \frac{1}{137}$$

2.18. Limitations of the Bohr's Theory

In spite of the extraordinary success of the Bohr's theory there were some serious limitations :

(1) It could not account for the spectra of complex atoms.

(2) Even simple atoms as neutral helium having more than one electron could not be dealt with by the Bohr's theory.

(3) Intensities of lines could not be calculated.

(4) It failed to give correct results in the complicated conditions of the splitting of lines in the presence of a magnetic field, known as the anomalous Zeeman effect.

(5) It did not account for the fine structure of the hydrogen lines and it did not account for the doublet structure of the alkali metal spectra.

EXERCISE II

1. What is a black body ?
2. Discuss briefly the inadequacy of classical mechanics.
3. Discuss the distribution of energy in the spectrum of a black body.
4. Derive Planck's radiation law.
5. Give briefly the Ritz combination principle in explaining the spectra of an atom.
6. Describe Einstein's photoelectric equation.
7. How was Millikan able to verify Einstein's photoelectric equation experimentally ?
8. How will you determine experimentally Planck's constant by cut off potential method ?
9. On the basis of Bohr's theory, deduce expressions for the radius, frequency and energy of an orbit of hydrogen atom.
10. What is meant by ionization potential ? Calculate its value in the case of hydrogen atom.
11. State and explain the fundamental postulates of Bohr's theory of the hydrogen atom. Derive

an expression for the energy of an electron in the *n*th orbit. Explain the various series in the hydrogen spectrum on the basis of Bohr's theory.

12. State Bohr's postulates for the theory of the hydrogen atom. Derive an expression for the energy of this atom when the electron is in the nth orbit. What is Binding energy ?

13. Explain the hydrogen spectrum on the basis of Bohr's theory.

14. Outline Bohr's theory of the hydrogen atom. What is the justification for assuming quantised stationary orbit ? Explain the formation of Balmer lines.

15. What are the limitations of the Bohr's theory of the hydrogen atom ?

16. Calculate the velocity of the electron in the first orbit of the hydrogen atom.

17. The wavelength of the photoelectric threshold for tungsten is 230 nm. Determine the energy of the electrons ejected from the surface by ultra-violet light of wavelength 180 nm.

18. Calculate the energy of a photon of wavelength 5890 Å. Express the result in (*i*) Joules and (*ii*) eV.

19. The longest wavelength of the incident radiation that can eject electrons from platinum is 1972 Å. Calculate the work function of platinum in electron volts.

20. Calculate the number of photons from the D_1 line of sodium ($\lambda = 5896 \times 10^{-10}$ m) required to do 6.624 joules of work. $h = 6.624 \times 10^{-34}$ Joules-second.

21. The wavelength of light falling on the surface of a metal of work function 2.3 eV is 4300 Å. With what velocity will the electons be emitted ?

22. The stopping potential is 4.6 volts for light of frequency 2×10^{15} Hz. When light of frequency 4×10^{15} Hz is used, the stopping potential is 12.9 volts. Calculate the value of Planck's constant.

23. When light of $\lambda = 4.0 \times 10^{-7}$ m falls on a sodium surface, the maximum energy of the emitted electrons is 0.8 eV. What is the work function of Na ?

24. The wavelength of the photoelectric threshold for silver is 3250 Å. Determine the maximum energy of the ejected electrons from a silver surface by light of wavelength 2537 Å.

25. The most energetic electrons emitted from a surface by 3500 Å photons are found to be of 1.5×10^{-15} Newton/m^2. Calcualte the work function of the material.

26. The wavelength of the second member of Balmer series of hydrogen is 4861 Å. Calculate the wavelength of its first member.

27. Calculate the wavelength of the shortest and longest wavelengths of the Balmer series of the hydrogen atom. $R = 1.097 \times 10^7$/m.

28. What are the series limit of Lyman, Paschen, Brackett and Pfund Series ?

$$R = 1.097 \times 10^7 \text{ m}^{-1}$$

3

MATTER WAVES

Wave-Particle Duality and Uncertainty Principle

3.1 Inadequacy of Classical Mechanics

Prior to 1900 most of the phenomena involving macroscopic systems could be explained satisfactorily with the accepted laws for classical Physics. In most of the cases the effects with which the classical theory is associated are either directly observable or capable of observation with the help of simple instruments. There is also a close link between the world of sensory perception and the realm of classical Physics.

With the discovery of an electron as a fundamental particle by J.J. Thomson in 1897, most of the studies in Physics concentrated to explore microscopic or atomic systems which are beyond the scope of direct observations. Many difficulties were encountered with phenomena viz. spectral distribution of heat radiations from black bodies, specific beats of solids at low temperature etc.

These difficulties were overcome by Planck in 1900 who suggested that the black body radiation (emission or absorption of electromagnetic radiations) takes place as discrete quanta and the energy equivalent of each quantum, $E = h\nu$.

The quantum concept was also used by Einstein to explain the observed photoelectric effect. Thus a dual character of electromagnetic radiation that it behaves as waves and particles became evident.

The concept of discrete energy levels has become very fundamental in the field of atomic and nuclear physics and it is a consequence of planck's hypothesis of quantam of energy.

3.2 Two Slit Experiment-Superposition Principle

A fine beam of electrons produces interference and diffraction patterns. For studying the two

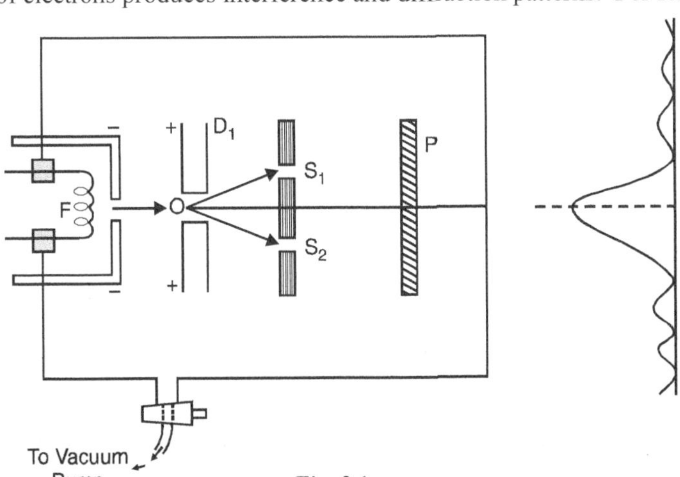

Fig. 3.1

slit interference pattern with electrons, the experimental arrangement consists of an electron gun and two slits (S_1 and S_2) (Fig. 3.1). The electrons emitted from the filament F are converged into a fine beam and are accelerated towards the diaphragm D_1 having a fine hole and maintained at a positive potential. The beam of electrons emerging out at O is incident on the slits S_1 and S_2. Any point on the photographic plate P receives electrons from both the slits S_1 and S_2. Similar to light waves, superposition of the two beams of electrons from S_1 and S_2 takes place and the distribution of intensity on the photographic plate shows an interference pattern with maxima and minima. The intensity distribution is shown in the Fig. 3.1.

Similarly employing a suitable experimental technique, the diffraction of electrons at a straight edge was studied and the distribution of intensity was found to be similar to that of ligh waves diffraction phenomenon.

These experiments clearly show that electrons behave as matter waves just similar to the light waves. The wavelength of the electron calculated from these experiments is the same as the value calculated from De-Broglie equation

$$\lambda = \frac{h}{m\upsilon} \quad ...(3.1)$$

3.3 Wave-Particle Dualism for Light and Matter

The earliest theory for explaining the nature of light radiations was the corpuscular theory of Newton. However, Huygen's wave theory could expalin satisfactorily the phenomena of interference, diffraction, double refraction and polarization. The quantum theory of radiation was helpful in explaining the phenomenon of photoelectric effect.

Bohr's quantization principle to explain the atom model and the spectral series of hydrogen was very satisfactory. However, Bohr's quantization theory had to be modified by Sommerfeld and other to explain the full structure of hydrogen spectrum. Though the theory was sucessful in case of hydrogen, it could explain satisfactorily the experimentally observed energy levels of helium or any atom with more than two electrons. There were other inconsistencies also to correlate the experiment and theory.

Einstein suggested that the waves merely serve as guides for photons carrying the energy $h\nu$ for a frequency ν. This suggestion, however brought a new and difficult concept to explain the various experimental data.

De-Broglie, in 1929, suggested that the material particles like photons should be associated with same form of wave motion. De-Broglie hypothesis suggests that such waves existed and he called them matter waves. This point of view was used to explain quantization principles employed in explaining the hydrogen atom. This was further verified by Davisson and Germer through their electron diffraction experiments. The experimental results positively conclude the existence of a system of waves associated with an electron beam. This new theory of the coexistence of *waves* and *particles* was fully developed subsequently by Schrodinger, Heisenburg, Dirac and Maxborn. This led to the foundation of *wave mechanics.*

De Broglie Waves

A photon of light of frequency υ has a momentum $p = \frac{h\nu}{c} = \frac{h}{\lambda}$ since $\lambda\nu = c$. Now, the wavelength of a photon is specified by the momentum by the relation

$$\lambda = \frac{h}{p}, \text{ where } h \text{ is Planck's constant.}$$

De Broglie suggested that the equation for the wavelength $\lambda = \frac{h}{p}$ should also be true for material particles. Therefore a particle of mass m and velocity v the momentum is mv and its **De-Broglie wavelength**

$$\lambda = \frac{h}{mv}.$$

3.4 De-Broglie Model of the Atom

The dual behaviour of photons inspired De-Broglie to suggest a model of the atom. The experiments on interference, diffraction and polarisation established the wave nature of light. The experiments on photoelectric effect established the corpuscular nature of light. Einstein's equation $E = mc^2$ where E is the energy, m is the mass and c is the velocity of light, gives the relation of conversion of mass into energy. De-Broglie noted that the corpuscular properties were more obvious for very energetic light *i.e.*, light of shorter wavelength. Thus, the light particle known as a *photon* has corpuscular as well as wave nature.

Thinking on the same lines, De-Broglie suggested that particles like electron have also a wave nature.

$$E = mc^2 = \frac{m_0c^2}{\sqrt{1-\left[\frac{\upsilon^2}{c^2}\right]}} \qquad ...(3.2)$$

Here m_0 is the rest mass.

The energy of a wave of frequency ν according to Planck is

$$E = h\nu \qquad ...(3.3)$$

From (3.1) and (3.2)

$$h\nu = mc^2 = \frac{m_0c^2}{\sqrt{1-\left[\frac{\upsilon^2}{c^2}\right]}}$$

But

$$\nu\lambda = c$$

$\therefore$

$$\frac{hc}{\lambda} = mc^2$$

or

$$\lambda = \frac{h}{mc} \qquad ..(3.4)$$

For an electron of velocity υ

$$\lambda = \frac{h}{m\upsilon} = \frac{h\sqrt{1-\left(\frac{\upsilon^2}{c^2}\right)}}{m_0\upsilon} \qquad ...(3.5)$$

For electrons, the wavelength is longest, while for proton it is smaller than the electrons. If the values of h, m and υ are substitued in equation (3.5), λ is found to be very small.

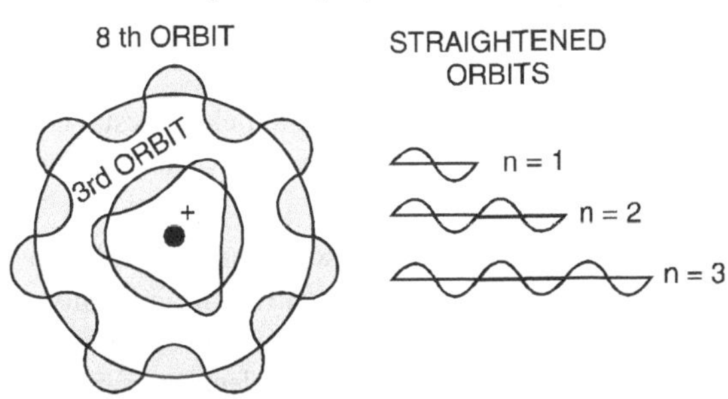

Fig. 3.2. De-Broglie Model of Atom.

Thus De-Broglie model of an atom has electron in various orbits and the electrons behave as matter waves of wavelength $\lambda = \frac{h}{m\upsilon}$. The electron exists as a standing wave in each orbit. The energy levels and 'orbits' of the Bohr model are retained. Moreover, this model explains the Bohr's postulate. The electron can be only in those orbits whose cirumference can contain the complete waves of the electron or the length of the orbit is a whole number multiple of the wavelength (Fig. 3.3).

The electron cannot be in an orbit whose length is not a whole number multiple of the wavelength. The necessary condition for the electron in a particular orbit can be calculated mathematically. The length of the orbit of radius $r = 2\pi r$ and $2\pi r = n\lambda$, where n is a whole number.

But $$\lambda = \frac{h}{m\upsilon} \qquad \text{[from equation (3.4)]}$$

$\therefore$ $$2\pi r = \frac{nh}{m\upsilon}$$

or $$m\upsilon r = \frac{nh}{2\pi} \qquad ...(3.6)$$

Equation (3.6) is in agreement with the Bohr's quantization hypothesis. Thus the De-Broglie model seems to be more exact and the electrons are matter waves in various orbits round the nucleus. Davisson and Germer experiment on the diffraction of electrons demonstrated the fact that the material particles like electrons exhibit a wavelength given by the De-Broglie equation, $\lambda = h/m\upsilon$.

3.5 Davisson-Germer Experiment

Davisson and Germer performed an experiment to study the wave-nature of electrons as suggested by De-Broglie. The experimental arrangement is shown in Fig. 3.3. Electrons from a filament F are rendered into a fine beam by applying a positive potential to the cylinder D.

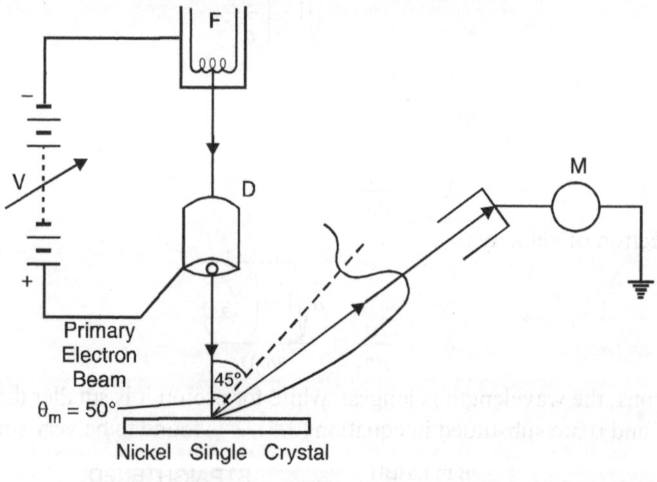

Fig. 3.3.

A fine narrow beam of electrons is incident on the nickel crystal C. The electrons are reflected and the intensity of the reflected electrons is measured at various angles with the help of a current meter M.

Different positive potential can be applied to the cylinder D and the velocity of the electrons can be varied. The curve represents a polar graph indicating the intensity of the reflected electrons at various angles. It is found that the intensity is maximum at 50° a critical energy of 54 eV. By altering the potential applied to the cylinder D, the energy of the electrons can be changed. It is found that the nature of the intensity graph is similar and the spur (maximum intensity) occurs always at 50°.

The intensity of the electrons scattered in any given direction is found by allowing the electrons to enter a chamber, set to an appropriate position and measuring the deflection produced by a meter connected to the chamber.

Results : When the difference of potential is 40 volts between F and D, the curve is fairly smooth. At 44 volts a distinct spur appears on the curve at an angle of 60 degrees. With increasing voltage the length of the spur increases unit it reaches a maximum at 54 volts at an angle of 50 degrees. After this the spur decreases and disappears completely at 68 volts.

The selective reflection of the '54 volt' electrons at 50 degrees angle can be explained due to the constructive interference. The waves from the successive layers of the crystal reinforce each other. Hence we conclude that waves must be associated with the electrons. The spur in its prominent stage is **conclusive evidence** that waves are associated with electrons and with matter in general.

From the De-Broglie relation the wavelength of 54-V electrons should be $\lambda_e = \dfrac{12.27}{\sqrt{54}} = 1.67$ Å.

Using Bragg's law we find the observed wavelength to be

$\lambda_e = D \sin \theta_m = 2.15 \sin 50° = 1.65$ Å $[D = 2.15$ Å$]$

The two values of λ_e are in excellent agreement.

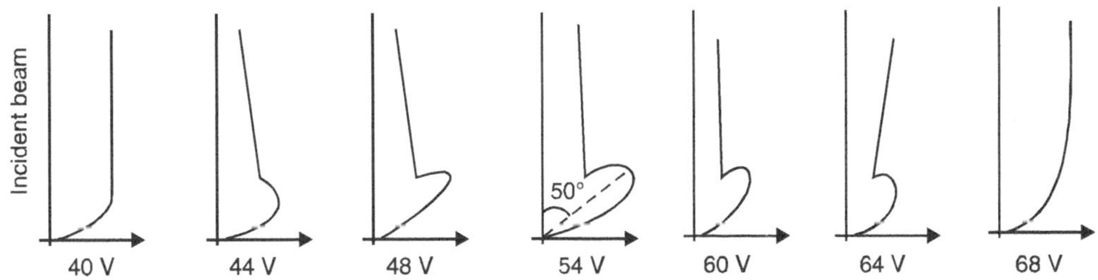

Fig. 3.4

The results obtained are similar to X-ray diffraction experiments. This experiments shows that the electrons behave like matter-waves as suggested by De-Broglie.

3.6 G.P. Thomson Experiment

The experimental arrangement is shown in Fig. 3.5.

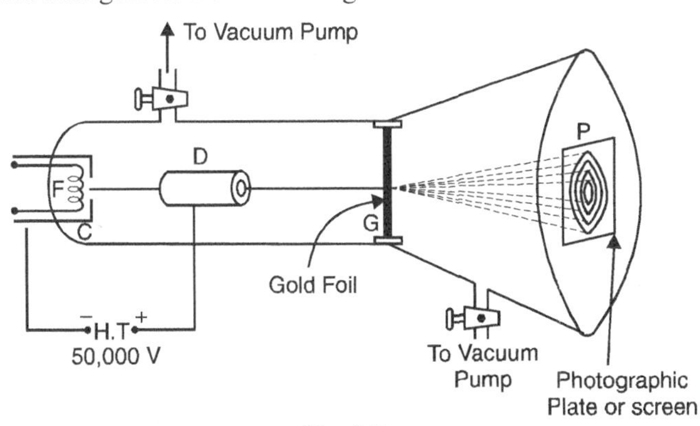

Fig. 3.5

The apparatus is evacuated with the help of evacuation pumps. The electrons from the filament *F* are accelerated through a potential difference of 50,000 volts by applying a positive potential to the *cylinder D*. The energy of each electron is 50,000 eV. This beam of electrons is incident on a thin gold foil *G* and after passing through the foil, it is incident on a photographic plate *P*. The pattern obtained in the photographic plate consists of concentric circular rings of varying intensity.

If the electrons behave only as corpuscles, the electrons passing through the foil should have been scattered through a wide angle. But the pattern obtained is similar to the Laue-pattern obtained with X-rays, which is possible only in the case of waves. Hence electrons also behave as waves.

3.7 Wave Velocity and Group Velocity

The equation for a plane progressive wave is given by

$$y = a \sin 2\pi \left[\frac{t}{T} - \frac{x}{\lambda}\right]$$

$$y = a \sin(\omega t - kx),$$

where $\omega = \frac{2\pi}{T} = 2\pi n$

and $k = \frac{2\pi}{\lambda}$.

Here ω is called the angular frequency of the wave and k is called the wave vector.

Also $\frac{\omega}{k} = \frac{2\pi n}{2\pi/\lambda} = n\lambda = \upsilon_p$

Here, υ_p is called the phase velocity or the wave velocity. In other words, $\upsilon_p = \frac{\omega}{k}$ is the velocity with which a plane progressive wavefront travels forward. It has a constant phase = $(\omega t - kx)$

$$\omega t - kx = \text{constant}$$

Differentiating this equation with respect to t and x,

$$\omega - k\frac{dx}{dt} = 0$$

or $\frac{dx}{dt} = \frac{\omega}{k} = \upsilon_p$ ...(3.7)

Group velocity. Consider a group of progressive waves with angular frequencies in the range ω

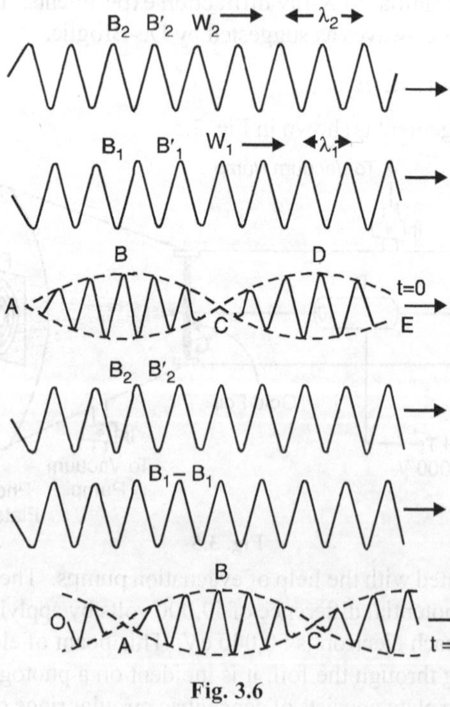

Fig. 3.6

and $\omega + \Delta\omega$ and wave vectors in the range k and $k + \Delta k$ superimposing on one another. The resultant wave pattern consists of points of maximum amplitude and points of minimum amplitude. In between any two consecutive maximum and minimum, the amplitude gradually decreases. Between any two consecutive minima, say A and C, there is a position of maximum amplitude midway between A and C *i.e.*, at B. The dotted loop represents a group of waves. The group of waves is called a packet and with time the packet moves forward in the medium with a velocity, called the group velocity. (Fig. 3.6).

Consider two plane progressive waves travelling in the positive direction of the X-axis. The equations of the two waves at any instant are,

$$y_1 = a_1 \sin(\omega_1 t - k_1 x) \quad ...(3.8)$$

and

$$y_2 = a_2 \sin(\omega_2 t - k_2 x) \quad ...(3.9)$$

Here

$$k_1 = \frac{2\pi}{\lambda_1} \quad ...(3.10)$$

and

$$k_2 = \frac{2\pi}{\lambda_2} \quad ...(3.11)$$

The wave velocities for the two waves are

$$\upsilon_1 = \frac{\omega_1}{k_1}$$

and

$$\upsilon_2 = \frac{\omega_2}{k_2}.$$

Initially the two peaks B_1 and B_2 are at the same position. Here $\lambda_2 > \lambda_1$ and $\upsilon_1 > \upsilon_2$. By the time the peak at B_2 travels to B_2, the difference in distance between the position of B_2' and B_1'

$$= (\lambda_2 - \lambda_1)$$

The difference in velocity between the two waves $= \upsilon_1 - \upsilon_2$. The time takes by B_1' to catch up with B_2' is given by

$$T = \frac{\lambda_2 - \lambda_1}{\upsilon_1 - \upsilon_2} \quad ...(3.12)$$

In this time T, the crest at B_1 moves forward a distance $= \upsilon_1 T$. But the point at which the two waves will reinforce moves along the first wave from B_1 to B_2' *i.e.*, a distance $= \lambda_1$.

Therefore, the total distance traversed by the point of reinforcement

$$= \upsilon_1 T + \lambda_1$$

$\therefore$ Group velocity,

$$\upsilon_g = \frac{\upsilon_1 T + \lambda_1}{T} \quad ...(3.13)$$

or

$$\upsilon_g = \upsilon_1 + \frac{\lambda_1}{T}$$

or

$$\upsilon_g = \upsilon_1 + \left[\frac{\lambda_1(\upsilon_1 - \upsilon_2)}{\lambda_2 - \lambda_1}\right]$$

$$\upsilon_g = \frac{\upsilon_1\lambda_2 - \upsilon_1\lambda_1 + \upsilon_1\lambda_1 - \upsilon_2\lambda_1}{(\lambda_2 - \lambda_1)}$$

$$\upsilon_g = \frac{\upsilon_1\lambda_2 - \upsilon_2\lambda_1}{(\lambda_2 - \lambda_1)}$$

Dividing the numerator and denominator by $\lambda_1 \lambda_2$,

$$\upsilon_g = \frac{\frac{\upsilon_1}{\lambda_1} - \frac{\upsilon_2}{\lambda_2}}{\frac{1}{\lambda_1} - \frac{1}{\lambda_2}}$$

Multiplying the numerator and denominator by 2π

$$\upsilon_g = \frac{\left(\frac{2\pi}{\lambda_1}\right)\upsilon_1 - \left(\frac{2\pi}{\lambda_2}\right)\upsilon_2}{\left(\frac{2\pi}{\lambda_1}\right) - \left(\frac{2\pi}{\lambda_2}\right)}$$

or

$$\upsilon_g = \frac{k_1\upsilon_1 - k_2\upsilon_2}{k_1 - k_2} \quad \text{...(3.14)}$$

$$\upsilon_1 = \frac{\omega_1}{k_1} \text{ and } \upsilon^2 = \frac{\omega_2}{k_2}$$

$$\therefore \quad \upsilon_g = \frac{\omega_1 - \omega_2}{k_1 - k_2} \quad \text{...(3.15)}$$

If the difference in angular frequency and difference in wave vector is extremely small,

$$\upsilon_g = \frac{d\omega}{dk} \quad \text{...(3.16)}$$

The group velocity υ_g is the speed with which the point of reinforcement of the resultant wave travels. The group velocity is different from the individual wave velocity. υ_g can be higher or lower than υ_p depending upon the relative direction of the individual waves. The modulated wave due to the superimposition will be in the form of wave packets as shown in Fig. 3.7. The group of waves packet in each dotted envelope travels with the group velocity and not with wave velocity.

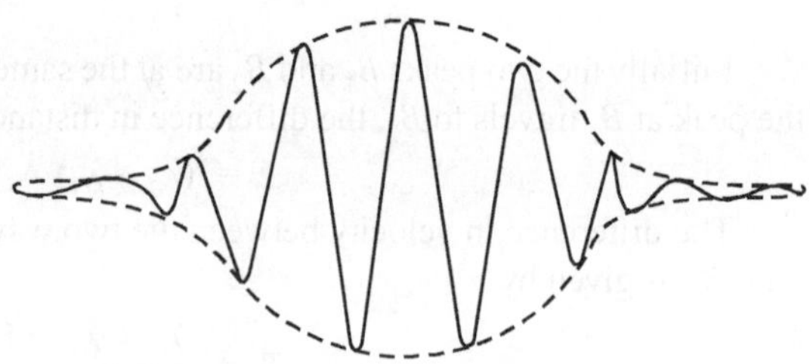

Fig. 3.7

The angular frequency ω is given by

$$\omega = 2\pi\, v = \frac{2\pi\, E}{h}$$

But

$$E = mc^2$$

$$\omega = \frac{2\pi\, mc^2}{h} \quad \text{...(3.17)}$$

$$m = \frac{m_0}{\left[\sqrt{1 - \frac{\upsilon^2}{c^2}}\right]}$$

$$\omega = \frac{2\pi\, c^2}{h}\left[\frac{m_0}{\left[\sqrt{1 - \frac{\upsilon^2}{c^2}}\right]}\right] \quad \text{...(3.18)}$$

The propagation constant k is given by

$$k = \frac{2\pi}{\lambda}$$

$$\lambda = \frac{h}{m\upsilon}$$

$$\therefore \qquad k = \frac{2\pi\, m\upsilon}{h}$$

$$k = \frac{2\pi\,\upsilon}{h}\left[\frac{m_0}{\sqrt{1-\frac{\upsilon^2}{c^2}}}\right] \qquad ...(3.19)$$

The group velocity

$$\upsilon_g = \frac{d\omega}{dk} = \frac{d\omega/d\upsilon}{dk/d\upsilon} \qquad ...(3.20)$$

Differentiating equation (3.18)

$$\frac{d\omega}{d\upsilon} = \frac{2\pi\, m_0}{h}\left[\frac{\upsilon}{\left[1-\frac{\upsilon^2}{c^2}\right]^{3/2}}\right] \qquad ...(3.21)$$

Differentiating equation (3.19)

$$\frac{dk}{d\upsilon} = \frac{2\pi\, m_0}{h}\left[\frac{1}{\left[1-\frac{\upsilon^2}{c^2}\right]^{\frac{1}{2}}} + \frac{\upsilon^2}{c^2\left[1-\frac{\upsilon^2}{c^2}\right]^{\frac{3}{2}}}\right]$$

$$= \frac{2\pi\, m_0}{h}\left[\frac{1}{\left[1-\frac{\upsilon^2}{c^2}\right]^{\frac{3}{2}}}\right] \qquad ...(3.22)$$

Substituting the values of $\frac{d\omega}{d\upsilon}$ and $\frac{dk}{d\upsilon}$ in equation (3.20)

$$\upsilon_g = \upsilon$$

Hence υ represents the particle velocity υ_g.

$$\therefore \qquad \upsilon_g = \upsilon_p \qquad ...(3.23)$$

It is clear from equation (3.23) that the electron can behave as a particle and as a wave at the same time.

The wave properties are indicated by the structure of the wave packet. The properties of the particle are determined by the position of the envelope of the wave packet. It is important to note the significant velocity in the description of a particle is the group velocity υ_g.

Therefore, the De-Broglie wave packet associated with a moving body (particle) travels with the same velocity as that of the body. Thus the particle inside a wave packet can be considered to be belonging to the entire packet travelling with the particle velocity υ_p.

Example 3.1. *If the De-Broglie wavelength of an electron is equal to 9×10^{-10} m, calculate its kinetic energy.* [*Delhi*]

Solution. Here

$$E = \frac{1}{2}m\upsilon^2 \text{ or } \upsilon = \sqrt{\frac{2E}{m}}$$

$$\lambda = \frac{h}{m\upsilon}$$

$$\lambda = \frac{h}{m\sqrt{\frac{2E}{m}}} = \frac{h}{[2mE]^{1/2}}$$

or $$E = \frac{h^2}{2m\lambda^2}$$

Here $h = 6.6 \times 10^{-34}$ J-s, m = 9.1×10^{-31} kg, $\lambda = 9 \times 10^{-10}$ m

$$\therefore \quad E = \frac{(6.6 \times 10^{-34})^2}{2 \times 9.1 \times 10^{-31} \times (9 \times 10^{-10})^2}$$

$$E = 2.955 \times 10^{-19} \text{ J}$$

$$E = \mathbf{1.8467 \text{ eV.}}$$

Example 3.2. *Calculate the De-Broglie wavelength of a beam of electrons whose energy is 100 eV. Take h = 6.6 × 10⁻³⁴ Joule-second and*

Solution. $$m = 9.1 \times 10^{-31} \text{ kg.}$$

Energy $$E = \frac{1}{2}m\upsilon^2 = 100 \text{ eV} = 100 \times 1.6 \times 10^{-19} \text{ Joule}$$

$$\upsilon^2 = \frac{2E}{m} \text{ or } \upsilon = \sqrt{\frac{2E}{m}}$$

Momentum, $$m\upsilon = m\sqrt{\frac{2E}{m}} = \sqrt{2mE}$$

$$\lambda = \frac{h}{m\upsilon} = \frac{h}{\sqrt{2mE}} \text{ metres}$$

Hence, $$h = 6.6 \times 10^{-34} \text{ kg}$$

$$\therefore \quad \lambda = \frac{6.6 \times 10^{-34}}{(2 \times 9.1 \times 10^{-31} \times 100 \times 1.6 \times 10^{-19})^{1/2}}$$

$$= 1.23 \times 10^{-19} \text{ metre}$$

$$= \mathbf{1.23 \text{ Å}}$$

Example 3.3. *What is the momentum of a photon of wavelength 6×10^{-7} m ?*

Solution.

$$\text{Momentum of photon} = \frac{hv}{c} = \frac{h}{c/v}$$

$$= \frac{h}{\lambda}$$

$$= \frac{6.624 \times 10^{-34}}{6 \times 10^{-7}}$$

$$= \mathbf{1.104 \times 10^{-27} \text{ kg-m/s.}}$$

Example 3.4. *Deduce the momentum of an electron having kinetic energy 1 Be V.* [*I.A.S.*]

Solution. $$E = \frac{p^2}{2m}$$

or $$p = \sqrt{2mE}$$

Here $$m = 9.1 \times 10^{-31} \text{ kg}$$

$$E = 1 \text{ BeV} = 10^9 \text{ eV} = 10^9 \times 1.6 \times 10^{-19} \text{ J}$$

$$E = 1.6 \times 10^{-10}\text{ J}$$
$$p = [2 \times 9.1 \times 10^{-31} \times 1.6 \times 10^{-10}]^{1/2}$$
$$= \mathbf{1.7 \times 10^{-20}\ kg\text{-}m/s.}$$

Example 3.5. *A particle of mass 0.51* MeV/c^2 *has kinetic energy 100* eV. *Find its De-Broglie wavelength.* [*I.A.S.*]

Solution.
$$U = 100\text{ eV}$$
$$= 100 \times 1.6 \times 10^{-19}\text{ J}$$
$$= 1.6 \times 10^{-17}\text{ J}$$
$$\text{mass, } m = \frac{0.51\text{ MeV}}{c^2} = \frac{0.51 \times 10^6 \times 1.6 \times 10^{-19}}{(3 \times 10^8)^2}$$
$$= 9 \times 10^{-31}\text{ kg}$$
$$U = \frac{1}{2}m\upsilon^2$$
$$\upsilon = \sqrt{\frac{2U}{m}}$$
$$\lambda = \frac{h}{m\upsilon} = \frac{h}{m\sqrt{2U/m}}$$
$$\lambda = \frac{h}{\sqrt{2mU}}$$
$$\lambda = \frac{6.624 \times 10^{-34}}{[2.9 \times 10^{-31}] \times 1.6 \times 10^{-17}]^{1/2}}$$
$$= \mathbf{1.234 \times 10^{-10}\ m = 1.234\ \text{Å}.}$$

Example 3.6. *Compare the energy of a photon with that of a neutron when both are associated with wavelength 1 Å. Mass of neutron = 1.67 × 10^{-27} kg.* [*I.A.S. 1987*]

$$1\text{Å} = 10^{-10}\text{ m}$$

Solution. (*i*) Energy of Photon

$$U_1 = hv = \frac{hc}{\lambda}$$
$$U_1 = \frac{6.624 \times 10^{-34} \times 3 \times 10^8}{10^{-10}}$$
$$U_1 = \mathbf{19.872 \times 10^{-16}\ J}$$

(*i*) Energy of neutron,

$$\lambda = \frac{h}{m\upsilon} \quad \text{or} \quad v = \frac{h}{m\lambda}$$
$$U_2 = \frac{1}{2}m\upsilon^2$$
$$U_2 = \frac{h^2}{2m\lambda^2}$$
$$U_2 = \frac{(6.624 \times 10^{-34})^2}{2 \times 1.67 \times 10^{-27} \times (10^{-10})^2}$$
$$U_2 = 13.14 \times 10^{-21}\text{ J}$$

$$\frac{U_1}{U_2} = \frac{19.872\times10^{-16}}{13.14\times10^{-21}}$$
$$= \mathbf{1.51 \times 10^5}$$

Example 3.7. *Photoelectrons are liberated by ultraviolet light of wavelength 3000 Å from a metallic surface for which the photoelectric threshold is 4000 Å. Calculate the De-Broglie wavelength of electrons emitted with maximum kinetic energy.*

Solution. Here,
$$\lambda = 3000 \text{ Å} = 3 \times 10^{-7} \text{ m}$$
$$\lambda = 4000 \text{ Å} = 4 \times 10^{-7} \text{ m}$$
$$hv - hv_0 = U$$
$$\frac{hc}{\lambda} - \frac{hc}{\lambda_0} = U$$
$$U = hc\left[\frac{1}{\lambda} - \frac{1}{\lambda_0}\right]$$
$$U = 6.624 \times 10^{-34} \times 3 \times 10^8 \left[\frac{1}{3\times10^{-7}} - \frac{1}{4\times10^{-7}}\right]$$
$$U = \mathbf{1.656 \times 10^{-19}\ J}$$

De-Broglie wavelength

$$\lambda = \frac{h}{m\upsilon}$$
$$U = \frac{1}{2}m\upsilon^2$$
$$\lambda = \frac{h}{\sqrt{2mU}} = \left[\frac{h^2}{2mU}\right]^{\frac{1}{2}}$$
$$\lambda = \left[\frac{(6.624\times10^{-34})^2}{2\times9.1\times10^{-31}\times1.656\times10^{-19}}\right]^{\frac{1}{2}}$$
$$= \mathbf{1.2 \times 10^{-9}\ m = 12Å}$$

Example 3.8. *Show that the De-Broglie wavelength of electrons accelerated through V volts is very nearly given by*

$$\lambda(\text{in Å}) = \left[\frac{150}{v\ (\text{volts})}\right]^{\frac{1}{2}}$$ [*Kanpur, 1991*]

Solution. Here,
$$\text{eV} = \frac{1}{2}m\upsilon^2$$
$$\upsilon = \left[\frac{2\ \text{eV}}{m}\right]^{\frac{1}{2}}$$
$$\lambda = \frac{h}{m\upsilon}$$
$$= \frac{h}{m\left[\frac{2\ \text{eV}}{\text{m}}\right]^{1/2}}$$
$$= \left[\frac{h^2}{2m\ \text{eV}}\right]^{\frac{1}{2}}$$

Here

$$h = 6.624 \times 10^{-34} \text{ J-s}$$
$$m = 9.1 \times 10^{-31} \text{ kg}$$
$$e = 1.6 \times 10^{-19} \text{ C}$$

$$\lambda = \left[\frac{(6.624 \times 10^{-34})^2}{2 \times 9.1 \times 10^{-31} \times 1.6 \times 10^{-19} \times V}\right]^{\frac{1}{2}}$$

$$\boldsymbol{\lambda = \left[\frac{150}{V}\right]^{\frac{1}{2}} \times 10^{-10} \text{ m}}$$

$$\boldsymbol{= \left[\frac{150}{V}\right]^{\frac{1}{2}} \text{ Å}}$$

Example 3.9. *Calculate the wavelength of matter-waves associated with an electron which has been accelerated from rest, through a potential difference of 1.25* kV.

Solution. Here

$$E = \text{eV} = 1.6 \times 10^{-19} \times 1.25 \times 10^3 \text{ J}$$
$$E = 2 \times 10^{-16} \text{ J}$$

Also

$$E = \frac{1}{2} m\upsilon^2$$
$$\upsilon = \sqrt{\frac{2E}{m}}$$
$$\lambda = \frac{h}{m\upsilon}$$
$$\lambda = \frac{h}{m\sqrt{\frac{2E}{m}}}$$
$$\lambda = \frac{h}{\sqrt{2mE}}$$

Here

$$E = 2 \times 10^{-16} \text{ J}$$
$$h = 6.624 \times 10^{-34} \text{ J-s}$$
$$m = 9.1 \times 10^{-31} \text{ kg}$$

$$\therefore \quad \lambda = \frac{6.624 \times 10^{-34}}{(2 \times 9.1 \times 10^{-31} \times 2 \times 10^{-16})^{1/2}}$$

$$\boldsymbol{\lambda = 0.348 \times 10^{-10} \text{ m}}$$
$$\boldsymbol{= 0.348 \text{ Å}}$$

Example 3.10. *What is the momentum of a proton having kinetic energy 1 B eV ?*

[*G.N.D.U., 1991*]

Solution. Here

$$U = 1 \text{ B eV} = 10^9 \text{ eV} = 10^9 \times 1.6 \times 10^{-19} \text{ J}$$
$$U = 1.6 \times 10^{-10} \text{ J}$$
$$m = 1.67 \times 10^{-27} \text{ kg}$$
$$U = \frac{1}{2} m\upsilon^2$$
$$p = mv = m\left[\frac{2U}{m}\right]^{\frac{1}{2}}$$
$$p = [2mU]^{1/2}$$

$$p = [2 \times 1.67 \times 10^{-27} \times 1.6 \times 10^{-10}]^{1/2}$$

$$p = \mathbf{7.31 \times 10^{-19} \text{ kg m/s}}$$

Example 3.11. *Calculate the wavelength of an electron of 1 Me V (non-relativistic).*

[*Kanpur, 1990*]

Solution.

$$U = 10^6 \text{ eV} = 10^6 \times 1.6 \times 10^{-19} \text{ J}$$

$$U = 1.6 \times 10^{-13} \text{ J}$$

$$m = 9.1 \times 10^{-31} \text{ kg}$$

$$U = \frac{1}{2} m\upsilon^2$$

$$\upsilon = \left[\frac{2U}{m}\right]^{\frac{1}{2}}$$

$$\lambda = \frac{h}{m\upsilon}$$

$$= \frac{h}{m\left[\frac{2U}{m}\right]^{1/2}} = \frac{h}{[2mU]^{1/2}}$$

$$= \frac{6.624 \times 10^{-34}}{[2 \times 9.1 \times 10^{-31} \times 1.6 \times 10^{-13})^{1/2}}$$

$$= \mathbf{1.227 \times 10^{-12} \text{ m}}$$

$$= \mathbf{1.227 \times 10^{-2} \text{ Å}.}$$

Example 3.12. *Calculate the most probable De-Broglie wavelength associated with thermal neutrons. Given, temperature = 27 °C, k = 1.38 × 10^{-21} J/K, mass of neutron = 1.6749 × 10^{-27} kg.*

[*Kolkata, 1992*]

Solution. For neutron,

$$U = \frac{3}{2} kT$$

or

$$\frac{1}{2} m\upsilon^2 = \frac{3}{2} kT$$

or

$$\upsilon = \left[\frac{3kT}{m}\right]^{\frac{1}{2}}$$

$$\lambda = \frac{h}{mv} = \frac{h}{m \times \left[\frac{3kT}{m}\right]^{\frac{1}{2}}}$$

$$\lambda = \frac{h}{[3kTm]^{1/2}}$$

Here

$$h = 6.624 \times 10^{-34} \text{ J.S}, \qquad m = 1.6749 \times 10^{-27} \text{ kg}$$

$$k = 1.38 \times 10^{-21} \text{ J/kg}, \qquad T = 27 + 273 = 300\,K$$

$$\lambda = \frac{6.624 \times 10^{-34}}{[3 \times 1.38 \times 10^{-21}] \times 300 \times 1.6749 \times 10^{-27}]^{1/2}}$$

$$\lambda = 1.45 \times 10^{-27} \text{ m} = \mathbf{0.145 \text{ Å}.}$$

Example 3.13. *The De-Broglie wavelength of a proton is 1 Å. Find its kinetic energy in eV.*

[*Delhi, 1992*]

Solution. Mass of proton $m = 1.67 \times 10^{-27}$ kg

$$\lambda = 1\ \text{Å} = 10^{-10}\ \text{m}$$

$$\lambda = \frac{h}{mv} \quad \text{or} \quad m = \frac{h}{m\lambda}$$

$$\upsilon_k = \frac{1}{2}m\upsilon^2$$

$$= \frac{1}{2}m\left[\frac{h}{m\lambda}\right]^2$$

$$= \frac{h^2}{2\lambda^2 m}$$

$$= \frac{(6.624 \times 10^{-34})^2}{2 \times (10^{-10})^2 \times 1.67 \times 10^{-27}}$$

$$= 1.34 \times 10^{-20}\ \text{J}$$

$$= \frac{1.314 \times 10^{-20}}{1.6 \times 10^{-19}} = \mathbf{8.2 \times 10^{-2}\ eV.}$$

3.8 Particle in a Box

Consider a particle of mass m in a box (Fig. 3.8). The particle is restrained to move only in one dimension say along the x-axis. At the positions $x = 0$ and $x = a$ the particle undergoes perfect elastic collisions with the walls of the box. The particle does not experience any other force. The momentum of the particle $= p_x = p$. It remains constant but only changes in sign when it collides against the wall.

The phase integral for the periodic motion of the particle is given by

$$J = \oint p_x dx$$

$$J = \oint p\, dx \qquad \text{...(3.24)}$$

$$J = p \times 2a = 2pa \qquad \text{...(3.25)}$$

According to Wilson-Sommerfeld quantization rule

$$\oint p\, dx = nh \qquad \text{...(3.26)}$$

or $$2pa = nh$$

or $$p = \frac{nh}{2a}$$

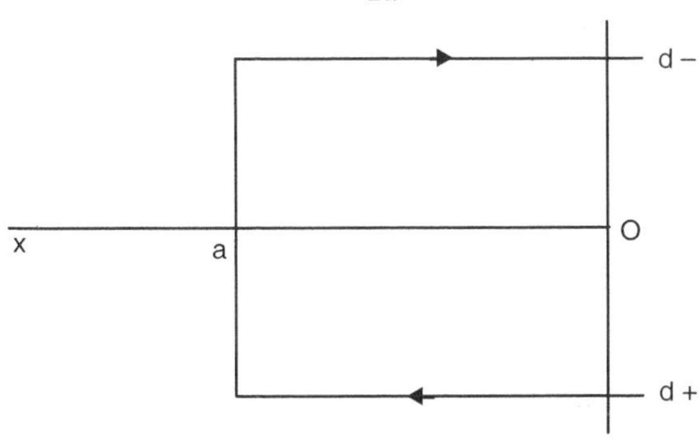

Fig. 3.8

The total energy of the system is only due to its kinetic energy.

$$\therefore \quad E_n = \frac{1}{2}m\upsilon^2 = \frac{p^2}{2m} \qquad ...(3.27)$$

Substituting the value of p, we get

$$E_n = \frac{n^2h^2}{8ma^2} \qquad ...(3.28)$$

Here $n = 0, 1, 2, 3...$ etc.
For $n = 1$

$$E_1 = \frac{h^2}{8ma^2} \qquad ...(3.29)$$

For $n = 2$

$$E_2 = 4\left[\frac{h^2}{8ma^2}\right] \qquad ...(3.30)$$

For $n = 3$

$$E_3 = 9\left[\frac{h^2}{8ma^2}\right]$$

Therefore energy values of the particle are discrete and not continuous. Moreover, they are not equally spaced as shown in Fig. 3.9.

From equation (3.28), the energy E_n depends on m and a^2.

Here $E \propto \frac{1}{m}$

and $E \propto \frac{1}{a^2}$.

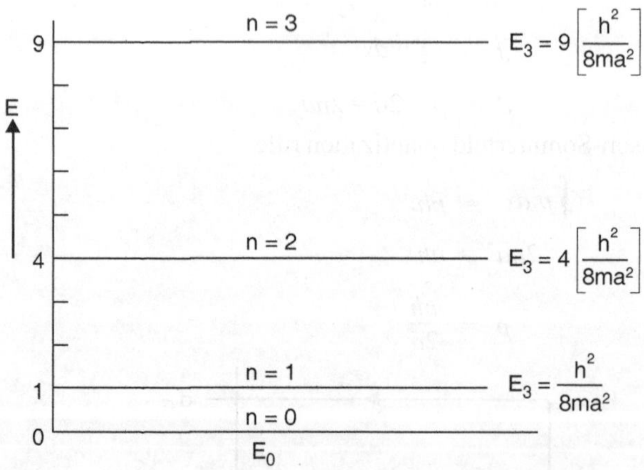

Fig. 3.9

Macroscopic system. In the case of macroscopic system say large m and large a, suppose

$$m = 1 \text{ kg}$$

and $a = 1$ m,

Then $E_n = \frac{n^2h^2}{8}$.

Here E_n is of the order of h^2 *i.e.*

E_n is the order of 10^{-68} J.

For $n = 1, 2, 3...$ etc.

The difference in any two energy states is extremely small and not detectable. Hence for macroscopic systems discrete energy values are not detectable and appear to be continuous.

Microscopic system. In the case of microscopic systems for an electron moving in a box, the values of a and m are very small. For an electron a is of the order of 10^{-10} m and m is of the order of 10^{-30} kg.

Hence the value E_1 and E_2 given by

$$E_n = \frac{n^2h^2}{8ma^2}$$

$$E_1 = \frac{(6.624 \times 10^{-34})^2}{8 \times (9.1) \times 10^{-31} \times (10^{-10})^2}$$

$$E_1 = 0.6 \times 10^{-17} \text{ J}$$

Similarly $$E_2 = 2.4 \times 10^{-17} \text{ J}$$

$$E_3 = 5.4 \times 10^{-17} \text{ J}$$

Here the difference between E_2 and E_1

$$= 2.4 \times 10^{-17} - 0.6 \times 10^{-17} = 1.8 \times 10^{-17} \text{ J}$$

$$= 112.5 \text{ eV}$$

Similarly $E_3 - E_2 = 3 \times 10^{-17}$ J = 187.5 eV

These values are quite appreciable.

Hence in microscopic systems the energy values are significant, discrete and non-continuous.

Example 3.14. *Consider a one-dimensional box of length a. There are two electrons in the box. State their quantum numbers and calculate the lowest energy of the system.*

Solution. The quantum numbers of the two electrons are

(1) $n = 1, l = 0, m_l = 0, m_s = +\frac{1}{2}$

(2) $n = 1, l = 0, m_l = 0, m_s = -\frac{1}{2}$

For lowest energy the electrons are in the ground state. As both the electrons are in the ground state, the total energy of the system

$$E = 2 \times \frac{n^2 h^2}{8\ ma^2}$$

Here $n = 1$

$$E = \frac{h^2}{4ma^2}$$

Example 3.15. *Consider that a one-dimensional box of length a has three electrons. State the quantum numbers and the lowest energy of the system.*

Solution. Here 2 electrons will be in ground state and 1 electron in the excited state ($n = 2$). (According to Pauli's exclusion principle) "No two electrons can be in the same quantum state".

(1) $n = 1, l = 0, m_l = 0, m_s = +\frac{1}{2}$

(2) $n = 1, l = 0, m_l = 0, m_s = -\frac{1}{2}$

(3) $n = 2, l = 0, m_l = 0, m_s = +\frac{1}{2}$

The energy of the system

$$E = \left[\frac{h^2}{8\,ma^2} + \frac{h^2}{8\,ma^2}\right] + \left(\frac{2^2 h^2}{8\,ma^2}\right)$$

$$E = \left[\frac{6h^2}{8\,ma^2}\right]$$

Example 3.16. *Deduce the zero point energy if the length of the box be 10^{-10} m and there are 10 electrons in it.*

(The Coulomb interaction may be disregarded.) [*I.A.S. 1991*]

Solution. The energy for a particle in a box

$$E = \frac{n^2 h^2}{8\,ma^2}$$

(*i*) There will be two electrons in the lowest energy level for $n = 1$.

$$E_1 = 2 \times \left[\frac{h^2}{8\,ma^2}\right] = \frac{h^2}{4\,ma^2} \qquad ...(i)$$

(*ii*) There are in all 10 electrons, the next 8 electrons will have $n = 2$. This is according to the Pauli's exclusion principle. Here $n = 2$

and
$$E_2 = 8 \times \left[\frac{2^2 h^2}{8\,ma^2}\right] = \frac{4h^2}{ma^2} \qquad ...(ii)$$

Therefore total lowest energy of the system,

$$E = E_1 + E_2 = \frac{h^2}{4\,ma^2} + \frac{4h^2}{ma^2} = \frac{17h^2}{4\,ma^2}$$

Here $h = 6.624 \times 10^{-34}$ J-s

$$m = 9.1 \times 10^{-31} \text{ kg}$$

$$a = 10^{-10} \text{ m}$$

$$E = \frac{17 \times (6.624 \times 10^{-34})^2}{4 \times 9.1 \times 10^{-31} \times (10^{-10})^2}$$

$$= \mathbf{2.05 \times 10^{-16} \; J}$$

Example 3.17. *Calculate the mean energy per electron at 0 K if electrons are enclosed in a long-chain molecule of length 50 Å.* [*I.A.S. 1990*]

Solution. Here energy per electron

$$E = \frac{h^2}{8\,ma^2}$$

$$h = 6.624 \times 10^{-34} \text{ J-s}$$

$$m = 9.1 \times 10^{-31} \text{ kg}$$

$$a = 50 \text{ Å} = 50 \times 10^{-10} \text{ m}$$

$$E = \frac{(6.624 \times 10^{-34})^2}{8 \times 9.1 \times 10^{-31} \times (50 \times 10^{-10})^2}$$

$$E = \mathbf{2.4 \times 10^{-21}\ J}$$

$$E = \frac{2.4 \times 10^{-21}}{1.6 \times 10^{-19}}$$

$$E = \mathbf{1.5 \times 10^{-2}\ eV.}$$

Example 3.18. *Consider that two electrons are confined to a box of length 10^{-10} m. Calculate the lowest energy of the system.*

Solution. Here both the electrons are in the ground state for $n = 1$.

Energy of the system

$$E = 2\left[\frac{n^2 h^2}{8\ ma^2}\right]$$

$$E = \frac{2h^2}{8\ ma^2}$$

Here

$$h = 6.624 \times 10^{-34}\ \text{J-s}$$

$$m = 9.1 \times 10^{-31}\ \text{kg}$$

$$a = 10^{-10}\ \text{m}$$

$$E = \frac{2 \times (6.624 \times 10^{-34})^2}{8 \times 9.1 \times 10^{-31} \times 10^{-10})^2}$$

$$E = \mathbf{1.205 \times 10^{-17}\ J}$$

$$E = \frac{1.205 \times 10^{-17}}{1.6 \times 10^{-19}}\ \text{eV}$$

$$E = \mathbf{75.3\ eV.}$$

Example 3.19. *Consider that there electrons are confined to a one dimensional box of length 1 Å. Calculate the lowest energy of the system.*

Solution. According to Pauli's exclusion principle, for first two electrons $n = 1$ and for the third electron $n = 2$.

$$a = 1\ \text{Å} = 10^{-10}\ \text{m}$$

Total energy of the system

$$E = 2\left(\frac{h^2}{8\ ma^2}\right) + \left(\frac{2^2 h^2}{8\ ma^2}\right)$$

$$E = \frac{6h^2}{8\ ma^2}$$

$$= \frac{6 \times (6.624 \times 10^{-34})^2}{8 \times 9.1 \times 10^{-31} \times (10^{-10})^2}$$

$$E = \mathbf{3.615 \times 10^{-17}\ J}$$

$$E = \frac{3.615 \times 10^{-17}}{1.6 \times 10^{-19}}\ \text{eV}$$

$$E = \mathbf{225.93\ eV.}$$

3.9 Heisenberg's Uncertainty Principle

In the experiments for the determination of specific charge of an electron, it was shown that electron gets deflected by the application of electron and magnetic fields. It means that the electron behaves as a particle and it has definite mass, momentum and energy which can be measured with desired accuracy. But the experiment of Davisson and Germer, and Thomson showed clearly the wave nature of electron and the wavelength of the electron is given by

$$\lambda = \frac{h}{m\upsilon} = \frac{h}{p}.$$

Here h is the Planck's constant and $m\upsilon$ is the momentum of the electron. A wave extends through space and it is difficult to locate the exact position of an electron behaving as a wave, at any given instant of time.

This dual behaviour of electron as a particle as a wave presents difficulty in locating the exact position and momentum at the same time. This difficulty is overcome according to *Heisenberg's uncertainty principle.*

This principle states that the exact position and momentum of a particle (say electron) cannot be determined simultaneously with a desired accuracy. Taking Δx as the error in determining its position and Δp the error in determining its momentum at the same instant, these quantities are related as follows :

$$\Delta x\, \Delta p \sim \frac{1}{2} \times \frac{h}{2\pi} \qquad ...(3.31)$$

The product of the two errors is approximately of the order of Plank's constant.

Multiplying and dividing the left hand side of equation (1) by υ, the velocity of the particle

$$\frac{\Delta x}{\upsilon}\Delta p\upsilon \sim \frac{1}{2} \times \frac{h}{2\pi}$$

or

$$\Delta t\, \Delta E \sim \frac{1}{2} \times \frac{h}{2\pi} \qquad ...(3.32)$$

Here ΔE represents the uncertainty in the measurement of the energy of a particle and Δt the uncertainty in the measurement of time.

In equation (3.31), the product $\Delta x\, \Delta p$ is of the order of $h/2\pi$. If Δx is small, Δp will be large and *vice versa.* It means that if one quantity is measured accurately, the other quantity becomes less accurate. A similar argument holds good for $\Delta E\, \Delta t$ also.

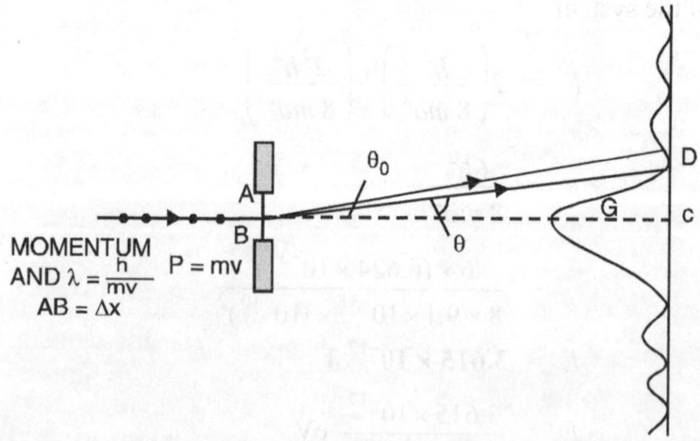

Fig. 3.10

Consider a beam of electrons travelling in the direction shown in Fig. 3.10. The slit AB of width Δx is perpendicular to the path of the electrons. Before entering the slit, the electron has a definite momentum $p = m\upsilon$. After passing through the slit the electron gets diffracted and acquires a momentum along OG. The angular deflection θ depends upon the component of the momentum parallel to the slit, *i.e.*, $\Delta p = p \sin\theta = p\theta$ for small angular deflection. The angle θ_0 corresponding to the direction of first minimum of the diffraction pattern is

$$\theta_0 = \frac{\lambda}{\Delta x} \text{ for small value of } \theta_0$$

or
$$\Delta x = \frac{\lambda}{0_0}$$

$\therefore$
$$\Delta p \Delta x \sim p\lambda \times \frac{\theta}{\theta_0}$$

But
$$p\lambda = h$$

$\therefore$
$$\Delta p\, \Delta x = h \frac{\theta}{\theta_0}$$

Taking,
$$\theta = \theta_0 \text{ approximately,}$$

$\therefore$
$$\Delta p \Delta x \sim h$$

The probable deflection θ of the electron is less than θ_0 and according to Heisenberg, the uncertainty relationship is

$$\Delta p \Delta x \geq \frac{1}{2} \times \frac{h}{2\pi}$$

In general any experiment that measures *position* or *momentum* accurately, introduces an uncertainty in the other. The uncertainty is never less than the predicted value according to Heisenberg's relation. Thus any instrument cannot measure the quantities more accurately than predicted by *Heisenberg's principle of uncertainty or indeterminancy.*

3.9A Applications of Uncertainty Principle

According to classical mechanics position and momentum can be predicated exactly. But according to uncertainty principle position and momentum of a particle cannot be measured accurately. Some of the applications of uncertainty principle are discussed.

(1) Energy and Radius of the Bohr's First Orbit

If Δx and Δp are the uncertainties in the position and momentum of an electron in the first orbit, then

$$\Delta x\, \Delta p \simeq \hbar \qquad \text{... (3.33)}$$

$$\Delta p \simeq \frac{\hbar}{\Delta x}$$

Total energy of electron in the Bohr's first orbit

$$E = \left[\frac{1}{2} m \upsilon^2\right] - \frac{Ze^2}{4\pi \epsilon_0 r} \qquad \text{... (3.34)}$$

$$\frac{1}{2} m \upsilon^2 = \frac{p^2}{2m}, \text{ Here } p \text{ is momentum.}$$

$$E = \left(\frac{p^2}{2m}\right) - \frac{Ze^2}{4\pi \epsilon_0 r} \qquad \text{...(3.35)}$$

Δx is the uncertainty in position and can be taken equal to r,

$\therefore$
$$\Delta p = \frac{\hbar}{r}$$

$\therefore$
$$E = \left(\frac{h^2}{2mr^2}\right) - \frac{Ze^2}{4\pi \epsilon_0 r} \qquad \text{...(3.36)}$$

Differentiating (3.34) with respect to r

$$\frac{dE}{dr} = -\frac{\hbar^2}{mr^3} + \frac{Ze^2}{4\pi \epsilon_0 r^2}$$

For minimum value of E, $\frac{dE}{dr} = 0$

$$-\frac{\hbar^2}{mr^3} + \frac{Ze^2}{4\pi \epsilon_0 r^2} = 0$$

$$\therefore \quad r = \frac{(4\pi \epsilon_0)\hbar^2}{mZe^2} \qquad \text{...(3.37)}$$

From hydrogen atom, $Z = 1$

$$m = 9.1 \times 10^{-31} \text{ kg}$$

$$\frac{1}{4\pi \epsilon_0} = 9 \times 10^9 \text{ N-m}^2/\text{C}^2$$

$$e = 1.6 \times 10^{-19} \text{ C}$$

$$\hbar = \frac{h}{2\pi} = \frac{6.624 \times 10^{-34}}{2\pi}$$

$$= 1.054 \times 10^{-34} \text{ J-s}$$

$$\therefore \quad r = \frac{(1.054 \times 10^{-34})^2}{9 \times 10^9 \times 9.1 \times 10^{-31} \times 1 \times (1.6 \times 10^{-19})^2}$$

$$r = \mathbf{5.29 \times 10^{-11} \text{ m}}$$

$$= \mathbf{0.529 \text{ Å}.}$$

This value agrees with the value of the Bohr's first orbit.

Energy $$E \simeq \frac{\hbar^2}{2m r^2} - \frac{Ze^2}{4\pi \epsilon_0 r}$$

$$E \simeq \frac{\hbar^2}{2m\left[\dfrac{4\pi \epsilon_0 \hbar^2}{mZe^2}\right]} - \frac{Ze^2}{4\pi \epsilon_0 \left[\dfrac{4\pi \epsilon_0 \hbar^2}{mZe^2}\right]}$$

$$E \simeq \frac{mZ^2e^4}{2(4\pi \epsilon_0)^2\hbar^2} - \frac{mZ^2e^4}{(4\pi \epsilon_0)^2\hbar^2}$$

$$E \simeq -\left[\frac{mZ^2e^4}{2(4\pi \epsilon_0)^2\hbar^2}\right]$$

Taking $Z = 1$ and substituting the values

$$E \simeq -\frac{(9.1 \times 10^{-31})(1.6 \times 10^{-19})^4 \times (9 \times 10^9)^2}{2 \times (1.054 \times 10^{-34})^2} \text{ J}$$

$$E \simeq -\frac{(9.1 \times 10^{-31})(1.6 \times 10^{-19})^4 \times (9 \times 10^9)^2}{2 \times (1.054 \times 10^{-34}) \times 1.6 \times 10^{-19}} \text{ eV}$$

$$\simeq \mathbf{-1351 \text{ eV}}$$

The energy of electron in the ground state (first orbit) of hydrogen atom is –13.51 eV. This value agrees with the value derived from Bohr's theory.

(2) Particle in a Box

Consider a particle confined to a box of length l. The uncertainty Δx in the position is l. Take $\Delta x = l$

$$\Delta x\,\Delta p \simeq \hbar$$

$$\Delta p \simeq \frac{\hbar}{\Delta x}$$

$$\Delta p \simeq \frac{\hbar}{l}$$

Energy,
$$E \simeq \frac{p^2}{2m} = \frac{(\Delta p)^2}{2m}$$

$$E \simeq \frac{\left(\frac{\hbar}{l}\right)^2}{2ml}$$

$$E \simeq \frac{\hbar^2}{2ml^2}$$

This result agrees with the result derived through Schodinger equation.

(3) Why electron cannot be present in the Nucleus ?

The radius of the nucleus is of order of the 10^{-14} m and maximum uncertainty in position is 2×10^{-14} m.

$$\Delta x\,\Delta p \simeq \hbar$$

$$\Delta p \simeq \frac{\hbar}{\Delta x}$$

$$\simeq \frac{1.054 \times 10^{-34}}{2 \times 10^{-14}}$$

$$\simeq 5.27 \times 10^{-21} \text{ kg-m/s.}$$

The minimum uncertainty in momentum can be taken as the momentum of electron

$$p = 5.27 \times 10^{-27} \text{ kg-m/s}$$

$$m = 9.1 \times 10^{-31} \text{ kg}$$

$\therefore$ Energy,

$$E = \frac{p^2}{2m}$$

$$= \frac{(5.27 \times 10^{-21})^2}{2 \times 9.1 \times 10^{-31}} \text{ J}$$

$$= \left[\frac{(5.27 \times 10^{-21})^2}{2 \times 9.1 \times 10^{-31} \times 1.6 \times 10^{-19}}\right] \text{eV}$$

$$= 95.37 \times 10^6 \text{ eV}$$

$$= \mathbf{95.37\ MeV.}$$

It means if electron exists inside the nucleus, it kinetic energy is of the order of **95.37** MeV.

But experimental results have shown that no electron in the atom possesses energy greater than 4 MeV. Hence electron is not present in the nucleus.

3.10 Energy and Time

We can apply the uncertainty principle to energy and time. Let E be the energy emitted in an atomic process, during the time interval Δt secs. If the energy is emitted as electro-magnetic waves, due to the limited time interval the accuracy to which its frequency can be calculated is limited. If we assume that the minimum uncertainty in the number of waves we can count in a wave group is one wave.

Then since frequency is no. of waves we can count divided by the time interval. Then the uncertainty Δv in the frequency measurement is Δv.

$$\therefore \qquad \Delta v \geq \frac{1}{\Delta t}$$

$$\because \qquad E = hv \quad \therefore \Delta E = h\Delta v$$

$$\text{or} \qquad \Delta E = \frac{h}{\Delta t} \quad \text{or} \quad \Delta E \,.\, \Delta t \geq h$$

This is the relation between energy and time.

A better result gives

$$\Delta E\, \Delta t \geq \frac{\hbar}{2} \qquad \text{...(3.38)}$$

$$\text{where} \qquad \hbar = \frac{h}{2\pi}$$

Example 3.20. *An electron is confined to a box of length 10^{-10} m. Calculate the minimum uncertainty in its velocity.*

Solution. Here

$$\Delta x = 10^{-10}\text{m}$$

$$\Delta \upsilon = ?$$

$$m = 9.1 \times 10^{-31} \text{ kg}$$

$$\hbar = 1.054 \times 10^{-34} \text{ J-s}$$

$$\Delta x\, \Delta p = \hbar$$

$$\Delta p = \frac{\hbar}{\Delta x}$$

$$m\Delta \upsilon = \frac{\hbar}{m\, \Delta x}$$

$$= \frac{1.054 \times 10^{-34}}{9.1 \times 10^{-31} \times 10^{-10}}$$

$$= \mathbf{1.58 \times 10^{6}\ m/s.}$$

Example 3.21. *A proton is confined to a nucleus of radius 5×10^{-15} m. Calculate the minimum uncertainty (i) momentum and (ii) kinetic energy of the proton.*

Solution.

$$m = 1.67 \times 10^{-27} \text{ kg}$$

$$\Delta x = 2r = 2 \times 5 \times 10^{-15} = 10^{-14} \text{ m}$$

$$h = 1.054 \times 10^{-34} \text{ J-s}$$

$$\Delta x\, \Delta p = \hbar$$

$$(i) \qquad \Delta p_{\min} \simeq \frac{\hbar}{\Delta x}$$

$$\simeq \frac{1.054 \times 10^{-34}}{10^{-14}}$$

$$\simeq 1.054 \times 10^{-20} \text{ kg-m/s}$$

$$(iii) \qquad E_{\min} = \Delta E = \frac{\Delta p^2}{2m}$$

$$= \frac{(1.054 \times 10^{-20})^2}{2 \times 1.67 \times 10^{-27}}$$

$$= \mathbf{3.326 \times 10^{-14}\ J}$$

$$= \frac{3.326 \times 10^{-14}}{1.6 \times 10^{-19}} \text{ eV}$$

$$= \mathbf{2.079 \times 10^5\ eV.}$$

Example 3.22. *An electron has a speed of 600 m/s with an accuracy of 0.005%. Calculate the certainty with which we can locate the position of the electron.*

$$h = 6.6 \times 10^{-34} \text{ Joule-second, } m = 9.1 \times 10^{-31} \text{ kg}$$

Solution. Here $\upsilon = 600$ m/second, $m = 9.1 \times 10^{-31}$ kg

$$h = 6.6 \times 10^{-34} \text{ joule-second}$$

$$\text{Momentum of electron} = m\upsilon = 9.1 \times 10^{-31} \times 600 \text{ kg-m/s}$$

$$\Delta p = \left(\frac{0.005}{100}\right) m\upsilon$$

$$= (5 \times 10^{-5} \times 9.1 \times 10^{-31} \times 600) \text{ kg-m/s}$$

$$\Delta x\, \Delta p \geq \frac{1}{2}\frac{h}{2\pi} \geq \frac{h}{4\pi}$$

$$\Delta x \geq \frac{h}{4\pi \times \Delta p}$$

$$\Delta x \geq \frac{6.6 \times 10^{-34}}{4\pi \times 5 \times 10^{-5} \times 9.1 \times 10^{-31} \times 600}$$

$$\Delta x \geq 0.001923 \text{ m}$$

$$\Delta x \geq \mathbf{1.923\ mm}$$

If the momentum of the electron is determined to the given accuracy, then the position of the electron cannot be measured to an accuracy, less than 2 mm. It means that the concept of an electron as a tiny particle does not hold good.

Example 3.23. *A bullet of mass 0.025 kg is moving with a speed of 400 m/s. The speed is measured accurate to 0.02%. Calculate the certainty with which we can locate the position of the bullet. Take $h = 6.6 \times 10^{-34}$ Joule-second.*

Solution. Here $\upsilon = 400$ m/s, $m = 0.025$ kg

Momentum of the bullet,

$$p = m\upsilon$$

$$= 0.025 \times 400 = 10 \text{ kg-m/s}$$

$$\Delta p = \left(\frac{0.02}{100}\right) \times p$$

$$\Delta p = (2 \times 10^{-4})(10) = 2 \times 10^{-3} \text{ kg-m/s}$$

$$\Delta x\, \Delta p \geq > \frac{h}{4\pi}$$

$$\Delta x \geq \frac{h}{4\pi \times \Delta p}$$

$$\Delta x \geq \frac{6.6 \times 10^{-34} \times 7}{4 \times 22 \times 2 \times 10^{-3}}$$

$$\Delta x \geq \mathbf{2.625 \times 10^{-32} \ m.}$$

Here Δx is a very small quantity and cannot be measured. Therefore, in heavy bodies the particle concept prevails.

Example 3.24. *An electron has a speed of 300 m/s accurate to 0.01%. Within what range at the best can we locate the position of the electron ?* [*I.A.S.*]

Solution. Here,

$$h = 6.6 \times 10^{-34} \text{ J-s}, m = 9.1 \times 10^{-31} \text{ kg}$$

$$u = 300 \text{ m/s}$$

Momentum of electron

$$p = m\upsilon = 9.1 \times 10^{-31} \times 300$$

$$p = 27.3 \times 10^{-29} \text{ kg-m/s}$$

$$\Delta p = \left(\frac{0.01}{100}\right) m\upsilon$$

$$= \frac{0.01}{100} \times 27.3 \times 10^{-29} = 27.3 \times 10^{-33}$$

$$\Delta x \, \Delta p \geq \frac{1}{2}\left(\frac{h}{2\pi}\right)$$

$$\Delta x \geq \frac{h}{4\pi \times \Delta p}$$

$$\Delta x \geq \frac{6.6 \times 10^{-34}}{4\pi \times 27.3 \times 10^{-33}}$$

$$\Delta x \geq 1.923 \times 10^{-3} \text{ m}$$

$$\Delta x \geq \mathbf{1.923 \ mm.}$$

Example 3.25. *An electron is confined to a box of length 10^{-10} m. Calculate the minimum uncertainty in the measurement of its velocity.* [*Kolkata, 1992*]

Solution.

$$\Delta x_{max} \, \Delta p_{min} \approx \frac{h}{2\pi}$$

$$\Delta x = 10^{-10} \text{ m}$$

$$\Delta p = m\Delta\upsilon$$

$$\therefore \quad \Delta\upsilon_{min} = \left[\frac{h}{2\pi \, \Delta x \times m}\right]$$

Here

$$h = 6.624 \times 10^{-34} \text{ J-s}$$

$$m = 9.1 \times 10^{-31} \text{ kg}$$

$$\Delta\upsilon_{min} = \frac{6.624 \times 10^{-34}}{2\pi \times 10^{-10} \times 9.1 \times 10^{-31}}$$

$$= \mathbf{1.16 \times 10^{6} \ m/s.}$$

Example 3.26. *The life time of an excited state of an atom is about 10^{-8} s. Calculate the minimum uncertainty in the energy of the excited state.* [*Delhi, 1992*]

Solution. Here

$$\Delta t = 10^{-8} \text{ S}$$

$$\Delta t \ \Delta \ E \sim \hbar$$

$$\Delta E = \frac{\hbar}{\Delta t} = \frac{1.054 \times 10^{-34}}{10^{-8}}$$

$$= \mathbf{1.054 \times 10^{-26} \ J}$$

3.11 Linear Harmonic Oscillator

(One Dimensional)

Consider a particle of mass m vibrating along the x-axis and executing simple harmonic motion. The restoring force on the particle is F and is directed towards the mean position F depends upon the displacement x.

$$\therefore \quad F = -Kx \qquad \text{...(3.39)}$$

where K is force constant.

But
$$F = m\frac{d^2x}{dt^2}$$

$$\therefore \quad m\frac{d^2x}{dt^2} = -Kx \qquad \text{...(3.40)}$$

or
$$m\frac{d^2x}{dt^2} + kx = 0 \qquad \text{...(3.41)}$$

This is the general equation of S.H.M.

The general equation is

$$\frac{d^2x}{dt^2} + \omega^2 x = 0 \qquad \text{...(3.4)}$$

Comparing equations (3.41) and (3.42), we get

$$\omega^2 = \frac{K}{m}$$

$$= \omega = \sqrt{K/_{\omega}}$$

Also
$$K = m\omega^2$$

But
$$\omega = 2\pi v$$

$$\therefore \quad K = 4\pi^2 mv^2 \qquad \text{...(3.43)}$$

Here v is the frequency of oscillation and ω is the angular frequency. The potential energy of the harmonic oscillator is given by

$$V = -\int F dx$$

But
$$F = -Kx$$

$$\therefore \quad V = \int Kx\, dx$$

$$V = \frac{1}{2}Kx^2 + C \qquad \text{...(3.44)}$$

Here C is the constant of integration.

3.12 Width of Atomic Spectral Lines

When an atom is raised to an excited state it generally decays to the ground state in about 10^{-8} secs. The wavelength of the emitted photon is

$$\lambda = \frac{hc}{E_i - E_f}, \text{ where } E_i \text{ is the energy of} \qquad \text{...(3.45)}$$

the initial, excited state and E_f the energy of the final ground state.

Now
$$\Delta E \geq \frac{\hbar}{2\Delta t}$$

Differentiating eq. (*i*) w.r.t. E_i, we get

$$\frac{d\lambda}{dE_i} = \frac{hc}{(E_i - E_f)^2} = \frac{-\lambda}{E_i - E_f} \quad ...(3.46)$$

From (*i*) and (*ii*) $$\frac{\Delta\lambda}{\lambda} = \frac{-\Delta E}{E_i - E_f}$$

ΔE is of the order of 5×10^{-8} eV about 8 orders smaller than the energy difference of the atomic transitions, which are a few eV. Hence the uncertainty in the line width is quite small and is overshadowed by other sources of line broadening such as the Doppler effect.

The life times of excited nuclear states are very much shorter, typically 10^{-20} to 10^{-18} seconds, the γ-rays from most nuclear transitions have natural line widths of the order of a few *K* eV.

Gama-ray Microscope

Heisenberg's uncertainty relation can be illustrated using an imaginary γ-rays microscope of high resolving power. We wish to determine the position and momentum of an electron using this microscope. The electron can be observed if at least one photon is scattered by it into the objective lens of the microscope. The resolving power of the microscope is

$$\Delta x = \frac{\lambda}{2 \sin \lambda},$$

where Δx is the smallest distance that can be resolved.

λ = wavelength of radiation used and α is the semi-vertical angle as shown in Fig. 3.11.

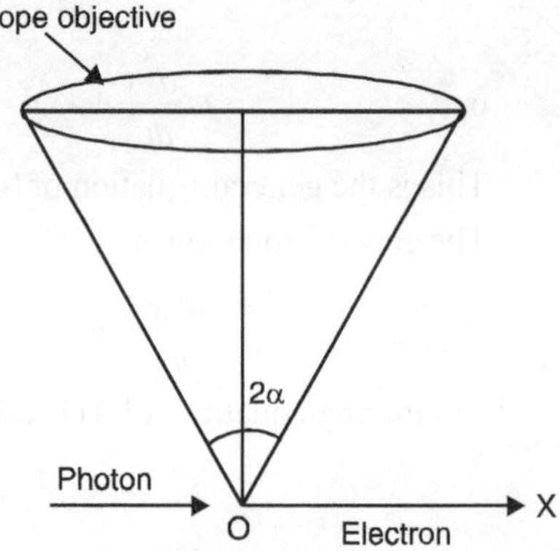

Fig. 3.11

The momentum given by the photon to the electron during the collision is $\frac{h}{\lambda}$ or $\frac{h\nu}{c}$ which therefore recoils. Since the scattered photon can enter the objective of the microscope any where between α and $-\alpha$, the momentum of the electron is uncertain to this extent.

Now, the X-component of the momentum lies between

$$+\frac{h\nu \sin\alpha}{c} \text{ and } -\frac{h\nu \sin\alpha}{c}.$$

$\therefore$ The uncertainty in X-component of momentum is

$$\Delta p_x = \frac{2h\nu}{c}\sin\alpha = 2\frac{h}{\lambda}\sin\alpha$$

Now Δp_x increases with decrease in λ (inverse relation).

$\therefore$ Any attempt to reduce Δx means an increase in Δp_x.

Now $$\Delta x\, \Delta p_x = \frac{\lambda}{2\sin\alpha}\cdot\frac{\alpha}{\lambda} h \sin\alpha = h \quad ...(3.47)$$

which is the uncertainty relation.

A more refined deduction shows that

$$\Delta x\, \Delta p_x \geq \frac{h}{2\pi}. \quad ...(3.48)$$

EXERCISE – III

1. What are De-Broglie waves ? Describe Davisson-Germer experiment. [*Delhi, 1983*]
2. Describe and explain Heisenburg's uncertainty principle. [*Delhi, 1982*]
3. Explain De-Broglie concept of matter waves. What is meant by velocity of De-Broglie waves ? [*Delhi, 1982*]
4. Describe the two-slit experiment to show that a beam of electrons behaves as a wave.
5. Describe wave-particle dualism for light and matter.
6. Discuss the behaviour of particle in a box constrained to move only along the x-axis.
7. Show that for a De-Broglie wave group associated with a moving particle, the group velocity is equal to the particle velocity. [*Delhi, (Hons.) 1985*]
8. Describe how Davisson Germer experiment verified De-Broglie hypothesis. [*Delhi (Hons.) 1985*]
9. Describe De-Broglie concept of matter waves. Derive De-Broglie wavelength. Describe Davission and Germer's experiment to study the diffraction of electrons. [*Kolkata, 1992*]
10. (*a*) Distinguish between the phase velocity and the group velocity. Show that the group velocity of the De-Broglie waves is the same as the velocity of the particle.

    (*b*) Give an account of an experiment that establishes the wave nature of electrons. [*Kolkata, 1992*]
11. (*a*) Discuss Heisenberg's uncertainty principle. What is Bohr's principle of complementarity ? [*Kolkata, 1992*]
12. State the De-Broglie postulate of matter waves. Describe the experiment establishing its validity. [*Osmania, 1992*]

    Describe the implications of applying the principles of quantum mechanics to a particle in one dimensional box of length '*L*'. [*Osmania, 1992*]
13. Calculate the wavelength associated with a thermal neutron of energy 0.025 eV. [**Ans.** 1.807×10^{-10} m]

4 QUANTUM MECHANICS

4.1 Schrödinger's Equation

In 1956 Schrodinger derived an equation for the motion of matter waves. He made the following assumptions :

(*i*) He assumed that De-Broglie wavelength holds good for any particle moving in a field of force with potential energy V. The sum of the kinetic and potential energy is given by

$$E = \text{Kinetic energy} + \text{Potential energy.}$$

i.e.

$$E = \frac{1}{2} m\upsilon^2 + V$$

$$E = \frac{p^2}{2m} + V \qquad ...(4.1)$$

$$\therefore \quad p^2 = 2m(E - V)$$

$$p = [2m(E - V)]^{1/2}$$

Therefore De-Broglie wavelength

$$\lambda = \frac{h}{p} = \frac{h}{[(2m(E - V)]^{\frac{1}{2}}} \qquad ...(4.2)$$

For large value of V, λ is large and for a minimum value of V, λ is also minimum.

(*ii*) He assumed that the wave function ψ is governed by the equation :

$$\psi = \psi_0 e^{-i\omega t}$$

The wave function ψ is also called probability amplitude for the position of the particle. The usual wave equation in Cartesian coordinates is given by

$$\frac{\partial^2\psi}{\partial x^2} + \frac{\partial^2\psi}{\partial y^2} + \frac{\partial^2\psi}{\partial z^2} = \frac{1}{u^2}\frac{\partial^2\psi}{\partial t^2} \qquad ...(4.3)$$

Here

$$\psi = \psi_0 e^{-i\omega t} \qquad ...(4.4)$$

and ψ is a function of (x, y, z, and t) and represents the amplitude of the wave associated with the particle and υ is the velocity.

Differentiating equation (4.4) twice with respect to t we get

$$\frac{\partial^2\psi}{\partial t^2} = -\omega^2\psi_0 e^{-i\omega t} = -\omega^2\psi$$

Substituting this value in equation (4.3)

$$\frac{\partial^2\psi}{\partial x^2} + \frac{\partial^2\psi}{\partial y^2} + \frac{\partial^2\psi}{\partial z^2} = -\left(\frac{\omega^2}{u^2}\right)\psi$$

where

$$\frac{\omega^2}{u^2} = \frac{(2\pi v^2)}{(v\lambda)^2} = \frac{4\pi^2}{\lambda^2}$$

$$\therefore \qquad \frac{\partial^2\psi}{\partial x^2}+\frac{\partial^2\psi}{\partial y^2}+\frac{\partial^2\psi}{\partial z^2} = -\left(\frac{4\pi^2}{\lambda^2}\right)\psi$$

Taking

$$\frac{\partial^2\psi}{\partial x^2}+\frac{\partial^2\psi}{\partial y^2}+\frac{\partial^2\psi}{\partial z^2} = \nabla^2\psi,$$

$$\nabla^2\psi+\frac{4\pi^2}{\lambda^2}\psi = 0 \qquad ...(4.5)$$

Equation (4.5) is a general equation and is independent of time.

Introducing the concept of wave mechanics, from De-Broglie wavelength, and from equation (4.2)

$$\lambda = \frac{h}{m\upsilon}=\frac{h}{[2m\,(E-V)]^{\frac{1}{2}}}$$

Substituting this value of λ in equation (4.5)

$$\nabla^2\psi+\frac{4\pi^2\,[2m\,(E-V)]}{h^2}\psi = 0$$

$$\nabla^2\psi+\left(\frac{8\pi^2 m}{h^2}\right)(E-V)\,\psi = 0 \qquad ...(4.6)$$

This represents Schrödinger time-independent wave equation. Here ψ is a wave function.

Also $$\hbar = \frac{h}{2\,\pi}$$

$$\therefore \qquad \nabla^2\psi+\left(\frac{2m}{\hbar^2}\right)(E-V)\,\psi = 0 \qquad ...(4.7)$$

Equation (4.7) shows that the wave function ψ is a function of coordinates. The potential V is also a function of coordinates.

Special case

In the case of a free particle (free from external force) $V = 0$ and the total energy is due to kinetic energy only.

$$\therefore \qquad \nabla^2\psi+\left(\frac{2m}{\hbar^2}\right)(E-V)\,\psi = 0 \qquad ...(4.8)$$

Equation (4.8) represents Schrödinger equation for a particle in the absence of external field.

4.2 Time Dependent Schrödinger Equation

The wave function

$$\psi = \psi_0 e^{-i\,\omega\,t} \qquad ...(4.9)$$

$$\frac{\partial\psi}{\partial t} = -i\,\omega\,\psi_0 e^{-i\,\omega\,t}$$

$$\frac{\partial\psi}{\partial t} = -i\,(2\,\pi\,v)\psi$$

But $$v = \left(\frac{E}{h}\right)$$

$$\frac{\partial\psi}{\partial t} = -i\,(2\,\pi)\left(\frac{E}{h}\right)\psi$$

$$E\,\psi = -\frac{1}{i}\left(\frac{h}{2\pi}\right)\frac{\partial\psi}{\partial t}$$

But

$$-1 = i^2$$

$$E\,\psi = \frac{i^2}{i}\left(\frac{h}{2\pi}\right)\frac{\partial\psi}{\partial t}$$

$$E\,\psi = i\hbar\,\frac{\partial\psi}{\partial t} = i\hbar\frac{\partial}{\partial t}(\psi) \qquad ...(4.10)$$

Schrodinger equation is given by

$$\nabla^2\psi + \left(\frac{2m}{\hbar^2}\right)(E - V)\,\psi = 0$$

$$\nabla^2\psi + \left(\frac{2m}{\hbar^2}\right)(E\,\psi) - \left(\frac{2m}{\hbar^2}\right)V\,\psi = 0 \qquad ...(4.11)$$

Substituting the value of $E\,\psi = i\hbar\,\frac{\partial\psi}{\partial t}$ in equation (4.11), we get

$$\nabla^2\psi + \left(\frac{2m}{\hbar^2}\right)i\hbar\,\frac{\partial\psi}{\partial t} - \frac{2m}{\hbar^2}V\,\psi = 0$$

$$\nabla^2\psi - \left(\frac{2m}{\hbar^2}\right)V\,\psi = -\left(\frac{2m}{\hbar^2}\right)i\hbar\,\frac{\partial\psi}{\partial t}$$

or

$$-\left(\frac{\hbar^2}{2m}\right)\nabla^2\psi + V\psi = i\hbar\,\frac{\partial\psi}{\partial t} \qquad ..(4.12)$$

This equation represents Schrödinger time-dependent equation because it depends upon time. Equation (4.12) can be written as

$$\left(-\frac{\hbar^2}{2m}\nabla^2 + V\right)\psi = i\hbar\,\frac{\partial\,\psi}{\partial r}$$

Taking $\left(-\frac{\hbar^2}{2m}\nabla^2 + V\right)$ as operator H called Hamiltonian, and from equation (4.10)

$$i\hbar\,\frac{\partial}{\partial t} = E, \text{ we get}$$

$$H\,\psi = E\,\psi \qquad ...(4.13)$$

Equation (4.13) represents the motion of a non-relativistic material particle.

4.3 Wave Function and Probability Density

The basic postulate of quantum mechanics is that the quality $\psi\psi^*\,dx\,dy\,dz$ is the probability that a particle represented by the wave function $\psi\,(x, y, z, t)$ is in the volume $dV = (dx\,dy\,dz)$ at the time t. ψ is the complex conjugate of ψ. The probability density of the particle is

$$P\,(t) = \psi\psi^* = |\,\psi\,(x, y, z, t)\,|^2$$

The wave function ψ is the probability amplitude for the position of the particle.

As the particle must be somewhere in the space, it is necessary that the probability density should satisfy the normalization condition

$$\int_{-\infty}^{+\infty}\int_{-\infty}^{+\infty}\int_{-\infty}^{+\infty} \psi\psi^*\,dx\,dy\,dz = 1 \qquad ...(4.14)$$

Here $\psi = \psi_0 e^{-i\omega t}$. But $\omega = \dfrac{2\pi E}{h}$

or $$\psi = \psi_0 e^{\frac{-i2\pi Et}{h}} \qquad ...(4.15)$$

or $$\psi = \psi_0 e^{-iEt/\hbar} \qquad ...(4.16)$$

The complex conjugate is

$$\psi^* = \psi_0^* e^{i2\pi Et/h} \qquad ...(4.17)$$

or $$\psi^* = \psi_0^* e^{iEt/\hbar} \qquad ...(4.18)$$

Substituting these values of ψ and ψ^* in equation (4.14)

$$\int_{-\infty}^{+\infty}\int_{-\infty}^{+\infty}\int_{-\infty}^{+\infty} \psi_0\psi_0^* \, dx\, dy\, dz = 1 \qquad ...(4.19)$$

Thus for a wave function satisfying the time independent Schrodinger equation, the probability is also independent of time.

1. The wave function ψ besides being normalizable must be single valued as $P(t)$ can only one value at a particular place at a particular time.
2. Wave function ψ and its partial derivatives $\dfrac{d\psi}{dx}, \dfrac{d\psi}{dy}$ and $\dfrac{d\psi}{dz}$ must be continuous every where.
3. ψ must be finite for all values of x, y and z.

4.4 Operators

For the equation

$$\Psi(x, t) = Ae^{i(kx - wt)}$$

where $$\vec{p} = \frac{h}{\lambda} = \hbar\vec{k} \quad \text{and} \quad k = \frac{2\pi}{\lambda}$$

and $\vec{k}$ is the wave vector;

We see that

$$ih\frac{\partial \Psi}{\partial t} = \hbar\omega\Psi = E\Psi \qquad ...(4.20)$$

Since $$E = \hbar\omega$$

and $$-ih\frac{\partial \Psi}{\partial x} = \hbar k\Psi = p\Psi \qquad ...(4.21)$$

Equations (4.20) and (4.21) are of the general form

(Differential Operator) Ψ = (Constant) Ψ

are called Eigenvalue equations. The constant is the Eigenvalue and Ψ is the corresponding Eigen function.

Since E and p are the energy and momentum; the differential operators

$$E_{op} = i\hbar\left(\frac{\partial}{\partial t}\right) = \hat{E}$$

and $$p_x\, op = i\hbar\left(\frac{\partial}{\partial x}\right) = \hat{p}_x$$

$$\vec{p}_{op} = i\hbar\nabla$$

where ∇ is the gradient.

$$\vec{\nabla} = -\hat{i}\left(\frac{\partial}{\partial x}\right)+\hat{j}\left(\frac{\partial}{\partial y}\right)+\hat{k}\left(\frac{\partial}{\partial z}\right)$$

where $\hat{i}$, $\hat{j}$ and $\hat{k}$ are unit vectors along the x, y and z directions.

An operator tells us what operation is to be carried out on the quantity that follows it.

4.1. For example $i\hbar\left(\frac{\partial}{\partial t}\right)$ instructs to take the partial derivative with respect to t and multiply it by $i\hbar$.

Now the Kinetic Energy = $K.E. = \frac{1}{2}m\upsilon^2 = \frac{p^2}{2m}$

$\therefore$ The kinetic energy operator is

$$\hat{K.E} = \frac{\hat{p}^2}{2m} = \frac{1}{2m}\left(\frac{\hbar}{i}\frac{\partial}{\partial x}\right)^2$$

$$= \frac{\hbar^2}{2m}\frac{\partial^2}{\partial x^2}$$

Table 4.1

Variable	Operator
x	$\hat{x}$
px	$\hat{p}_x = -i\hbar\frac{\partial}{\partial x}$
p	$\hat{p} = -i\hbar\nabla$
E	$\hat{E} = i\hbar\frac{\partial}{\partial t}$
$K.E.$	$\hat{K.E} = -\frac{\hbar^2}{2m}\frac{\partial^2}{\partial x^2}$

Commutative Property of operators

For two operators $\hat{A}$ and $\hat{B}$ if

$\left[\hat{A},\hat{B}\right] = \hat{A}\hat{B} - \hat{B}\hat{A} = 0$, then the operators are commutative.

Some results :

$$\left[\hat{x}, \hat{px}\right] = i\hbar \qquad \left[\hat{x}, \hat{py}\right] = 0$$

$$\left[\hat{px}, \hat{py}\right] = 0$$

4.5 Expectation Values

We know from the uncertainty principle that we cannot expect to obtain a definite value for the position or the momentum of a particle knowing its energy. Therefore, we must be content to know the probability that a measurement of these variables will yield a particular result. The average of a large number of such measurements of a variable is known as the expectation value of the variable and we can calculate this once we know the wave function $\psi(x)$.

$\therefore$ The expectation value or the average value of a physical quantity is defined as :

$$<\alpha> = \int_{-\infty}^{+\infty} \frac{\Psi^*(x)\hat{\alpha}\Psi(x)\,dx}{\int_{-\infty}^{+\infty}\Psi^*(x)\psi(x)\,dx} \quad \text{where } \Psi = Ae^{-(i\hbar)(Et - px)}$$

$\Psi(x)$ = states of system and Ψ^* is its complex conjugate. If however, $\psi(x)$ is normalised, *i.e.*, $\int_{-\infty}^{+\infty}\Psi^*(x)\Psi(x)\,dx = 1$

then $$<\alpha> = \int_{-\infty}^{+\infty}\Psi^*(x)\hat{\alpha}\Psi(x)\,dx \text{ only}$$

For 3-dimensional motion

$$<\alpha> = \frac{\int_\tau \psi^*(\vec{r})\,\hat{\alpha}\,\psi(\vec{r})\,d\tau}{\int_\tau \psi^*(\vec{r})\,\psi(\vec{r})\,d\tau}$$

Since p and E are replaced by the corresponding operators in eqn., the expectation value of

$$<P_x> = \int_{-\infty}^{+\infty}\Psi^*\hat{P}_x\Psi^*\,dx = \int_{-\infty}^{+\infty}\Psi^*\left(-\frac{i\hbar\partial}{\partial x}\right)\Psi\,dx$$

$$= \frac{\hbar}{i}\int_{-\infty}^{+\infty}\Psi^*\frac{\partial\Psi}{\partial x}dx$$

$$<x> = \int_{-\infty}^{+\infty}\Psi^*x\Psi\,dx$$

$$<p> = \int_{-\infty}^{+\infty}\Psi^*(-i\hbar\nabla)\Psi\,dx$$

$$<E> = \int_{-\infty}^{+\infty}\Psi^*\hat{E}\psi\,d\tau = \int_{-\infty}^{+\infty}\Psi^*\left(\frac{i\hbar\partial}{\partial t}\right)\Psi\,d\tau$$

$$<\frac{p^2}{2m}> = \int_{-\infty}^{+\infty}\Psi^*\left(-\frac{i\hbar^2}{2m}\nabla^2\right)\Psi\,d\tau$$

Operators in Quantum mechanics are linear operators. Linear operators commute and they also obey the distributive and associative properties.

4.6 Transition Probability

According to Bohr's theory of the hydrogen atom

$$v = \frac{E_m - E_n}{h}$$

υ is the frequency of radiation emitted when an atom goes from an energy level E_m to a lower level E_n.

The wave function ψ of an electron capable of existing in states n and m is

$$\Psi = a\Psi_n + b\Psi_m$$

where a^*a is the probability of an electron existing in state n. b^*b is the probability in the state m.

Note $\quad a^*a + b^*b = b$

Initially, $a = 1$ and $b = 0$, when the electron is in the excited state. $a = 0$ and $b = 1$ and finally it returns to $a = 1$ and $b = 0$ state.

There is no radiation emitted when the electron is in either state, but when a transition takes place from m to n radiation is emitted.

The expectation value for the composite function is

$$<x> = \int_{-\infty}^{+\infty} x\left(a^*\Psi_n{}^* + b^*\Psi_m{}^*\right)\left(a\Psi_n + b\Psi_m\right)dx$$

$$= \int_{-\infty}^{+\infty} x\left(a^2\Psi_n^*\Psi_n + b^*a\Psi_m^*\Psi_n + a^*b\Psi_n^*\Psi_m + b^2\Psi_m^*\Psi_m\right)dx$$

Now, $a^*a = a^2$ and $b^*b = b^2$

Note the first and last term do not vary with time. Hence only second and third terms contribute to the time variation of $<x>$.

$$<x> = a^2\int_{-\infty}^{+\infty} x\,\Psi_n^*\Psi_n\,dx + b^*a\int_{-\infty}^{+\infty} x\Psi_m e^{+(\frac{iEm}{\hbar})t}\Psi_n e^{-i(\frac{E_n}{\hbar})t}dx + a^*b\int_{-\infty}^{+\infty} x\Psi_n{}^* e^{+i(\frac{E_n}{\hbar})t}\Psi_m e^{-\left(\frac{iE_m}{\hbar}\right)t}dx + b^2\int_{-\infty}^{+\infty} x\Psi_m^*\Psi_m dx \qquad ...(4.22)$$

Now,

$$e^{i\theta} = \cos\theta + i\sin\theta$$

$$e^{-i\theta} = \cos\theta - i\sin\theta$$

$$\therefore \quad e^{i\theta} + e^{-i\theta} = 2\cos\theta$$

$\therefore$ Equation becomes

$$2a^*b\cos\left(\frac{E_m - E_n}{\hbar}\right)t\int_{-\infty}^{+\infty} x\Psi_n^*\psi_m^* dx$$

$$+\,i\sin\left(\frac{E_m - E_n}{\hbar}\right)t\int_{-\infty}^{+\infty}\left[b^*a\Psi_m^*\Psi_n - a^*b\Psi_n^*\psi_m\right]dx$$

The real part is

$$\cos\frac{(E_m - E_n)\,t}{\hbar} = \cos 2\pi\frac{(E_m - E_n)\,t}{h} = \cos 2\pi\,\upsilon t.$$

$$\therefore \quad E = h\upsilon.$$

$\therefore$ The electron's position oscillates sinusoidally at the frequency $\nu = \dfrac{E_m - E_n}{h}$

when the electron is in state n, $b^2 = 0$;

when the electron is in state m, $a^2 = 0$

and the electron is, on the average, stationary. When it undergoes a transition between states its position oscillates with a frequency υ. Such an electron is like an electric dipole and radiates energy of frequency υ. This is the result obtained by Bohr's theory also.

4.7 Application of Schrödinger Equation :

Particle in a Box

Consider a free particle in a one dimensional box. The particle has a mass m and is restricted to move in a straight line along the x-axis. The range of the particle is $0 < x < a$. The particle is reflected back whenever it reaches the end of the range. The general equation is

$$\nabla^2\psi + \left(\frac{2m}{\hbar^2}\right)(E - V)\psi = 0 \qquad ...(4.23)$$

For a free particle $\quad V = 0$ for $\quad 0 < x < a.$

Also the wave function ψ vanishes at $x = 0$, and $x = a$. For one dimension, only the modified equation is given by

$$\frac{\partial^2\psi}{\partial x^2} + \left(\frac{2m}{\hbar^2}\right)E\psi = 0 \qquad ...(4.24)$$

$$\frac{\partial^2\psi}{\partial x^2} + \left(\frac{8\pi^2 mE}{h^2}\right)\psi = 0$$

Taking $\quad \dfrac{8\pi^2 mE}{h^2} = k^2,$

$$E = \frac{k^2h^2}{8\pi^2 m}$$

$$\therefore \qquad \frac{\partial^2\psi}{\partial x^2} + k^2\psi = 0 \qquad ...(4.25)$$

The general solution for equation (4.25) is given by

$$\psi(x) = A\sin kx + B\cos kx \qquad ...(4.26)$$

$$\psi = 0 \text{ at } x = 0$$

Therefore from equation (4.21), B must be zero.

$$\psi(a) = A\sin ka = 0$$

Therefore A should be zero. But we cannot take A to be zero because it does not yield any solution. Hence $\sin ka = 0$

i.e. $\qquad ka = n\pi$

$$k = \frac{n\pi}{a} \text{ where } n = 1, 2, 3, \text{ etc.}$$

Hence the only permissible solutions of the wave equations are

$$\psi n(x) = A\sin\left(\frac{n\pi}{a}\right)x \qquad ...(4.27)$$

$$E = \frac{k^2h^2}{8\pi^2 m} \qquad ...(4.28)$$

Substituting the value of $k = \dfrac{n\pi}{a}$ in equation (4.28) we get

$$E_n = \frac{n^2h^2\pi^2}{8\pi^2 ma^2} = \frac{n^2h^2}{8ma^2}$$

$$\therefore \qquad E_n = \frac{n^2 h^2}{8\,ma^2} \qquad ...(4.29)$$

It means E_n depends upon the value of n. Each value of E_n is called Eigen value or proper value. Each value of ψ_n is called an Eigen function or proper function.

According to the requirement of the wave function,

$$\int_0^a |\psi_n(x)|^2\,dx = A^2 \int_0^a sin^2\left(\frac{n\pi x}{a}\right)dx = 1$$

or
$$\frac{A^2 a}{2} = 1$$

or
$$A = \sqrt{\frac{2}{a}}$$

$$\therefore \qquad \psi_n(x) = \sqrt{\frac{2}{a}}\,\sin\left(\frac{n\,\pi}{a}\right)x$$

For $n = 1$

$$\psi_1(x) = \sqrt{\frac{2}{a}}\,\sin\left(\frac{\pi}{a}\right)x$$

and
$$\psi_2(x) = \sqrt{\frac{2}{a}}\,\sin\left(2\,\frac{\pi}{a}\right)x$$

$$\psi_2(x) = \sqrt{\frac{2}{a}}\,\sin\left(3\,\frac{\pi}{a}\right)x$$

Plotting graphically, the first three functions are shown in Fig. 4.1.

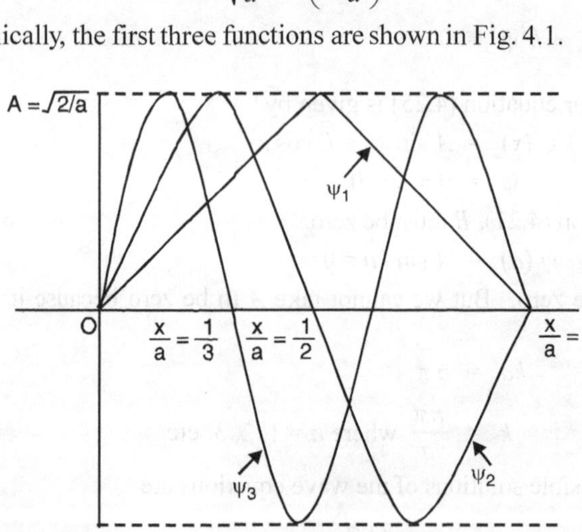

Fig. 4.1

Equation (4.2) represents that energy E_n of the particle is quantized. Similarly for the momentum of the particle,

$$E_n = \frac{P_n^2}{2m}$$

or
$$p_n = (2m\,E_n)^{1/2} = \left(\frac{2mn^2h^2}{8ma^2}\right)^{\frac{1}{2}}$$

or $$p = \frac{nh}{2a}$$

$$\therefore \quad p_n \propto n$$

It means the momentum of the particle is also quantized. Further, for each value of E_n, there is a corresponding value of p_n.

According to quantum mechanics the probability is given by

$$|\psi_n(x)|^2 \, dx = \frac{2}{a}\sin^2\left(\frac{n\pi x}{a}\right) dx$$

This is wave like function and
for $$n = 3$$

$$|\psi_3(x)|^2 \, dx = \frac{2}{a}\sin^2\left(\frac{3\pi x}{a}\right) dx$$

The graph between probability and x is shown in Fig 4.2.

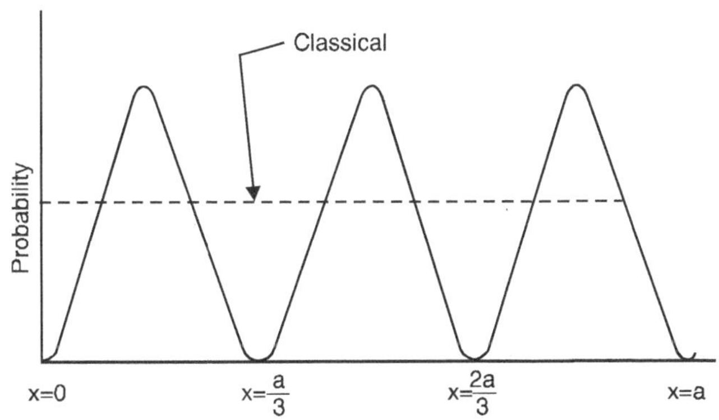

Fig. 4.2

In the graph the dotted line represents the constant probability according to classical mechanics. But the curve represents the probability variation with variation of x according to quantum mechanics. The probability is zero for values of

$$x = 0, x = \frac{a}{3}, x = \frac{2a}{3} \text{ and } x = a,$$

and so on.

4.8 Particle in three Dimensional Box

Consider a particle enclosed in a three dimensional box of length l, breadth b, height t whose walls are rigid and hard (Fig. 4.3). The potential energy is assumed to be zero inside the box. As no force acts on the particle inside the box, therefore in this region

$0 < x < a$

$0 < y < b$

$0 < z < c$

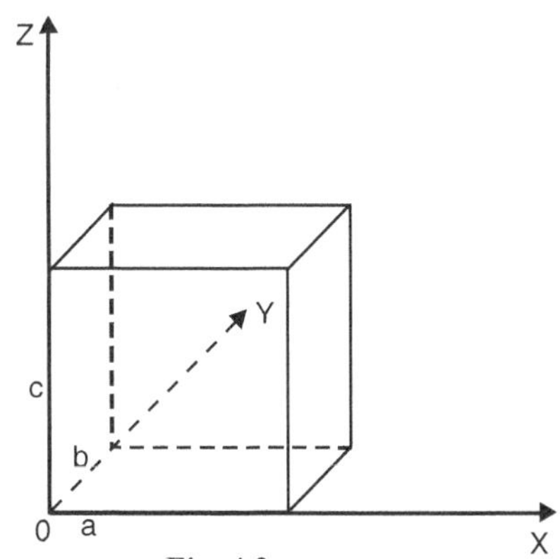

Fig. 4.3

The potential energy $V(x, y, z) = 0$

and for region outside the box, $V(x, y, z) = \infty$.

The Schrodinger equation is

$$\nabla^2\psi + \frac{2m}{\hbar^2}(E - V)\psi = 0 \qquad ...(4.30)$$

For a particle inside the box, $V = 0$

$$\nabla^2\psi + \left[\frac{2mE}{\hbar^2}\right]\psi = 0 \qquad ...(4.31)$$

or
$$\frac{\partial^2\psi}{\partial x^2} + \frac{\partial^2\psi}{\partial y^2} + \frac{\partial^2\psi}{\partial z^2} + \left[\frac{2mE}{\hbar^2}\right] = 0 \qquad ...(4.32)$$

Applying the method of separation of variables, we get

$$\psi(x, y, z) = X(x)\,Y(y)\,Z(z) \qquad ...(4.33)$$

It means wave function $\psi(x, y, z)$ is equal to the product of the functions X, Y and Z.

Take the second derivatives and substituting their values in equation (4.32), we get

$$YZ\frac{d^2X}{dx^2} + ZX\frac{d^2Y}{dy^2} + XY\frac{d^2Z}{dz^2} + \left[\frac{2mE}{\hbar^2}\right] \times XYZ = 0 \qquad ...(4.34)$$

Dividing by XYZ, we get

$$\left(\frac{1}{X}\right)\frac{d^2X}{dx^2} + \left(\frac{1}{Y}\right)\frac{d^2Y}{dy^2} + \left(\frac{1}{Z}\right)\frac{d^2Z}{dz^2} + \left(\frac{2mE}{\hbar^2}\right) = 0 \qquad ...(4.35)$$

For a particular value of kinetic energy the term $\frac{2mE}{\hbar^2}$ has a constant value. As velocity is a vector quantity and can be resolved into three components along the co-ordinate axes, therefore kinetic energy E can be resolved into three components such that

$$E = E_x + E_y + E_z \qquad ...(4.36)$$

Substituting this value in equation (4.35) and rearranging

$$\left[\left(\frac{1}{X}\right)\frac{d^2X}{dx^2} + \frac{2mE_x}{\hbar^2}\right] + \left[\left(\frac{1}{Y}\right)\frac{d^2Y}{dy^2} + \frac{2mE_y}{\hbar^2}\right] + \left[\left(\frac{1}{Z}\right)\frac{d^2Z}{dz^2} + \frac{2mE_z}{\hbar^2}\right] = 0$$

This equation gives three independent equations which are as follows :

$$\frac{d^2X}{dx^2} + \left(\frac{2mE_x}{\hbar^2}\right)X = 0 \qquad ...(4.37)$$

$$\frac{d^2Y}{dy^2} + \left(\frac{2mE_y}{\hbar^2}\right)Y = 0 \qquad ...(4.38)$$

and
$$\frac{d^2Z}{dz^2} + \left(\frac{2mE_z}{\hbar^2}\right)Z = 0 \qquad ...(4.39)$$

The boundary conditions applicable to the solution are

$X = 0$ at $x = 0, a$

$Y = 0$ at $x = 0, b$

$Z = 0$ at $z = 0, c$.

The general solution for X, Y and Z are

$$X = X_0 \sin\left[\left(\frac{2mE_x}{\hbar^2}\right)^{\frac{1}{2}} x + \phi_x\right] \qquad ...(4.40)$$

$$Y = Y_0 \sin\left[\left(\frac{2mE_y}{\hbar^2}\right)^{\frac{1}{2}} y + \phi_y\right] \qquad ...(4.41)$$

$$Z = Z_0 \sin\left[\left(\frac{2mE_z}{\hbar^2}\right)^{\frac{1}{2}} z + \phi_z\right] \quad ...(4.42)$$

The boundary conditions $X = 0$, $Y = 0$, $Z = 0$ are satisfied at $x = 0$, $y = 0$ and $z = 0$ if we take

$$\phi_x = 0, \phi_y = 0 \text{ and } \phi_z = 0.$$

In order to satisfy the boundary conditions

$$X = 0 \text{ at } x = 0, a$$

$$Y = 0 \text{ at } x = 0, b$$

and

$$Z = 0 \text{ at } z = 0, c.$$

It is necessary that energies assume Eigen values given by

$$E_x = \frac{\pi^2 \hbar^2 n_x^2}{2ma^2} \quad ...(4.43)$$

$$E_y = \frac{\pi^2 \hbar^2 n_y^2}{2mb^2} \quad ...(4.44)$$

$$E_z = \frac{\pi^2 \hbar^2 n_z^2}{2mc^2} \quad ...(4.45)$$

Here n_x, n_y and n_z are integers whose values can be from one to infinity. The total energy of the particle

$$E = E_x + E_y + E_z$$

$$E = \frac{\pi^2 \hbar^2}{2m}\left[\frac{n_x^2}{a^2} + \frac{n_y^2}{b^2} + \frac{n_z^2}{c^2}\right] \quad ...(4.46)$$

Here $n_x = 1, 2, 3....$

$$n_y = 1, 2, 3.....$$

$$n_z = 1, 2, 3....$$

The wave function ψ for a particle is obtained from equation (4.40), (4.41) and (4.42) with normalization over the volume of the box

$$\psi = \left(\frac{8}{abc}\right)^{\frac{1}{2}} \sin\left(\frac{n_x \pi x}{a}\right) \sin\left(\frac{n_y \pi y}{b}\right) \sin\left(\frac{n_z \pi z}{c}\right) \quad ...(4.47)$$

Results :

1. n_x, n_y and n_z are integers and are called quantum numbers, n_x, n_y, n_z have positive integral values.
2. The lowest possible energy occurs when

$$n_x = n_y = n_z = 1.$$

$$\therefore \quad E_{111} = \frac{\pi^2 \hbar^2}{2m}\left[\frac{1}{a^2} + \frac{1}{b^2} + \frac{1}{c^2}\right] \quad ...(4.48)$$

3. **Non-Degenerative Energy Levels**

For a particle contained in a cubical box of length l

$$E = \frac{\pi^2 \hbar^2}{2ml^2}\left[n_x^2 + n_y^2 + n_z^2\right] \quad ...(4.49)$$

The energy in the ground state is given by

$$E_{111} = \frac{3\pi^2 \hbar^2}{2ml^2} \qquad ...(4.50)$$

In this case, no other state will have this energy. This state has only one wave function. Therefore the ground state and this energy level is said to be **Non-Degenerate.**

4. **Degenerate Energy Levels**

As for a particle in a cubical box the energy depends upon the sum of the squares of the quantum numbers, the particles having same energy of an excited state can have different energy levels and different wave functions. Such states and energy levels are said to be degenerate.

Example : (1) *For first excited state the possible values of quantum numbers are*

n_x	n_y	n_z
2	1	1
1	2	1
1	1	2

Therefore there are three different energy states having same energy.

The energy in this case

$$E_{211} = E_{121} = E_{112} = \frac{6\pi^2 \hbar^2}{2ml^2}$$

As the number of different wave functions in this case is 3, therefore the degree of degeneracy of the level is said to be *three fold*.

(2) For second excited state,

n_x	n_y	n_z
2	2	1
2	1	2
1	2	2

$$E_{221m} = E_{212} = E_{122} = \frac{9\pi^2 \hbar^2}{2ml^2}$$

It is *three fold* degenerate.

(3) For $n_x = 2, n_y = 2, n_z = 2$

$$E_{122} = \frac{12\pi^2 \hbar^2}{2ml^2}$$

It is *non-degenerate*.

Similarly degree of degeneracy for any energy level can be calculated.

4.9 Linear Harmonic Oscillator

(One Dimensional)

Consider a particle of mass m vibrating along the x-axis and executing simple harmonic motion. The restoring force on the particle is F and is directed towards the mean position. F depends upon the displacement x.

$$\therefore \qquad F = -Kx \qquad ...(4.51)$$

where K is force constant.

But
$$F = m\frac{d^2x}{dt^2}$$

$$\therefore \qquad m\frac{d^2x}{dt^2} = -Kx \qquad ...(4.52)$$

or

$$m\frac{d^2x}{dt^2} + Kx = 0 \qquad ...(4.53)$$

This is the general equation of SHM.

The general equation is

$$\frac{d^2x}{dt^2} + \omega^2 x = 0 \qquad ...(4.54)$$

Comparing equations (4.53) and (4.54)

$$\omega^2 = \frac{K}{m}$$

or

$$\omega = \frac{K}{m}$$

Also

$$K = m\omega^2$$

But

$$\omega = 2\pi\nu$$

$$\therefore \qquad K = 4\pi^2 m\nu^2 \qquad ...(4.55)$$

Here ν is the frequency of oscillation and ω is the angular frequency. The potential energy of the harmonic oscillator is given by

$$V = -\int F\,dx$$

But

$$F = -Kx$$

$$V = \int Kx\,dx$$

$$V = \frac{1}{2}kx^2 + C \qquad ...(4.56)$$

Here C is the constant of integration.

When the particle is in the mean position, x is zero and its potential energy is also zero. Therefore from equation (4.56)

$$0 = 0 + C$$

$$\therefore \qquad C = 0$$

Hence from equation (4.56)

$$V = \frac{1}{2}Kx^2 \qquad ...(4.57)$$

Substituting the value of $K = 4\pi^2 m\nu^2$ in equation (4.57)

$$V = 2\pi^2 m\nu^2 x^2 \qquad ...(4.58)$$

This is the potential energy of the harmonic oscillator.

If a is the amplitude, then the maximum potential energy of the harmonic oscillator is given by

$$V = 4\pi^2 m\nu^2 a^2 \qquad ...(4.59)$$

$$\frac{d^2\psi}{dx^2} + \frac{2m}{\hbar^2}[E - V]\psi = 0 \qquad ...(4.60)$$

represents schrödinger equation.

Substituting the value of $V = \frac{1}{2}Kx^2$ in equation (4.60) we get

$$\frac{d^2\psi}{dx^2} + \frac{2m}{\hbar^2}\left[E - \frac{1}{2}Kx^2\right]\psi = 0 \quad ...(4.61)$$

Also $\frac{d^2\psi}{dx^2} + \frac{2m}{\hbar^2}\left[E - 2\pi^2 m\nu^2 x^2\right]\psi = 0$

or $$\frac{d^2\psi}{dx^2} + \left[\frac{2mE}{\hbar^2} - \frac{4\pi^2 m^2\nu^2 x^2}{\hbar^2}\right]\psi = 0 \quad ...(4.62)$$

Taking $$\lambda = \frac{2E}{\hbar\,(2\pi\nu)}$$

Here λ is the Eigen value, a pure dimensionless quantity, where

$$\lambda_n = (2n+1)$$

and $$\frac{2\pi m\nu}{\hbar} = \alpha^2.$$

$$\therefore \quad \alpha^2\lambda = \frac{2mE}{\hbar^2}$$

Substituting these value in equation (4.62)

$$\therefore \quad \frac{d^2\psi}{dx^2} + (\alpha^2\lambda - \alpha^4 x^2)\,\psi = 0 \quad ...(4.63)$$

Dividing by α^2

$$\frac{1}{\alpha^2}\frac{d^2\psi}{dx^2} + (\lambda - \alpha^2 x^2)\,\psi = 0 \quad ...(4.64)$$

Taking new variable

$$\xi = \alpha x$$

$$\therefore \quad \frac{d\xi}{dx} = a$$

where α is a constant.

$$\frac{d\psi}{dx} = \frac{d\psi}{d\xi} \times \frac{d\xi}{dx} = \alpha\,\frac{d\psi}{d\xi}$$

$$\frac{d^2\psi}{dx^2} = \frac{d}{dx}\left(\frac{d\psi}{dx}\right)$$

$$= \frac{d}{d\xi}\left(\frac{d\psi}{dx}\right)\frac{d\xi}{dx}$$

$$= \frac{d}{d\xi}\left(\frac{d\psi}{dx}\right)\alpha$$

Substituting $\frac{d\psi}{dx} = \alpha\,\frac{d\psi}{d\xi}$ in the above equation, we get

$$\frac{d^2\psi}{dx^2} = \frac{d}{d\xi}\left[\alpha^2\,\frac{d\psi}{d\xi}\right]$$

$$= \alpha^2 \frac{d^2\psi}{d\xi^2}$$

$$\frac{1}{\alpha^2}\frac{d^2\psi}{dx^2} = \frac{d^2\psi}{d\xi^2}$$

Substituting these values in equation (4.64)

$$\frac{d^2\psi}{d\xi^2} + \left(\lambda - \xi^2\right)\psi = 0 \qquad ...(4.65)$$

Solution

When $\xi^2 >> \lambda$, λ can be neglected in equation (4.65)

$$\therefore \qquad \frac{d^2\psi}{d\xi^2} - \xi^2\psi = 0 \qquad ...(4.66)$$

Its general solution is

$$\psi = e^{\pm\xi^2/2} \qquad ...(4.67)$$

There are two possible solutions

(*i*) $\psi = e^{\xi^2/2}$

and (*ii*) $\psi = e^{-\xi^2/2}$ ...(4.68)

Since $|\psi|^2$ is the probability of locating along x-axis, therefore it must decrease continuously to zero as x or ξ approaches $\pm\infty$.

Therefore, the first solution is not possible hence the second solution

$$\psi = e^{-\xi^2/2} \text{ holds goods.}$$

Recursion Formula

It is clear that equation (4.68) contains the term $e^{-\xi^2/2}$ as a factor. The solution of equation (4.68) can be written as

$$\psi = e^{-\xi^2/2}\,\phi(\xi) \qquad ...[4.68(a)]$$

Here $\phi(\xi)$ is a function of x and also therefore of x.

Differentiating equation [4.68(*a*)] twice with respect to ξ we get

$$\frac{d^2\psi}{d\xi^2} = e^{-\xi^2/2}\left[\frac{d^2\phi}{d\xi^2} - 2\xi\frac{d\psi}{d\xi} + (\xi^2 - 1)\,\phi\right] \qquad ...(4.69)$$

Substituting the values of ψ and $\frac{d^2\psi}{d\xi^2}$ in equation (4.65) and simplifying we get

$$\frac{d^2\phi}{d\xi^2} - 2\xi\frac{d\phi}{d\xi} + (\lambda - 1)\,\phi = 0 \qquad ...(4.70)$$

Now representing the solution of equation (4.70) in the form of power series of ξ.

$$\phi[\xi] = \sum_{n=0}^{\infty} a_n\,\xi^n \qquad ...[4.70(a)]$$

$$= a_0 + a_1\xi + a_2\xi^2 + a_3\xi^3 + ...$$

Differentiating above equation *w. t. r.* ξ

$$\frac{d\phi}{d\xi} = \sum_{n=1}^{\infty} n a_n\,\xi^{n-1}$$

$$= a_1 + 2a_2\xi + 3a_3\xi^2 + ...$$

Differentiating further *w.r.t.*ξ, we get

$$\frac{d^2\phi}{d\xi^2} = \sum_{n=2}^{\infty} n(n-1)\, a_n\, \xi^{n-2}$$

$$= 2a_2 + 6a_3\xi + 12a_4\xi^2 + ...$$

Substituting these values in equation (4.70) we get

$$\sum_{n=2}^{\infty} n(n-1)\, a_n\, \xi^{n-2} - 2\sum_{n=1}^{\infty} na_n\xi^n + (\lambda-1)\sum_{n=0}^{\infty} a_n\, \xi^n = 0.$$

$$\sum_{n=2}^{\infty} n(n-1)\, a_n\, \xi^{(n-2)} + (\lambda - 2n - 1)\sum_{n=0}^{\infty} a_n\, \xi^n = 0 \qquad ...(4.71)$$

This equation is satisfied for all values of ξ, if the coefficients of each power of x vanishes separately [*i.e.*, here for the first term *n* should be taken as (*n* + 2).]

$$(n+2)(n+1)\, a_{n+2} + (\lambda - 2_n - 1)\, a_n = 0$$

$$\therefore \qquad a_{n+2} = \frac{2n+1-\lambda}{(n+2)(n+1)}\, a_n \qquad ...(4.72)$$

This is known as *Recursion formula.*

4.10 Eigen Values of Harmonic Oscillator

In order to obtain a satisfactory wave function the series given in equation [4.70(*a*)] should break off after a finite number of terms. In equation (4.72) the term $2n + 1 - \lambda$ must be zero. So that the series break off after the *n*th *i.e.* the finite term.

$$\therefore \qquad 2n + 1 - \lambda = 0$$

$$\text{or} \qquad \lambda = 2n + 1 \qquad ...(4.73)$$

Here λ is dimensionless Eigen value.

$$\text{But} \qquad \lambda = \frac{2E}{\hbar(2\pi\nu)} \qquad ...(4.74)$$

∴ Equating (4.73) and (4.74)

$$2n + 1 = \frac{2E}{\hbar(2\pi\nu)}$$

$$\text{or} \qquad E = \left[n + \frac{1}{2}\right]\hbar(2\pi\nu)$$

$$E = \left[n + \frac{1}{2}\right]\hbar\omega$$

$$\text{E} = \left[n + \frac{1}{2}\right]h\nu$$

$$\text{In general} \qquad E_n = \left[n + \frac{1}{2}\right]\hbar\omega$$

$$\text{or} \qquad E_n = \left[n + \frac{1}{2}\right]h\nu \qquad ...(4.75)$$

$$\text{where} \qquad n = 0, 1, 2, 3, ...$$

It means the energy of the harmonic oscillator has discrete values depending upon the values of ν. The difference in energies between two adjacent states is *h*ν.

4.11 Zero Point Energy

The energy of the linear harmonic oscillator is given by the equation

$$E_n = \left[n + \frac{1}{2}\right] h\nu.$$

4.12 Potential Step

The general Schrodinger equation is

$$\nabla^2\psi + \left(\frac{2m}{\hbar^2}\right)(E - V)\,\psi = 0 \qquad ...(4.76)$$

For one dimension only the modified equation is given by (particle constrained to move along the x-axis)

$$\frac{\partial^2\psi}{\partial x^2} + \left(\frac{2m}{\hbar^2}\right)(E - V)\,\psi = 0 \qquad ...(4.77)$$

If the total energy of the particle has a fixed value E, the time dependent equation is given by

$$(x, t)\psi = y_0 e^{-i\omega t} = \psi_0 e^{-i/h\, Et} \qquad ...(4.78)$$

Therefore ψ also satisfies the time -independent Schrodinger equation

$$\frac{\partial^2\psi}{\partial x^2} + \frac{2m}{\hbar^2}(E - V)\,\psi = 0$$

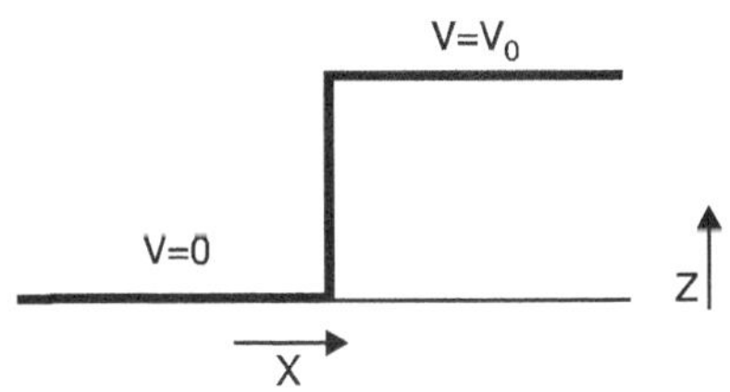

Fig. 4.4

The square potential step is shown in Fig. 4.4.

Here $V_n = 0$ for $x < 0$

and $V_n = V_0$ for $x > 0$

Therefore time independent Schrodinger equation

1. For $x < 0$ is given by

$$\frac{d^2\psi}{dx^2} + \frac{2m}{\hbar^2}E\,\psi = 0$$

Taking $K_0^2 = \dfrac{2mE}{\hbar^2}$

$$\frac{d^2\psi}{dx^2} + K_0^2\,\psi = 0 \qquad ...(4.79)$$

2. For $x > 0$, $V_n = V_0$

$$\therefore \quad \frac{d^2\psi}{dx^2} + \frac{2m}{\hbar^2}(E - V_0)\,\psi = 0$$

Taking $k^2 = \dfrac{2m}{\hbar^2}(E - V_0)$

$$\frac{d^2\psi}{dx^2} + k^2\psi = 0 \qquad ...(4.80)$$

The solutions of equations (4.79) and (4.80) are

$$\psi = Ae^{ik_0x} + Be^{-ik_0x} \text{ (for } x < 0) \qquad ...(4.81)$$

and

$$\psi = Ce^{ikx} + De^{-ikx} \text{ (for } x > 0) \qquad ...(4.82)$$

Here A, B, C and D are integration constants. There is discontinuity of V at $x = 0$.

The Schrödinger equation changes discontinuously at $x = 0$.

The constant of integrations A, B, C and D must be so chosen that, y and its derivatives are continuous at this points. It means wave functions must be equal at x = 0.

From equation (4.81) and (4.82)

When $x = 0$

$$\psi = A + B \text{ and } \psi = C + D$$

$$\therefore \quad A + B = C + D \qquad ...(4.83)$$

When derivatives of equations (4.81) and (4.82) are made equal at $x = 0$, in that case

$$\frac{d\psi}{dx} = k_0(A - B)$$

and

$$\frac{d\psi}{dx} = k(C - D)$$

$$\therefore \quad k_0(A - B) = k(C - D) \qquad ...(4.84)$$

or

$$A - B = \left(\frac{k}{k_0}\right)(C - D) \qquad ...(4.85)$$

From equations (4.83) and (4.84)

$$B = \left(\frac{k_0 - k}{k_0 + k}\right)A + \left(\frac{2k}{k_0 + k}\right)D \qquad ...(4.86)$$

and

$$C = \left(\frac{2k_0}{k_0 + k}\right)A - \left(\frac{k_0 - k}{k_0 + k}\right)D \qquad ...(4.87)$$

Substituting the value of B in equation (4.81) for $x < 0$

$$\psi = Ae^{ik_0x} + \left[\left(\frac{k_0 - k}{k_0 + k}\right)A + \left(\frac{2k}{k_0 + k}\right)D\right]e^{-ik_0x}$$

$$\psi = A\left[e^{ik_0x} + \left(\frac{k_0 - k}{k_0 + k}\right)e^{-ik_0x}\right] + \left[\left(\frac{2k}{k_0 + k}\right)e^{-ik_0x}\right]D$$

Taking

$$\left.\begin{aligned} e^{-ik_0x} + \left(\frac{k_0 - k}{k_0 + k}\right)e^{-ik_0x} &= \psi_1 \\ \text{and} \left(\frac{2k}{k_0 + k}\right)e^{-ik_0x} &= \psi_2 \end{aligned}\right] \text{for } x < 0$$

$$\psi = A\psi_1 + D\psi_2 \qquad ...(4.88)$$

Substituting the value for C in equation (4.82) for $x > 0$

$$\psi = \left[\left(\frac{2k_0}{k_0 + k}\right)A - \left(\frac{k_0 - k}{k_0 + k}\right)D\right]e^{-ikx} + De^{-ikx}$$

$$\psi = \left[\left(\frac{2k_0}{k_0 + k}\right)e^{-ikx}\right]A + \left[e^{-ikx} - \left(\frac{k_0 - k}{k + k}\right)e^{-ikx}\right]D$$

Taking

$$\left(\frac{2k_0}{k_0 + k}\right)e^{-ik_0k} = \psi_1$$

and

$$\left[e^{-ikx} - \left(\frac{k_0 - k}{k_0 + k}\right)e^{ikx}\right] = \psi_2$$

$$\psi = A\psi_1 + D\psi_2 \qquad ...(4.89)$$

The equations (4.88) and (4.89) are the same and conditions of continuity at $x = 0$ are satisfied by both ψ_1 and ψ_2. Also another condition to be imposed upon the wave function is that ψ must remain finite as $|x|$ becomes finite.

4.13 Rectangular Potential Barrier

Consider a rectangular potential barrier of width a and height V_0 as shown in Fig. 4.5.

In regions *I* and *III*, the value of $V = 0$ and in region *II* for $x = 0$ to $x = a$, the potential energy V of the particle is constant and is equal to V_0.

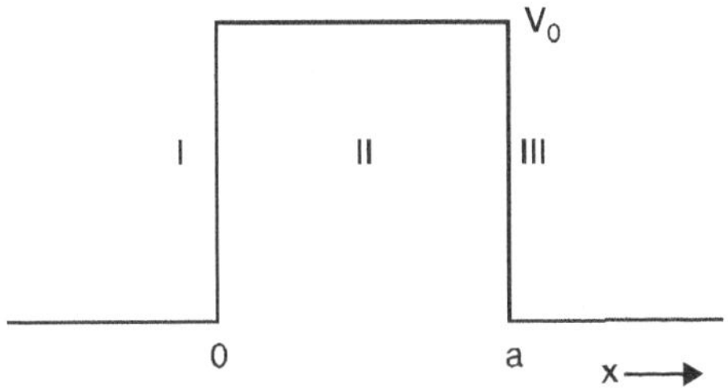

Fig. 4.5

Consider that a beam of particle is travelling from left to right. It is incident on the potential barrier. In region *I* the total energy of the particle is kinetic energy. In region *III* also the total energy is kinetic energy. In region *II*, the total energy is partly kinetic energy and partly potential energy.

If $E < V_0$ then according to quantum mechanics the probability of transmission is small but it has a definite value. The effect is called tunnel effect. However according to classical mechanics the particle cannot reach region *III* after crossing region *II*.

The time independent Schrodinger equation for a particle constrained to move along x-axis is given by

$$\frac{d^2\psi_1}{dx^2} + \frac{2m}{\hbar^2}(E - V)\psi = 0.$$

Taking wave function as ψ_1, ψ_2 and ψ_3 in regions *I*, *II* and *III* respectively,

1. In region I, $-\infty < x < 0$

$$V = 0$$

$$\therefore \quad \frac{d^2\psi}{dx^2} + \frac{2mE}{\hbar^2}\psi_1 = 0$$

Taking $$\frac{2mE}{\hbar^2} = k_0^2$$

$$\frac{d^2\psi_1}{dx^2} + k_0^2\psi_1 = 0 \qquad \text{...(4.90)}$$

2. For region II, $0 < x < a$, and $E < V_0$

$$V = V^0$$

$$\frac{d^2\psi}{dx^2} + \frac{2m}{\hbar^2}(E - V_0)\,\psi_2 = 0$$

$$\frac{d^2\psi_2}{dx^2} - \frac{2m}{\hbar^2}(V_0 - E)\psi_2 = 0$$

Taking $$\frac{2m\,(V_0 - E)}{\hbar^2} = k^2$$

$$\frac{d^2\psi_2}{dx^2} - k^2\psi_2 = 0 \qquad ...(4.91)$$

3. For region *III*, $a < x < \infty$

$$V = 0$$

$$\frac{d^2\psi_3}{dx^2} + \left(\frac{2mE}{\hbar^2}\right)\psi_3 = 0$$

$$\frac{d^2\psi_3}{dx^2} + k_0^2\psi_3 = 0 \qquad ...(4.92)$$

The solutions of equations (4.90), (4.91) and (4.92) are

$$\psi_1 = Ae^{ik_0x} + Be^{-ik_0x} \qquad ...(4.93)$$

$$\psi_2 = Ce^{kx} + De^{-kx} \qquad ...(4.94)$$

$$\psi_3 = Le^{ik_0x} + Me^{-ik_0x} \qquad ...(4.95)$$

As no particle comes from left in region III, M = 0

$$\therefore \qquad \psi_3 = Le^{ik_0x}$$

In equation (4.93) Ae^{ik_0x} represents the wave travelling along the *x*-axis with amplitude *A* and Be^{-ik_0x} represents the reflected wave along negative *x*-axis with amplitude *B*.

In equation (4.94) De^{-kx} represents exponentially decreasing wave moving along x-axis in the barrier and Ce^{kx} is the reflected wave within the barrier.

In equation (4.95) Le^{ik_0x} represents the transmitted wave moving along *x*-axis in region *III*.

Determination of values of coefficients

1. At $x = 0$

$$\psi_1 = \psi_2$$

From equations (4.93) and (4.94)

$$A + B = C + D \qquad ...(4.96)$$

Also their derivatives are equal at $x = 0$.

$$\therefore \qquad \frac{d\psi_1}{dx} = \frac{d\psi_2}{dx}$$

$$\therefore \qquad ik_0A - ik_0B = kC - kD$$

$$A - B = \left(\frac{k}{ik_0}\right)(C - D) \qquad ...(4.97)$$

From equations (4.96) and (4.97)

$$A = \frac{1}{2}\left[\left(1 + \frac{k}{ik_0}\right)C + \left(1 - \frac{i}{ik_0}\right)D\right] \qquad ...(4.98)$$

and

$$B = \frac{1}{2}\left[\left(1 - \frac{k}{ik_0}\right)C + \left(1 + \frac{k}{ik_0}\right)D\right] \qquad ...(4.99)$$

2. At $x = a$

$$\psi_2 = \psi_3$$

$$Ce^{ka} + De^{-ka} = Le^{ik_0a} \qquad ...(4.100)$$

Also their derivatives are equal at $x = a$.

$$\therefore \quad \frac{d\psi_2}{dx} = \frac{d\psi_3}{dx}$$

$$Cke^{ka} - Dk\,e^{-ka} = L\,ik_0\,e^{ik_0 a}$$

or
$$Ce^{ka} - De^{-ka} = \left(\frac{ik_0}{k}\right) L\,e^{ik_0 a} \qquad ...(4.101)$$

From equations (4.100) and (4.101)

$$C = \frac{1}{2}\left[\left(1 + \frac{ik_0}{k}\right) e^{-ka}\right]\left[Le^{ik_0 a}\right] \qquad ...(4.102)$$

$$D = \frac{1}{2}\left[\left(1 - \frac{ik_0}{k}\right) e^{-ka}\right]\left[Le^{ik_0 a}\right] \qquad ...(4.103)$$

Substituting these values of C and D in equation (4.98)

$$A = \frac{1}{2}\left[\left(1 + \frac{ik_0}{k}\right)\frac{1}{2}\left(1 + \frac{ik_0}{k}\right) e^{-ka}(Le^{ik_0 a})\right.$$
$$\left. + \left(1 + \frac{k}{ik_0}\right)\frac{1}{2}\left(1 - \frac{ik_0}{k}\right) e^{-ka}\,(Le^{ik_0 a})\right]$$

Rearranging and simplifying we get

$$A = \left[\left(\frac{e^{ka} + e^{-ka}}{2}\right) - \frac{1}{2}\left(\frac{k}{ik_0} + \frac{ik_0}{k}\right)\left(\frac{e^{ka} - e^{-ka}}{2}\right)\right](Le^{ik_0 a})$$

But
$$\frac{e^{ka} + e^{-ka}}{2} = \cosh\,(ka)$$

and
$$\frac{e^{ka} - e^{-ka}}{2} = \sinh\,(ka)$$

$$\therefore \quad A = [\cosh\,(ka) - \frac{1}{2}\left(\frac{k}{ik_0} + \frac{ik_0}{k}\right)\sinh\,(ka)]\,(Le^{ik_0 a})$$

$$A = [\cosh\,(ka) - \frac{1}{2}\left(\frac{k}{k_0} - \frac{k_0}{m}\right)\sinh\,(ka)]\,(Le^{ik_0 a}) \qquad ...(4.104)$$

Similarly, by substituting the values of C and D in equation (x) and rearranging and simplifying we get

$$B = -\frac{1}{2}\left(\frac{k}{k_0} + \frac{k_0}{k}\right)\sinh\,(ka)\,(Le^{ika_0}) \qquad ...(4.105)$$

4.14 Transmission Probability (Transmission Coefficient)

The transmission coefficient or probability T is given by

$$T = \left(\frac{L}{A}\right)\left(\frac{L}{A}\right)^*$$

Here $\left(\frac{L}{A}\right)^*$ is the complex conjugate of $\left(\frac{L}{A}\right)$.

From equation (4.104)

$$\left(\frac{A}{L}\right) = \left[\cos h\,(ka) + \frac{1}{2}\left(\frac{k}{k_0} - \frac{k_0}{k}\right)\sin h\,(ka)\right] e^{ik_0 a}$$

$$\left[\frac{A}{L}\right]^* = \left[\cosh(ka) - \frac{1}{2}\left(\frac{k}{k_0} - \frac{k_0}{k}\right)\sinh(ka)\right] e^{-ik_0 a}$$

$$\therefore \quad \left[\frac{A}{L}\right]\left[\frac{A}{L}\right]^* = \cosh^2(ka) + \frac{1}{4}\left(\frac{k}{k_0} - \frac{k_0}{k}\right)^2 \sinh^2(ka)$$

But $\cosh^2(ka) = 1 + \sinh^2(ka)$

$$\therefore \quad \left[\frac{A}{L}\right]\left[\frac{A}{L}\right]^* = 1 + \sinh^2(ka) + \frac{1}{4}\left(\frac{k}{k_0} - \frac{k_0}{k}\right)^2 \sinh^2(ka)$$

or

$$\frac{1}{T} = 1 + \frac{1}{4}\left[\left(\frac{k}{k_0} - \frac{k_0}{k}\right)^2 + 4\right]\sinh^2(ka)$$

$$\frac{1}{T} = 1 + \frac{1}{4}\left(\frac{k}{k_0} + \frac{k_0}{k}\right)^2 \sinh^2(ka) \qquad \text{...(4.106)}$$

But

$$\frac{k}{k_0} = \left[\left(\frac{2m(V_0 - E)}{\hbar^2}\right) \times \left(\frac{\hbar^2}{2mE}\right)\right]^{\frac{1}{2}}$$

$$\therefore \quad \frac{k}{k_0} = \left[\frac{V_0 - E}{\hbar^2}\right]^{\frac{1}{2}}$$

and

$$\frac{k_0}{k} = \left[\frac{E}{V_0 - E}\right]^{\frac{1}{2}}$$

Substituting these values in equation (4.106)

$$\frac{1}{T} = 1 + \frac{1}{4}\left[\left(\frac{V_0 - E}{E}\right)^{\frac{1}{2}} + \left(\frac{E}{V_0 - E}\right)^{\frac{1}{2}}\right]^2 \sinh^2(ka)$$

$$\frac{1}{T} = 1 + \frac{1}{4}\left[\frac{V_0^2}{E(V_0 - E)}\right]\sinh^2(ka)$$

or

$$T = \frac{1}{1 + \left[\frac{V_0^2}{4E(V_0 - E)}\right]\sinh^2(ka)} \qquad \text{...(4.107)}$$

Results :

1. Equation (4.107) shows that there is finite probability for a particle to cross the potential barrier or height V_0 even when $E < V_0$.

2. When $ka \geq 1$, in that case e^{ka} is large and e^{-ka} is extremely small as compared to 1,

$$\therefore \quad \sinh^2(ka) = 1 + \left(\frac{e^{ka} - e^{-ka}}{2}\right)^2 \approx \left(\frac{e^{2ka}}{4}\right)$$

$$\frac{1}{T} = 1 + \left[\frac{V_0^2}{4E(V_0 - E)}\right]\left[\frac{e^{2ka}}{4}\right]$$

Neglecting 1

$$\frac{1}{T} = \frac{V_0^2 e^{2ka}}{16E(V_0 - E)}$$

or
$$T = \frac{16E(V_0 - E)}{V_0^2} e^{-2ka}$$

or
$$T = \left(\frac{16E}{V_0}\right)\left(1 \frac{E}{V_0}\right) e^{-2ka} \qquad ...(4.108)$$

It means when $E < V_0$ and $ka \geq 1$, the value T decreases exponentially with increase in the value of a. Equation (4.108) also shows that if a is constant and $\frac{E}{V_0}$ decreases exponentially. The curve for T and $\frac{E}{V_0}$ is shown in Fig. 4.6.

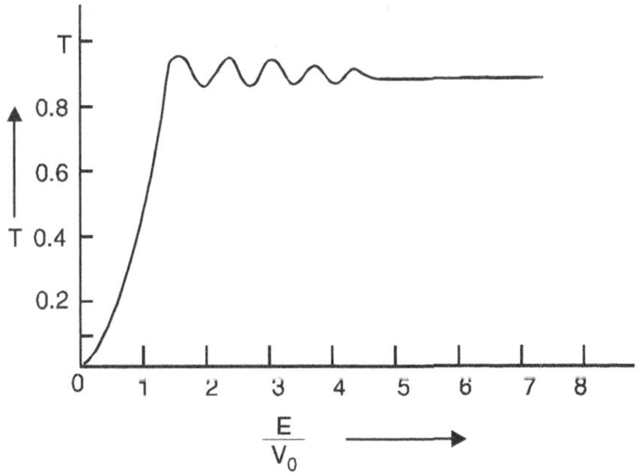

Fig. 4.6.

3. When the energy of the particle E approaches the potential energy V_0, in that case

$$ka = a\sqrt{\frac{2m(V_0 - E)}{\hbar^2}}$$

and its value is ≤ 1.

$\therefore$
$$\sinh(ka) = a\sqrt{\frac{2m(V_0 - E)}{\hbar^2}}$$

or
$$\sinh^2(ka) = \frac{2m(V_0 - E)\,a^2}{\hbar^2}$$

Substituting this value in equation (4.107)

$$T = \frac{1}{1 + \left(\frac{V_0^2}{4E(V_0 - E)}\right)\left(\frac{2m(V_0 - E)\,a^2}{\hbar^2}\right)}$$

$$T = \frac{1}{1 + \frac{m V_0^2 a^2}{2E\hbar^2}}$$

As E approaches V_0

$$T = \frac{1}{1 + \frac{m V_0 a^2}{2\hbar^2}}$$

Therefore when E approaches V_0 the value of T increases. However T is less than 1.

Special Case

When $E > V_0$

$$k_2 = \frac{2m(V_0 - E)}{\hbar^2} = -\frac{2m(E - V_0)}{\hbar^2}$$

or
$$k = i\sqrt{\frac{2m(E - V_0)}{\hbar^2}}$$

Taking
$$\sqrt{\frac{2m(E - V_0)}{\hbar^2}} = \alpha$$

$$k = i\,\alpha$$

$$\sinh(ka) = \frac{e^{ka} - e^{-ka}}{2} = \left[\frac{e^{i\alpha a} - e^{-i\alpha a}}{2}\right]$$

$$\sinh(ka) = i\left[\frac{e^{i\alpha a} - e^{-i\alpha a}}{2}\right]$$

$$\sinh(ka) = i\sin\alpha\, a$$

Squaring we get

$$\sinh^2(ka) = -\sinh^2(\alpha a)$$

Substituting this value of $\sinh^2(ka)$ in equation (4.107)

$$T = \frac{1}{1 - \left[\frac{V_0^2}{4E(V_0 - E)}\sin^2(\alpha\, a)\right]}$$

$$T = \frac{1}{1 + \left[\frac{V_0^2}{4E(E - V_0)}\sin^2(\alpha\, a)\right]} \qquad ...(4.109)$$

Equation (4.109) gives the value of *transmission coefficient for* $E > V_0$.

Results

1. When E approaches V_0

$$\alpha\, a = \sqrt{\frac{2m(E - V_0)}{\hbar^2}} \text{ will be far less than 1.}$$

$$\therefore \quad \sin(\alpha\, a) = \alpha\, a \text{ and } \sin^2(\alpha\, a) = (\alpha\, a)^2$$

$$\therefore \quad \sin^2(\alpha\, a) = \frac{2m(E - V_0)\,a^2}{\hbar^2}$$

Substituting this value in equation (4.109)

$$T = \frac{1}{1 + \left(\frac{V_0^2}{4E(E - V_0)}\right)\left(\frac{2m(E - V_0)\,a^2}{\hbar^2}\right)}$$

Taking $E \approx V_0$, and simplifying

$$T = \frac{1}{1 + \left(\frac{m V_0 a^2}{2\hbar^2}\right)} \qquad ...(4.110)$$

Equation (4.110) is same as equation (4.108) for $E < V_0$.

2. For $E > V_0$ in equation (4.109)

$$T = 1 \text{ when } \sin^2(\alpha a) = 0$$

i.e., when $\alpha a = n\pi$ where $n = 1, 2, 3, \ldots.$, etc.

or
$$a = \frac{n\pi}{\alpha} = \frac{n\pi}{\left[\dfrac{2m(E - V_0)}{\hbar^2}\right]^{\frac{1}{2}}} = \frac{n\pi\hbar}{[2m(E - V_0)]^{\frac{1}{2}}}$$

$$a = \frac{n\pi(h/2\pi)}{[2m(E - V_0)]^{\frac{1}{2}}}$$

$$a = \left(\frac{n}{2}\right)\left[\frac{h}{[2m(E - V_0)]^{\frac{1}{2}}}\right]$$

But
$$\frac{h}{[2m(E - V_0)]^{\frac{1}{2}}} = 0 \text{ (De-Broglie wavelength)}$$

or
$$a = n\left(\frac{\lambda}{2}\right)$$

Therefore
$$T = 1 \text{ when}$$
$$a = n\left(\frac{\lambda}{2}\right)$$

It means when the width of the potential barrier is an integral multiple of half wavelength, there is 100% transmission of the incident beam of the particles.

4.15 Reflection Coefficient (Reflection Probability)

The reflection coefficient

$$R = \left(\frac{B}{A}\right)\left(\frac{B}{A}\right)^*$$

From equations (4.104) and (4.105)
Here

$$A = \left[\cosh(ka) + \frac{1}{2}\left(\frac{k}{k_0} - \frac{k_0}{k}\right)\sinh(ka)\right]\left(Le^{ik_0 a}\right)$$

and
$$B = -\frac{1}{2}\left(\frac{k}{k_0} + \frac{k_0}{k}\right)\sinh(ka)\left(Le^{ik_0 a}\right)$$

$\therefore$
$$\left(\frac{B}{A}\right) = \frac{-\dfrac{1}{2}\left(\dfrac{k}{k_0} + \dfrac{k_0}{k}\right)\sinh(ka)}{\cosh(ka) + \dfrac{1}{2}\left(\dfrac{k}{k_0} - \dfrac{k_0}{k}\right)\sinh(ka)}$$

and
$$\left(\frac{B}{A}\right)^* = \frac{\dfrac{1}{2}\left(\dfrac{k}{k_0} - \dfrac{k_0}{k}\right)\sinh(ka)}{\cosh(ka) - \dfrac{1}{2}\left(\dfrac{k}{k_0} - \dfrac{k_0}{k}\right)\sinh(ka)}$$

$\therefore$
$$R = \left(\frac{B}{A}\right)\left(\frac{B}{A}\right)^*$$

$$R = \frac{\frac{1}{4}\left[\frac{k}{k_0} + \frac{k_0}{k}\right]^2 \sinh^2(ka)}{\cosh^2(ka) + \frac{1}{4}\left[\frac{k}{k_0} - \frac{k_0}{k}\right] \sinh^2(ka)}$$

But $\cosh^2(ka) = 1 + \sinh^2(ka)$

$$\therefore \quad R = \frac{\frac{1}{4}\left[\frac{k}{k_0} + \frac{k_0}{k}\right]^2 \sinh^2(ka)}{\left[1 + \frac{1}{4}\left(\frac{k}{k_0} - \frac{k_0}{k}\right)^2\right] \sinh^2(ka)}$$

$$R = \frac{\frac{1}{4}\left[\frac{k}{k_0} + \frac{k_0}{k}\right]^2 \sinh^2(ka)}{\left[1 + \frac{1}{4}\left(\frac{k}{k_0} + \frac{k_0}{k}\right)^2\right] \sinh^2(ka)} \quad \text{...(4.111)}$$

From equation (4.106) and

$$T = \frac{1}{1 + \frac{1}{4}\left(\frac{k}{k_0} + \frac{k_0}{k}\right)^2 \sinh^2(ka)} \quad \text{...(4.112)}$$

Adding equations (4.111) and (4.112)

$$R + T = \frac{1 + \frac{1}{4}\left(\frac{k}{k_0} + \frac{k_0}{k}\right)^2 \sinh^2(ka)}{1 + \frac{1}{4}\left(\frac{k}{k_0} + \frac{k_0}{k}\right)^2 \sinh^2(ka)}$$

$$\therefore \quad R + T = 1$$

$$\text{or} \quad R = 1 - T \quad \text{...(4.113)}$$

The following effects can be easily explained with the help of tunnel effect.

1. Emission of Alpha particles from a radioactive element.
2. Switching action of tunnel diode.
3. Emission of electrons from cold metallic surfaces.

Example 4.1. *A beam of electrons each of energy* $E = 4$ eV *is incident on a potential barrier of height* $V_0 = 6$ eV. *The width of the potential barrier is 10 Å. Calculate the percentage transmission of the beam through its barrier.*

Solution.

$$T = \left[\frac{16E}{V_0}\right]\left[1 - \frac{E}{V_0}\right] e^{-2ka} \quad \text{...(1)}$$

$$k = \left[\frac{2m(V_0 - E)}{\hbar^2}\right]$$

$$2ka = \frac{2a}{\hbar}[2m(V_0 - E)]^{\frac{1}{2}}$$

Here

$$a = 10\text{Å} = 10^{-9} \text{ m}$$

$$E = 4 \text{ eV} = 4 \times 1.6 \times 10^{-19} = 6.4 \times 10^{-19} \text{ J}$$

$$V_0 = 6 \text{ eV} = 6 \times 1.6 \times 10^{-19} = 9.6 \times 10^{-19} \text{ J}$$

$$\hbar = 1.054 \times 10^{-31} \text{ J–s}$$

$$\therefore \qquad 2ka = \left[\frac{2\times10^{-9}}{1.054\times10^{-34}}\right][2\times9.6\times10^{-31}\times3.2\times10^{-19}]^{\frac{1}{2}}$$

$$= 14.48$$

Substituting the values in equation (1)

$$T = \left[\frac{16\times6.4\times10^{-19}}{9.6\times10^{-19}}\right]\left[1-\frac{6.4\times10^{-19}}{9.6\times10^{-19}}\right]e^{-14.48}$$

$$T = \left[\frac{32}{3}\right]\left[\frac{1}{3}\right]\left[5.145\times10^{-7}\right]$$

$$T = \mathbf{1.829\times10^{-6}}$$

Percentage Transmission = $\mathbf{1.829 \times 10^{-6} \times 100 = 1.829 \times 10^{-4}\%}$

Example 4.2. *Calculate the width of the potential barrier of an α particle emitted with kinetic energy 4* MeV *from a radioactive atom of atomic weight A = 222 and atomic number Z = 86.*

Solution. Here $r_0 = (1.5 \times 10^{-15})A^{1/3}$, But $A = 222$

$$r_0 = (1.5\times10^{-15})(222)^{1/3}$$

$$r = 9\times10^{-15}\text{ m}$$

Let r_1 be the distance from the centre of the nucleus where potential energy of the α particle is equal to the kinetic energy.

$$E = \frac{2(Z-2)e^2}{4\pi\varepsilon_0 r_1}$$

$$r_1 = \frac{2(Z-2)e^2}{4\pi\varepsilon_0 E}$$

Here $Z = 86$

$$\frac{1}{4\pi\varepsilon_0} = 9\times10^9$$

$$e = 1.6\times10^{-19}\text{ C}$$

$$E = 4\text{ MeV} = 4\times10^{-6}\times1.6\times10^{-19}\text{ J}$$

or $E = 6.4\times10^{-13}$ J

$$\therefore \qquad r_1 = \frac{2\times84\times(1.6\times10^{-19})^2\times9\times10^9}{6.4\times10^{-13}}$$

$$r_1 = 60.48\times10^{-15}\text{m}$$

Width of the potential barrier

$$a = r_1 - r_0$$

or $a = 60.48\times10^{-15} - 9\times10^{-15}$

$$= \mathbf{51.48\times10^{-15}}\text{ m}$$

Example 4.3. *A beam of electrons is incident on a potential barrier 5* eV *high, 0.5 nm wide. What energy should they have if half of them are to get through the barrier ?* [*G.N.D.U., 1992*]

$$h = 1.054\times10^{-34}\text{ J-s}$$

Solution. Here $V_0 = 5\text{ eV} = 5\times1.6\times10^{-19}$ J

$$V_0 = 8\times10^{-19}\text{ J}$$

$$m = 9.1\times10^{-31}\text{ kg}$$

$$a = 0.5\text{ nm} = 0.5\times10^{-9}\text{ m}$$

$$a = 5 \times 10^{-10} \text{ m}$$

$$T = \frac{1}{2}$$

$$E = ?$$

$$T = \frac{1}{1 + \dfrac{m V_0^2 a^2}{2 E \hbar^2}}$$

$$1 + \left(\frac{m V_0^2 a^2}{2 E \hbar^2} \right) = \frac{1}{T}$$

$$\frac{m V_0^2 a^2}{2 E \hbar^2} = \frac{1}{T} - 1 = \left(\frac{1}{1/2} \right) - 1 = 1$$

$$\therefore \quad E = \frac{m V_0^2 a^2}{2 E \hbar^2}$$

$$E = \frac{(9.1 \times 10^{-31})(8 \times 10^{-19})^2 \times (5 \times 10^{-10})^2}{2 \times (1.054 \times 10^{-34})^2}$$

$$E = 65.53 \times 10^{-19} \text{ J}$$

$$= \frac{65.53 \times 10^{-19}}{1.6 \times 10^{-19}} \text{ eV}$$

$$= \mathbf{40.95 \text{ eV}}$$

Example 4.4. *Calculate the zero point energy of a system, constrained to move only in one direction, consisting of mass 1 gram connected to fixed point by a spring which is stretched 1 cm by a force of 1,000 dynes.* [*Delhi*]

Solution. Displacement $x = A \sin 2 \pi \nu t$

Restoring Force $= 4 \pi^2 r^2 mx$

Here

$$m = 10^{-3} \text{ kg}$$

$$x = 10^{-2} \text{ m}$$

$$F = 10^{-2} \text{ N}$$

$$F = 4 \pi^2 \nu^2 (10^{-3}) \times (10^{-2})$$

$$10^{-2} = 4 \pi^2 \nu^2 (10^{-5})$$

$$4 \pi^2 \nu^2 = 1000$$

or

$$\nu = \frac{(1000)^{\frac{1}{2}}}{2\pi}$$

Zero point energy

$$U = \frac{1}{2} h\nu$$

$$U = \frac{6.6 \times 10^{-34} \times (1000)^{\frac{1}{2}}}{2 \times 2 \times 3.14}$$

$$U = \mathbf{1.66 \times 10^{-33} \text{ J}}$$

4.16 Solution of the Schrödinger Equation in Spherical Co-ordinates

In atoms, the potential has Spherical symmetry. Hence we write the Schrödinger equation in spherical co-ordinates using the relations :

$x = r \sin \theta \cos \phi$

$y = r \sin \theta \sin \phi$

and $z = r \cos \theta$.

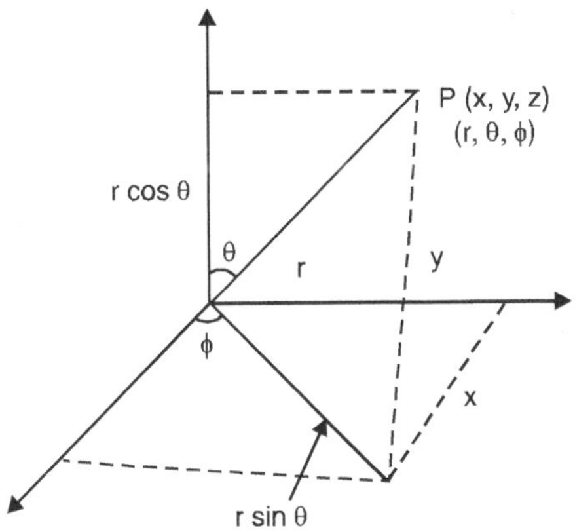

Fig. 4.7

$$\therefore \quad \nabla^2 = \frac{1}{r^2}\frac{\partial}{\partial r}\left(r^2\frac{\partial}{\partial r}\right) + \frac{1}{r \sin\theta}\frac{\partial}{\partial \theta}\left(\sin\theta\frac{\partial}{\partial\theta}\right) + \frac{1}{r^2 \sin^2\phi}\frac{\partial^2}{\partial\phi^2} \qquad ...(4.114)$$

Now, for the Hydrogen atom the Potential energy is given by

$$V(r) = -\frac{\hbar e^2}{4\pi\epsilon_0 r}$$

The wave function Ψ is a function of polar co-ordinates (r, θ, ϕ).

$\therefore$ The Schrödinger's equation is

$$\frac{1}{r^2}\frac{\partial}{\partial r}\left(r^2\frac{\partial\Psi}{\partial r}\right) + \frac{1}{r^2 \sin\theta}\frac{\partial}{\partial\theta}\left(\sin\theta\frac{\partial\Psi}{\partial\theta}\right) + \frac{1}{r^2\sin^2\theta}\frac{\partial^2\Psi}{\partial\phi^2} + \frac{2m}{\hbar^2}(E - V)\Psi = 0$$

Multiplying $r^2 \sin^2\theta$ we get

$$\sin^2\theta\frac{\partial}{\partial r}\left(r^2\frac{\partial\Psi}{\partial r}\right) + \sin\theta\frac{\partial}{\partial\theta}\left(\sin\theta\frac{\partial\Psi}{\partial\theta}\right) + \frac{\partial^2\Psi}{\partial\phi^2} + \frac{2mr^2\sin^2\theta}{\hbar^2}\left(\frac{E + e^2}{4\pi t_0 r}\right)\Psi = 0 \qquad ...(4.115)$$

Let the wave function $\psi(r, \theta, \phi)$ be written as the product of $R(r)$ function of r only, $\Theta(\theta)$ function of θ only, and $\Phi(\phi)$ function of ϕ only.

$$\therefore \quad \Psi(r, \theta, \Phi) = R(r), \Theta(\theta), \Phi(\phi)$$

Substituting in equation and writing as three terms and divide by ψ we get,

$$\frac{\partial\Psi}{\partial r} = \Theta\phi\frac{dR}{dr}, \frac{\partial\Psi}{\partial\theta} = R\Theta\frac{d\theta}{d\theta}, \frac{\partial^2\Psi}{\partial\phi^2} = R\Theta\frac{d^2 I}{d\phi^2}$$

Note the use of differential equations instead of partial derivatives because each function depends only on one variable.

$$\therefore \quad \frac{\sin^2\theta}{R}\frac{\partial}{\partial r}\left(r^2\frac{dR}{dr}\right) + \frac{\sin\theta}{\Theta}\frac{\partial}{\partial\theta}\left(\sin\theta\frac{d\Theta}{d\theta}\right) + \frac{1}{\Phi}\frac{d^2\Phi}{d\theta^2} + \frac{2mr^2\sin^2\theta}{\hbar^2}\left(\frac{E + e^2}{4\pi t_0 r}\right) = 0 \qquad ...(4.116)$$

$$\therefore \quad \frac{\sin^2\theta}{R}\frac{\partial}{\partial r}\left(r^2\frac{dR}{dr}\right) + \frac{\sin\theta}{\Theta}\frac{\partial}{\partial\theta}\left(\sin\theta\frac{d\Theta}{d\theta}\right) + \frac{2mr^2\sin^2\theta}{\hbar^2}\left[E + \frac{e^2}{4\pi t_0 r}\right] = -\frac{1}{\Theta}\frac{d^2\Phi}{d\phi^2} \qquad ...(4.117)$$

Since R.H.S. is a function of ϕ only and L.H.S. is r and ϕ.

We write each equal to a constant $= m_l^2$

$$\therefore \quad \frac{d^2\Phi}{d\phi^2} = -m_l^2\Phi \quad ...(4.118)$$

and $$\frac{\sin^2\theta}{R}\frac{\partial}{\partial r}\left(r^2\frac{\partial R}{\partial r}\right)+\frac{\sin\theta}{\Theta}\frac{\partial}{\partial\theta}\left(\sin\theta\frac{d\Theta}{d\theta}\right)+\frac{2mr^2\sin^2\theta}{\hbar^2}\left[E+\frac{e^2}{4\pi\epsilon_0 r}\right] = m_l^2 \quad ...(4.119)$$

Also we can divide by $\sin^2\theta$ and transferring, the terms depending on r to the left in Eq(2)

$$\frac{1}{R}\frac{d}{dr}\left(r^2\frac{dR}{dr}\right)+\frac{2m}{\hbar^2}\left(E+\frac{e^2}{4\pi\varepsilon_0 r}\right) = \frac{m_l^2}{\sin^2\theta}-\frac{1}{\Theta\sin\theta}\frac{\partial}{\partial\theta}\left(\sin\theta\frac{d\Theta}{d\theta}\right)$$

$$\therefore \frac{1}{R}\frac{d}{dr}\left(r^2\frac{dR}{dr}\right)+\frac{2m}{\hbar^2}\left(E+\frac{e^2}{4\pi\epsilon_0 r}\right) = b = l\,(l+1) \quad ...(4.120)$$

and $$\frac{m_l^2}{\sin^2\theta}-\frac{1}{\Theta\sin\theta}\frac{d}{d\theta}\left(\sin\theta\frac{d\Theta}{d\theta}\right) = b = l\,(l+1) \quad ...(4.121)$$

Quantum Numbers

Hence we have three Eqs. for three variables.

Equation (1) $\frac{d^2\Phi}{d\phi^2}+m_l^2\,\Phi = 0$ is easiest to solve

and has sol. $\phi = Ae^{im_l\phi}$ where A = Const.

The function Φ must be periodic *i.e.*,

$$\Phi\,(\phi+2\pi) = \Phi(\phi)$$

$$\therefore \quad e^{im_l(\psi+2\pi)} = e^{im_l\phi}$$

Divide by $e^{im_l\phi}$, we get

$$e^{2\pi i m_l} = 1$$

$$\therefore \quad e^{ix} = \cos x + i\sin x = 1$$

$$\therefore \quad \cos 2\pi m_l = 1$$

i.e., $m_l = 0, \pm 1, \pm 2, \pm 3 .... \pm l$.

Hence we see that magnetic quantum number m_l has arisen from the Schrödinger's equation.

Normalising Φ *i.e.*,

$$\int_0^{2\pi}\Phi^*(\phi)\,\Phi(\phi)\,d\phi = 1 \qquad \therefore A = \frac{1}{\sqrt{2\pi}}$$

Solutions of Eqs. (4.120) and (4.121) are more involved. The eqn. for θ takes the form of Legendres's differential equation. All solutions of equation become infinite except those for which $b = l\,(l+1)$ where $l \geq 0$ and also $l > |m_l|$.

Since Eq. for r is the only one which contains E the energy we see that $E = -\frac{me^2}{32\pi^2\epsilon_0^2\hbar^2}\frac{1}{n^2}$ are the only ones which are finite.

$$E = -\frac{13.6}{n^2}\text{eV} \quad \text{where } n = 1, 2, 3, 4,......$$

and n is called the principal quantum number.

Now $n \geq l+1$ $\quad\therefore n = 0, 1, 2, 3......(n-1)$

l is the orbital quantum number.

Now Eq. (4.120) after multiplying by R becomes

$$r^2 \frac{dR}{dr^2} + 2r \frac{dR}{dr} - \frac{2mr^2}{\hbar^2}(E - V) = 0 \qquad ...(4.122)$$

The solution is of form $R = \exp(-r/a)$

Substituting these in Eq. (4.122) we get,

$$\frac{1}{a^2} + \frac{2m}{\hbar^2} E = 0$$

and $$\frac{-2}{a} + \frac{2m}{\hbar^2} \frac{Ze^2}{4\pi \in_0} = 0$$

Since coefficients of r and r^2 must vanish separately

Solving we get $a = \dfrac{\hbar^2 4\pi \in_0}{mZe^2}$ which is the first Bohr radius.

$$a_0 = 0.529 \times 10^{-10} \text{ m.}$$

The Eigen functions are obtained by Normalising the wave function.

The dependence of R, Θ and Φ on the quantum numbers n, l, m_l we write

$$\Psi = R_{nl} \Theta_{lm_l} \phi m_l$$

The wave functions R, Θ and Φ with Ψ are given for a few Quantum numbers.

n	l	m_l	ϕ	θ	R	Ψ
1	0	0	$\frac{1}{\sqrt{2\pi}}$	$\frac{1}{\sqrt{2}}$	$\frac{2}{a_0^{\frac{3}{2}}} e^{-r/a_0}$	$\frac{1}{\sqrt{\pi}\, a_0^{\frac{3}{2}}} e^{-\frac{r}{a_0}}$
2	0	0	$\frac{1}{\sqrt{2\pi}}$	$\frac{1}{\sqrt{2}}$	$\frac{1}{2\sqrt{2}\, a_0^{\frac{3}{2}}} \left(\frac{2-r}{a_0}\right) e^{\frac{-r}{2a_0}}$	$\frac{1}{4\sqrt{2}\, a_0^{\frac{3}{2}}} \left(\frac{2-r}{a_0}\right) e^{-\frac{r}{2a_0}}$

EXERCISE IV

1. State the Physical significance of the wave function. What is a well-behaved wave function ?
2. Obtain the time dependent Schrödinger's equation.
3. Obtain the time independent Schrödinger's equation.
4. Introduce time-independent Schrödinger's equation for one dimensional potential barrier and solve it.
5. What is a harmonic oscillator ? Write down the Schrodinger's equation for a quantum mechanical harmonic oscillator. Use this equation to explain the formation of energy levels in a quantum mechanical oscillator. How do you explain the existence of zero point energy ? Also write the first three normalized wave functions of the harmonic oscillator.
6. Discuss the behaviour of a particle in a box constrained to move only along the x-axis.
7. Derive one-dimensional time-dependent Schrödinger's equation for a particle with potential energy $V(x, t)$.
8. Derive an expression for the transmission coefficient of a particle through a rectangular potential barrier for an energy less than the barrier height. Discuss qualitatively how an α -decay is explained in the light of the above derivation.

9. Using quantum mechanics, derive Bohr's postulate :

$\upsilon = \frac{(E_m - E_n)}{h}$ where υ is the frequency of the radiation emitted by an atom dropping from an energy level E_m to a lower level E_n.

10. Discuss the motion of a particle in a three Dimensional box. Find the Eigen-values and Eigen functions. Explain non-degenerate and degenerate energy levels.

11. Write down the Schrödinger's wave equation for the Hydrogen atom. From this equation obtain differential equations for separate variables. Hence obtain solution for Φ.

12. Show that $\hat{x}\hat{p}_x - \hat{p}_x\hat{x} = i\hbar$ and $\left[\hat{x}\,\hat{p}_y\right] = 0$.

13. Obtain solutions of the Schrodinger's equation for the motion of an electron with an initial energy E approaching a potential barrier V (x) given by

$$V(x) = V \text{ for } x \geq 0$$
$$\text{and } V(x) = 0 \text{ for } x < 0, \text{ the cases,}$$
$$\text{for } E > V \text{ and } E < V.$$

14. Obtain the expressions for the transmission and reflection coefficient for a particle in a rectangular potential barrier.

15. What is meant by the expectation value of a dynamical variable. Find the operators for the linear momentum, Position and kinetic energy of a system ?

16. What is meant by normalization and orthogonality of a wave function ?

5

ATOMIC PHYSICS

5.1 Optical Spectra and Electronic Structure

In the solution of the Schrödinger's equation the main feature is the appearance of the three quantum numbers n, l and m_l.

The following rules have to be satisfied

$$n = 1, 2, 3, 4, 5 \ldots\ldots$$

$$l = 0, 1, 2 \ldots\ldots (n-1)$$

and $$m_l = 0, \pm 1, \pm 2 \ldots\ldots \pm l.$$

Hydrogen Spectra

Rydberg suggested the optical series as

$$\bar{\upsilon} = \bar{\upsilon}_\infty - \frac{RZ^2}{(n+\phi)^2}$$

where R is the Rydberg's constant, n is an integer and ϕ is a fraction less than 1.

In terms of Rydberg's formula the four intense series are :

For Principal Series $$\bar{\upsilon} = P_\infty - \frac{R}{(n+P)^2} \qquad n = 2, 3, 4 \ldots.$$

Sharp Series $$\bar{\upsilon} = S_\infty - \frac{R}{(n+S)^2} \qquad n = 2, 3, 4 \ldots.$$

Diffuse Series $$\bar{\upsilon} = D_\infty - \frac{R}{(n+D)^2} \qquad n = 3, 4, 5 \ldots.$$

Fundamental Series $$\bar{\upsilon} = F_\infty - \frac{R}{(n+F)^2} \qquad n = 4, 5, 6 \ldots.$$

Note, if we replace P_∞, S_∞, D_∞ and F_∞ for $\bar{\upsilon}_\infty$ in Rydberg's formula and the variable term ϕ by S, P, D and F.

i.e. $$P_\infty = \frac{R}{(1+S)^2},\; S_\infty = \frac{R}{(2+P)^2},\; D_\infty = \frac{R}{(2+P)^2} \text{ and } F_\infty = \frac{R}{(3+D)^2}$$

i.e., If nP is used to denote $\frac{R}{(n+P)^2}$, nS for $\frac{R}{(n+S)^2}$ etc. then Principal series is given by

$$\bar{\upsilon} = 1S - nP \qquad n = 2, 3, 4, 5$$

Sharp $$\bar{\upsilon} = 2P - nS \qquad n = 2, 3, 4, 5$$

Diffuse $$\bar{\upsilon} = 2P - nD \qquad n = 3, 4, 5, 6$$

Fundamental by $$\bar{\upsilon} = 3D - nF \qquad n = 4, 5, 6, 7$$

Orbital Quantum Number

The radial part of the equation for Ψ

i.e. $$\frac{1}{r^2}\frac{d}{dr}\left(r^2\frac{dR}{dr}\right)+\left[\frac{2m}{\hbar^2}\left(E+\frac{e^2}{4\pi\epsilon_0 r}\right)-\frac{l(l+1)}{r^2}\right]R=0 \qquad ...(5.1)$$

Contains E which has nothing to do with the radial motion, E being the total energy.

The Kinetic Energy = $K.E_{radial} + K.E_{orbital} = K.E.$

and the potential energy $V = \frac{-e^2}{4\pi\epsilon_0 r}$.

$\therefore$ Equation for $E = K.E_{radial} + K.E_{orbital} + \frac{-e^2}{4\pi\epsilon_0 r}$

Substitute for E and rewriting as

$$\frac{1}{r^2}\frac{d}{dr}\left(r^2\frac{dR}{dr}\right)+\frac{2m}{\hbar^2}\left[K.E_{radial}+K.E_{orbital}-\frac{\hbar^2\, l/l+1}{2mr^2}\right]R=0 \qquad ...(5.2)$$

If $K.E_{radial}$ alone is required we find that

$$K.E_{orbital} = \frac{\hbar^2\, l(l+1)}{2m\,r^2}. \qquad ...(5.3)$$

$\therefore$ $K.E_{orbital} = 1/2\, mv^2_{orbital}$ and angular momentum $L = mv_{orbital}\, r$

$$\therefore \qquad K.E_{orbtial} = \frac{L^2}{2mr^2} \qquad ...(5.4)$$

Hence we have $$\frac{L^2}{2mr^2} = \frac{\hbar l^2(l+1)}{2mr^2}. \qquad ...(5.5)$$

$\therefore$ The electron's angular momentum $L = \sqrt{l(l+1)}\hbar$

$$l = 0, 1, 2, ... (n-1) \text{ and } \hbar = 1.054\times10^{-34} \text{ J.S.}$$

Vector Atom Model

5.2 Magnetic Quantum Number

The angular momentum $\vec{L}$ is a vector whose $|\vec{L}| = L$ is given by the orbital quantum number l (direction is given by the right hand rule).

$$|\vec{L}| = \sqrt{l(l+1)\hbar}. \qquad ...(5.6)$$

When an electron with angular momentum $\vec{L}$ interacts with an external magnetic field $\vec{B}$, m_l gives the direction of $\vec{L}$ *i.e.* component in the field direction called the magnetic quantum number. This is referred to as space quantisation.

$\therefore$ $L_Z\, m_l\, \hbar$ $\qquad m_l = 0, \pm1, \pm2, ... \pm l,$

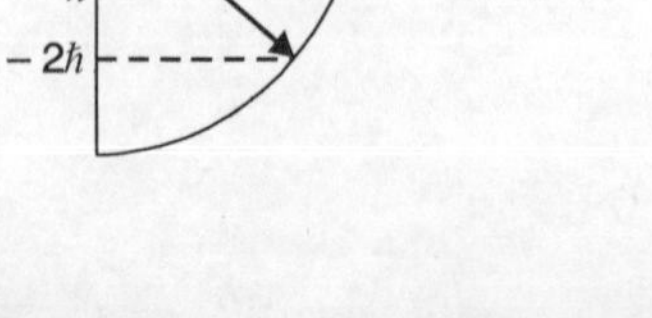

$\because \qquad m_l = l\cos\theta$

$\therefore \qquad \cos\theta = \frac{m_l}{\sqrt{l(l+1)}}$

Fig. 5.1

When	$l = 0$	$L_Z = 0$ only
For	$l = 1$	$L_Z = \hbar$, 0, and $-\hbar$
	$l = 2$	$L_Z = 2\hbar, \hbar, 0, -\hbar$ and $-2\hbar$

Note that in the absence of an external magnetic field, the direction of Z is arbitrary.

5.3 Electron Spin

Uhlenbeck and Goudsmit introduced the concept of electron spin in order to account for the fine structure of the lines in the spectral series and to explain the anomalous Zeeman effect.

It need not be assumed that the electron is spinning or turning on its axis. According to Dirac's theory the electron does behave as if it had an internal angular momentum and an associated magnetic moment.

$\therefore$ Every electron has an intrinsic angular momentum called spin. The quantum number S describes the spin angular momentum of the electron. The only value of S is 1/2. The magnitude of the spin quantum number $\vec{S}$ is $|\vec{S}| = \sqrt{S(S+1)}\hbar = \sqrt{3}/2\hbar$.

Vectorially $\vec{S}$ can be represented by a length 1/2 in units of $\hbar$ and its projection on the magnetic field direction gives m_s the magnetic spin quantum number.

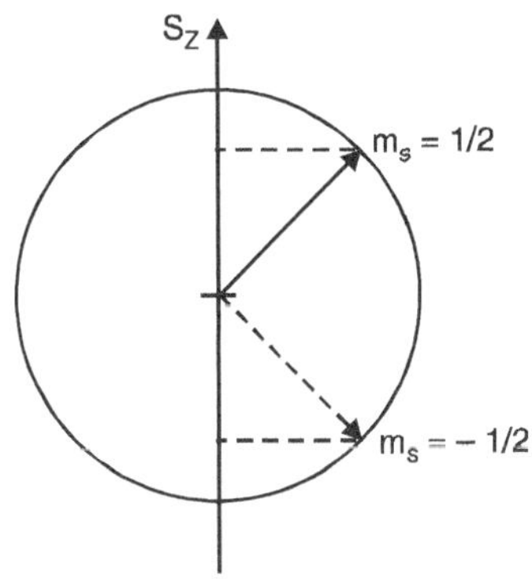

Fig. 5.2

$|S_Z| = m_s \hbar = 1/2\ \hbar$. m_s is called the magnetic spin quantum number.

5.4 Russel-Saunder's or *L-S* Coupling

For atoms with more than one electron; $\vec{l}$ represents the orbital angular momentum of a single electron and $\vec{L}$ for the total angular momentum of all the electrons.

$\therefore$ Total angular momentum $\vec{L} = \vec{l_1} \oplus \vec{l_2} \oplus \vec{l_3} + ...$

$\oplus$ implies quantised vector addition.

$$\text{Magnitude of} \quad \vec{L} = |\vec{L}| = \sqrt{L(L+1)}\ \hbar. \qquad ...(5.7)$$

In atoms that are not too heavy, containing many valance electrons the electrical interaction between the electrons is large and is due to the individual orbital angular momentum of the electrons which add to form a resultant $\vec{L}$ and all their spin angular momentum vectors combine to form a resultant $\vec{S}$. This type of coupling which occurs most frequently is Russel-Saunders or *L-S* type of coupling (also called normal coupling).

The process may be symbolically represented as $(s_1 \oplus s_2 \oplus s_3 + ...) + (l_1 \oplus l_2 \oplus l_3 + ...) = \vec{S} \oplus \vec{L} = \vec{J}$.

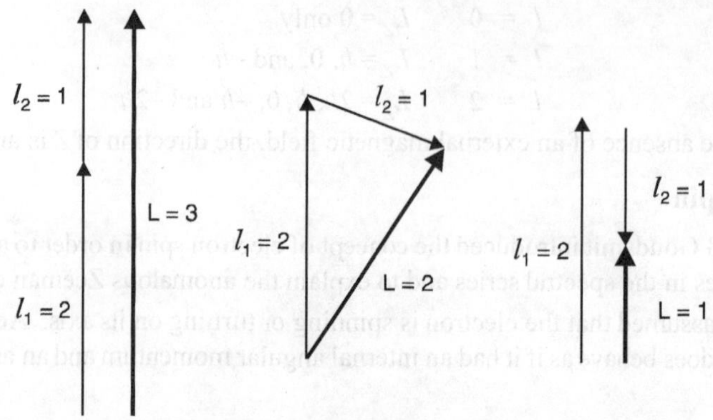

Fig. 5.3

$$\therefore \qquad |\vec{L}| = \sqrt{L(L+1)}\,\hbar = \sqrt{2}\,\hbar, \sqrt{6}\,\hbar \text{ and } \sqrt{12}\,\hbar$$

The total orbital angular momentum vector $\vec{L}$ can have only certain values in space. The discrete values are given by

$|\vec{L}_Z| = M_L \hbar$ where M_L is the total orbital quantum no. and takes values from $-L$ to $+L$ *i.e.*, $(2L+1)$ values.

Also the spin angular momentum vectors of the electrons combine to form a resultant spin angular momentum $\vec{S}$.

$$\vec{S} = \vec{s}_1 \oplus \vec{s}_2 \oplus \vec{s}_3 + ... \qquad ...(5.8)$$

where $|\vec{S}| = \sqrt{S(S+1)}\,\hbar$ and S = Total spin quantum number.

The vector sum $\vec{S}$ of several electrons is odd multiple of 1/2 for odd numbers of electrons and for even no. of electrons it is an integer.

Hence we see that the spin vectors are either parallel or anti-parallel.

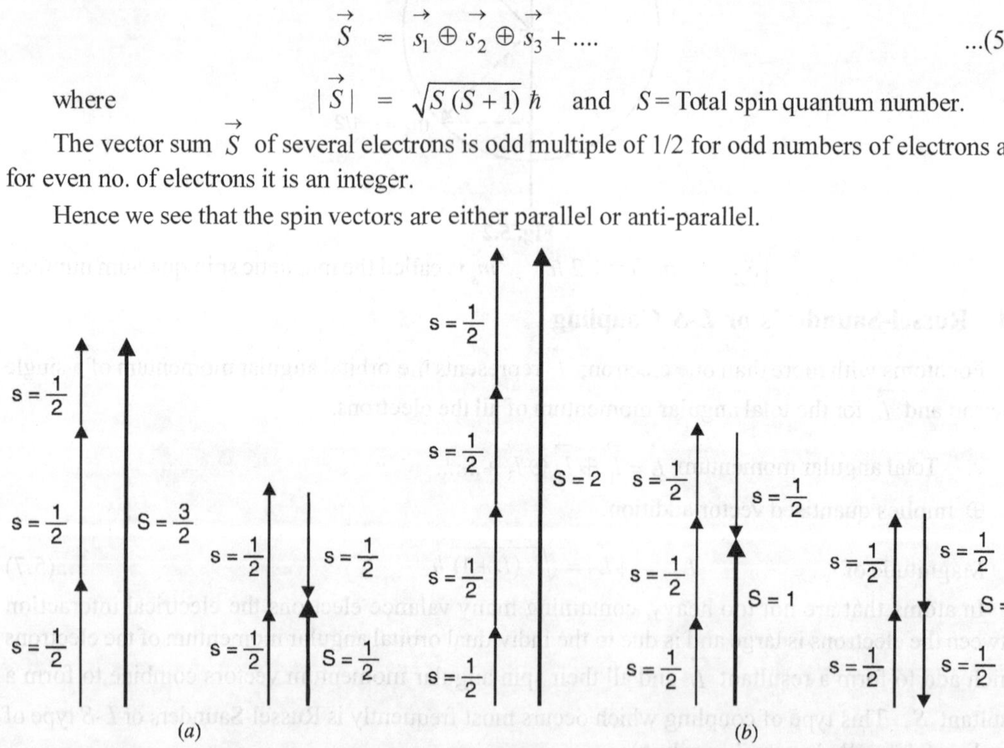

(*a*) (*b*)

Fig. 5.4 Addition of electron spin vectors (*a*) for three electrons (*b*) for four electrons.

The direction of $\vec{S}$ is quantised and its projection on any fixed direction has discrete values given by

$$|\vec{S}|_Z = M_S \hbar$$

$$M_S = \text{Total magnetic spin quantum number.}$$

For $\vec{S} = 3/2$ M_S has values $\pm 3/2, \pm 1/2$.

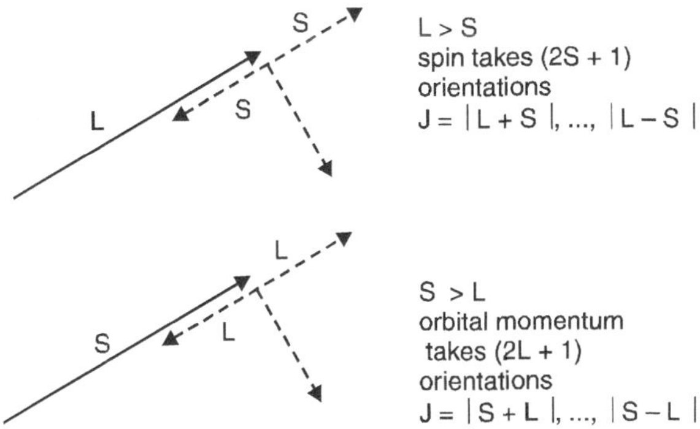

Fig. 5.5. In Russell-Saunders coupling, the total orbital momentum (denoted by number L) and the total spin (denoted by S) combine to form the total angular momentum J. Above : for $L > S$, the possible J values are given by the simple vector model as $J = | L + S |, ..., | L - S |$. Below : For $S > L$, the possible values are $J = | S + L |, ..., | S - L |$.

The total angular momentum of the atom is then given by

$$\vec{J} = \vec{L} \oplus \vec{S}.$$

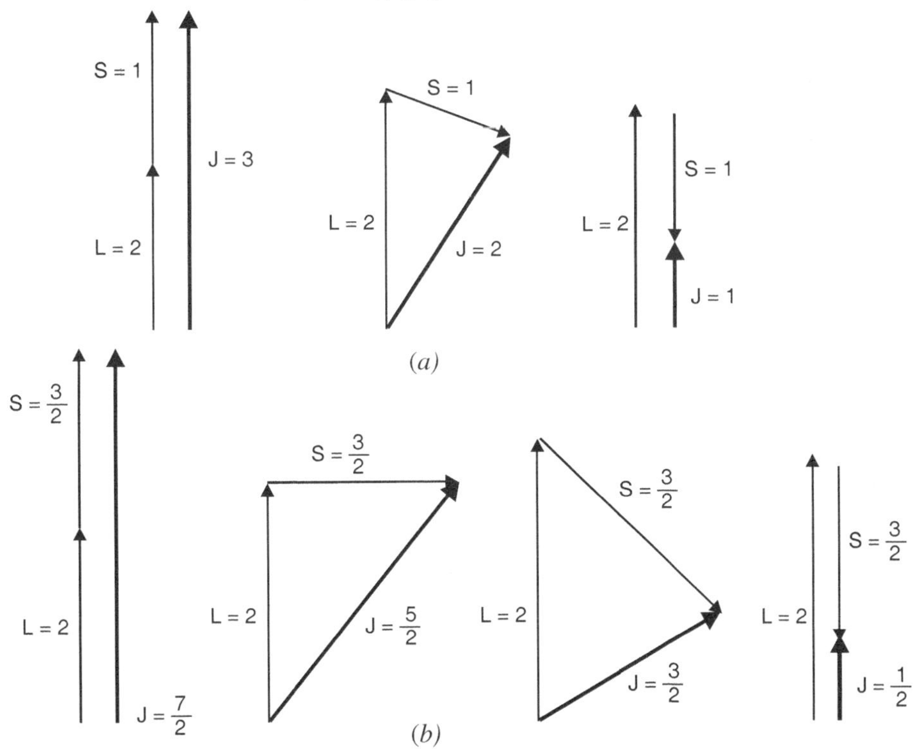

Fig. 5.6. (*a*) Addition of the vectors L and S to yield integral values of J. (*b*) Addition of the vectors L and S to yield odd half-integral values of J.

$\vec{J}$ must be an integer if $\vec{S}$ is an integer.

$\vec{J}$ is an odd multiple of 1/2 if $\vec{S}$ is odd multiple of 1/2.

Fig. shows for $L = 2, S = 1$ and $L = 2$ and $S = 3/2$

No. of possible values of $\vec{J}$ for $L > S$ is $2S + 1$

and for $\vec{J}$ $L < S$ $2L + 1$

If $\vec{L} = 0 \quad \vec{J} = \vec{S}$

$$|\vec{J}| = \sqrt{J(J+1)}\,\hbar$$

Also $$|J_Z| = M_J \hbar \qquad \text{. . . (5.9)}$$

M_J is the total magnetic quantum number of the atom.

j - j Coupling

In alkali elements, the changes in atomic configuration giving optical spectra are produced due to a single valence electron. Now the total angular momentum of this electron is the vector sum of orbit and spin angular momentum of the single electron.

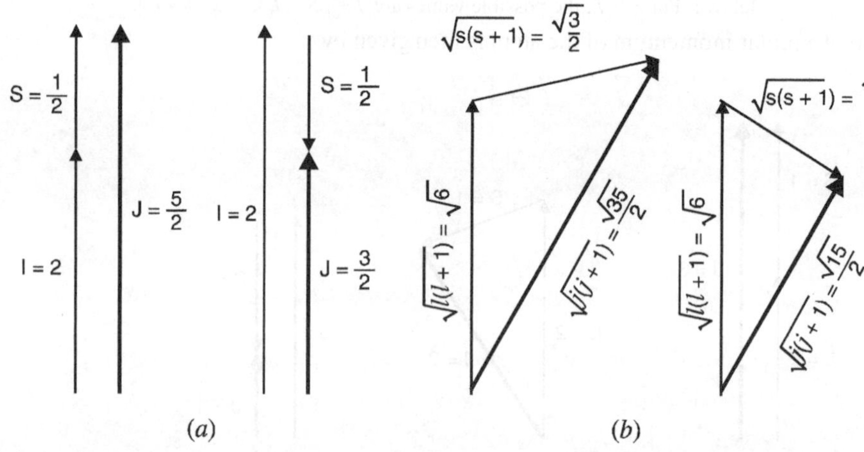

Fig. 5.7. (*a*) Addition of the vectors l and s to form j according to the vector model. (*b*) Addition to the vectors l and s to form j according to wave mechanics

Total angular momentum is $j\hbar$ where j = total angular quantum number.

$$\vec{j} = \vec{l} + \vec{s} \qquad \text{. . . (5.10)}$$

The vector sum of all the individual j-vectors of the different electrons gives the total angular momentum $\vec{J}$ of the atom. This sort of coupling may be symbolically represented as

$$[(s_1 + l_1) + (s_2 + l_2) + ...] = (j_1 + j_2 + ...) = \vec{J}.$$

$\vec{J}$ is always an odd multiple of 1/2

$\because$ $s = 1/2$ j can have only two values $l + 1/2$ and $l - 1/2$

When $l = 0, \quad j = 1/2$

In Fig. 5.7 $l = 2$ and $s = 1/2$ values are shown.

The angle between $\vec{l}$ and $\vec{s}$ in Fig. 5.7 is given by the cosine rule.

$$\cos(s, l) = \frac{j(j+1) - l(l+1) - s(s+1)}{2\sqrt{s(s+1)}\sqrt{l(l+1)}} \quad ...(5.11)$$

5.5 Pauli's Exclusion Principle

Pauli's Exclusion principle, sometimes termed also as the **equivalence principle** states that every completely defined quantum state in an atom can be occupied by only one electron. In other words, it is impossible for two electrons in an atom to be identical as regards all their quantum numbers, *i.e.*, in such a case one of the two will be excluded from entering into the constitution of the atom. Hence the name "exclusion principle".

The principle may be stated in yet another way : Two systems of quantum numbers which are deducible from each other by interchange of two electrons represent only one state. Thus understood, the principle enunciates the indistinguishability of electrons which have identical quantum numbers. Hence the name "equivalence principle".

It is clear that the principle defines a certain minimum individuality of the electrons in the atom. Among the several quantum numbers associated with the electron, four, *viz.* n, l, m_l and m_s or n, l, j and m_j are strictly required to specify completely the state of any particular electron, the former set being used in the presence of a strong magnetic field which breaks the coupling between l and s in the electron, whereas the latter in the rest of the cases.

Pro. Pauli

The principle was first introduced by Pauli to explain certain experimental facts, such as (*a*) the occurrence or non-occurrence of spectral lines in optical and X-ray spectra, it was found that for the missing energy states of the atoms all the four quantum numbers of the electron agreed, and (*b*) the complete scheme of the successive formation of the atoms as arranged in the periodic table, discovered by Bohr, Stoner and others-such an arrangement required the association of the above-stated four quantum numbers with each electron.

Naturally then, the principle finds its chief use in the elucidation of electronic structure and atomic spectra. It has been realised later that the principle is much more fundamental and universal, holding good for the totality of electrons in any arbitrary molecule, nay, even for the comprehensive system of conduction electrons that belong to a metal bulk state on the one hand and for the constituent particles of the nucleus in the ultramicroscopic state on the other.

No theoretical proof of the principle can be given as yet and for the present it must be regarded as something empirical added to and regulating the vector atom model.

The interval rule : Lande discovered a rule regarding the interval in frequency between the different levels constituting a multiple. It is called the *lande interval rule* and states that the frequency interval between two levels with total angular momenta $(J + 1)$ and J respectively is proportional to $(J + 1)$.

There are, however, many deviations from this internal rule, chiefly in cases when the coupling scheme follows neither the L-S type, nor the j-j type, but is intermediate between these extreme cases.

Spectral terms : Since a spectral line arises due to the transition of the atom from one energy state to another, in the interpretation of the fine structure of a spectral line, one must necessarily start with the determination of the different possible energy states. Then by the use of adequate selection rules the states between which transitions actually take place are picked out. Such transitions should account for the observed fine structure.

The different energy states of the atom between which transitions can take place have come to be known by the special name in spectroscopic study, *viz.*, *spectral terms*. The problem of the fine structure is thus essentially a problem of spectral terms. When the latter is solved, the rimer is readily understood. Hence in the application of the vector atom model to the fine structure phenomenon, it must, first of all be seen how the model can be used to fix up the spectral terms involved.

For the sake of clearness, the fine structure in optical and X-ray spectra will be considered separately, as the notations used in the two cases are different.

5.6 Optical Spectra

Spectral terms : In the classification of the spectral terms by the use of the vector atom model, it is convenient to divide atoms into two main categories, *viz. one-electron system and many-electron system.*

The atoms belonging to the first class have *only one valence or optical electron, e.g.*, hydrogen-like atoms, such as the alkalis. Even if the atom of this system actually contains several electrons all of these, except one, are interlocked in closed shells, which, in consequence, do not contribute to the total angular momentum of the atom, according to Pauli's principle. Hence the single free electron in the outermost shell is alone effective in determining the spectral properties of the atom.

The atoms of the second class have *more than one valence or optical electron, i.e.*, several free electrons not interlocked in closed shells. They will, therefore, contribute their share to the total angular momentum of the atom and hence become effective in fixing the spectral properties, *e.g.*, the alkaline earths, belonging to the two electron system, aluminium, scandium, etc., belonging to the three electron system, titanium to the four electron system and so on.

Considering the **one-electron system**, since the state of the atom as a whole is determined by that of a single free electron, the values of l, s and j referring to the electron are equal to those of L, S, J, defining the whole atom. Hence $S = s = \pm 1/2$ and the multiplicity of state given by $r = (2S + 1)$ is two. This system **will therefore give rise only to doublet terms with the exception of the ground terms** which corresponds to the normal state of the atom, *i.e.*, to $L = 0$. For, in present case, the possible values of J are only two *viz*, $J = (L + 1/2)$ and $(L - 1/2)$. Now, if $L = 0$, $J = \pm 1/2$; but since the net angular momentum of the atom is always a positive quantity, the only value of J in this case is $J = +1/2$, which means a *singlet*. If $L = 1$, $J = 1 \pm 1/2$. *i.e.*, 3/2, or 1/2 (*doublet*). For $L = 2$, $J = 2 \pm 1/2$, *i.e.*, 5/2 or 3/2 (*doublet again*) and so on.

In the **many-electron system**, on the other hand, the state of the atom will be determined by the states of several free electrons. In consequence, the vectors L, S and J of these electrons. Hence the value of S is not always 1/2, but may be 0, 1/2, 1, 3/2, 2, etc., according to the number of free electrons involved and their orientations, parallel or anti-parallel, and the multiplicity r is no longer two in all cases, but is l when $S = 0$, 2, when $S = 1/2$, 3 when $S = 1$, 4 when $S = 3/2$, 5 when $S = 2$ and so on. Thus the possible values of J for a given value of L may be one, two, three, four, etc., which means **each state may be a singlet or multiplet.**

In the *two-electron system*, since $S = 0$ or 1, the state can be only a *singlet* or *triplet*. In the *three-electron system*, as $S = 1/2$ or 3/2, each state can be a *doublet* or a *quartet*. In the *four-electron system* since $S = 0$, 1 or 2, *singlet triplet* and *quintent* are possible. In general, *odd electron systems* have even number of terms, while *even electron systems have odd number of terms*. Accordingly, the number of spectral terms of successive atoms in the periodic table of elements will alternate between even and odd.

But the ground terms is always a singlet in the system also. For, it corresponds to $L = 0$ and $J = S$. Now, since S can assume values 0, 1/2, 1, 3/2, etc. S is never less than L. For $S = L = O$, the possible value of J is one; when $S > L$, the multiplicity of the state is given by $(2L + 1)$ which leads again to the possible value of J as one.

Limitation of term multiplicity due to Pauli's principle : Pauli's exclusion principle causes, however, the multiplicity of terms not to attain always the maximum $(2S + 1)$ values which are to be expected from the number of free electrons involved. Taking, for instance, the case of an atom with 6 electrons in the L shell, at first sight, one expects the multiplicity $r = (2S + 1)$ to attain the maximum value of 7 on the assumption that the spins of all the six electrons are similarly directed there by leading to a value of 3 for S. But such an arrangement of spins cannot take place without violation of Pauli's principle. For, in this case, $n = 1$, $l = 0$ and 1; m_l can have only four values, *viz.*, 0 corresponding to $l = 0$, and 1, 0, −1, corresponding to $l - 1$. Now if we associate with these 4 values of m_1 *six like* values of m_l *i.e.*, all either +1/2 or −1/2, the result would be that more than one electron will have identical quantum numbers as shown below :

L_I $(n = 2, l = 0)$	L_{II} $(n = 2, l = 1)$
$m_l = 0$	+1, 0, −1.
$m_s = +1/2, +1/2$	+1/2, +1/2 +1/2 +1/2
The *correct arrangement is*	
$m_l = 0$	+ 1, 0, −1
$m_s = +1/2, -1/2$	+1/2, +1/2, +1/2, −1/2

Hence only four of the six electrons can have similarly directed spins (*i.e.*, same sign for m_s), while the other two must be directed oppositely. In consequence, the value of S reduced to 1, from which it follows that the actual multiplicity is only 3. Another inference is that, as a shell is gradually filled by the addition of successive electrons, the term multiplicity first rises and then falls, the variation being symmetrical with respect to the middle of the shell.

Notations of spectral terms : The states of the atom, in which the values of its L vector are 0, 1, 2, 3, 4, etc. are symbolically represented by the capital letters S, P, D, F, G, etc. respectively. The value of J is appended to this symbol as a subscript and the multiplicity of the state r written as a left superscript.

Thus the spectral terms corresponding to $L = 1$ and $S = 1/2$ are written as $^2P_{1/2}$ and $^2P_{3/2}$. For, since $L = 1$, the capital letter which represents the state is P, the multiplicity of the state $r = (2 \times 1/2 + 1) = 2$ which is put as a superscript; since $J = L \pm S = 1 \pm 1/2$, *i.e.*, 1/2 or 3/2, these are put as subscripts.

To take another examples, the notation $^4D_{5/2}$ specifies a spectral terms belonging to a state whose $L = 2, J = 5/2$ multiplicity $r = 4$ and hence $S = 3/2$ (from $r = 2S + 1$). Since the possible values of J are four, *viz.*, 7/2, 5/2, 3/2, 1/2, the above notation refers only to one of the four terms of the given state.

In the atoms belonging to the one-electron system, since every state is a doublet, except the ground state, the notations for the spectral terms in these atoms are

$$^2S_{1/2},\ ^2P_{1/2}\backslash ^2D_{5/2},\ ^2F_{5/2},\ ^2F_{7/2} \text{ etc.}$$

In the many-electron system atoms, the notations will be more complex, but they can be readily written, provided the necessary data are available.

It may be noted that the *multiplicity symbol of the system is used on all terms even if all of them are not present*. Thus the ground term of all alkali atoms is written as $^2S_{1/2}$ although the multiplicity is one and we should strictly speaking write $^1S_{1/2}$. But this method has the advantage of indicating to which system the ground term belongs; in the present case, for instance, to the one-electron system, giving rise to doublet terms in general.

It is likewise important to note that the vector model, where only the vectors L, S and J are involved, cannot determine the magnitude of the energies corresponding to the different spectral terms, since the total quantum number n, necessary for such a determination, does not directly enter in the fixing up of the spectral terms. Yet the three vectors L, S and J characterise the different energy states in such a definite manner that they may be used, without the intervention of n, to predict which transitions between the various energy states are permitted and hence which spectral lines may be expected.

5.7 Fine Structure

As spectral lines arise due to transitions of the atom from one spectral term to another, it is easy to understand, in a general way, how complex lines may appear when there are several spectral terms for a given atomic state defined by the vector L.

In the atoms belonging to the one-electron system all the states except the ground one are doublets. Transitions between these doublet states give rise to the fine structure. To account fully for the actually observed fine structure, as regards their number and intensity, the selection and intensity rules must be taken into consideration.

In the case of atoms pertaining to the many-electron system the fine structure may be expected to be much more complex on account of the greater term multiplicities involved. But here also, knowing the spectral terms and applying the selection and intensity rules, the observed fine structure can be accounted for. It is to be remarked that on account of the special property of closed shells the free electrons responsible for the spectra are limited to a small number. Hence the complexity of the fine structure does not go on increasing from one end to the other of the periodic table. The selection rules also, in their turn, put a limit on the complexity of fine structure. Likewise the limitations of term multiplicity arising from Pauli's exclusion principle may reduce the complexity still further.

These general principles may now be applied to the interpretation of the fine structure in the two simple cases, already mentioned, *viz.*, the sodium D line and the hydrogen H_x line, to serve as illustrations.

Fine structure of the H_α-line : As the hydrogen atom, in whose spectrum the H_α line appears, belongs to the one-electron system, $L = 1$, $S = s$, and $J = j$ and all the quantum states are doublets except the singlet ground state. According to the simple theory of Bohr, the H_x-line the first of the Balmer series, arises, due to transition from the third quantum state ($n = 3$) to the second ($n = 2$).

For the *upper state* ($n = 3$), L can take any of the values 0, 1, 2. Using the relation $J = L \pm S$, since S in this case has the value 1/2, the possible number of terms are five, *viz.*, ${}^2D_{5/2}$, ${}^2D_{3/2}$, ${}^2P_{3/2}$, ${}^2P_{1/2}$ and ${}^2S_{1/2}$.

For the lower state, ($n = 2$), L can have two values 0 and 1, which gives the possible number of terms as three, *viz.*, ${}^2P_{3/2}$, ${}^2P_{1/2}$ and ${}^2S_{1/2}$.

Fifteen transitions are theoretically possible between the five terms of the upper state and the three terms of the lower state. But the selection rules $\Delta L = \pm 1$, and $\Delta J = 0$ or ± 1 reduce the permitted transitions to the following seven :

$D_{5/2} \to P_{3/2}$, $D_{3/2} \to P_{3/2}$, $D_{3/2} \to P_{3/2}$, $D_{3/2}, \to P_{1/2}$, $P_{3/2}, \to S_{1/2}$, $P_{1/2} \to S_{1/2}$, $S_{1/2} \to P_{3/2}$ and $S_{1/2} \to P_{1/2}$.

Two pairs in separate cases of these seven allowed transitions are identical *viz.*

$$D_{3/2} \to P_{1/2} \text{ and } P_{3/2} \to S_{1/2}, P_{1/2} \to S_{1/2} \text{ and } S_{1/2} \to P_{1/2}$$

Since they represent transitions between *coincident levels, i.e.*, levels whose L values differ by units, but J values are the same.

Taking into account these identical lines, *the fine structure of the H_α line should have five components.*

Applying the intensity rules, the two strongest components are :

$$D_{5/2} \to P_{3/2} \text{ and } \begin{Bmatrix} D_{3/2} \to P_{1/2} \\ P_{3/2} \to S_{1/2} \end{Bmatrix}$$

since they are of the type $\Delta L = -1$, $\Delta J = -1$.

The two components $D_{3/2} \to P_{3/2}$ and $P_{1/2} \to S_{1/2}$ are less intense being of the type $\Delta L = -1$. $\Delta J = 0$. The component $S_{1/2} \to P_{3/2}$ is still weaker, since it is of the type $\Delta L = +1$, $\Delta J = +1$. The weakest line would be $S_{1/2} \to P_{1/2}$ ($\Delta L = +1$, $\Delta J = 0$), but it is fused with $P_{1/2} \to$ for reasons given above.

Energy level diagram (Fig. 5.8). The *coincident levels* are represented by horizontal lines drawn close together. There are two such levels in the upper state and one in the lower. The other lines represent, the *neighbouring* levels. If we call those vertical lines which terminate at the upper level of the lower state as group I and those which terminate at the lower level as group II and if we number the initial levels of the upper state as *a, b* and *c*, in no way as they do not coincide, then the two identical pairs are II_b and II_c. The five components are marked 1, 2, 3, 4 and 5, corresponding to I_a, I_b, I_c, II_b, and II_c respectively.

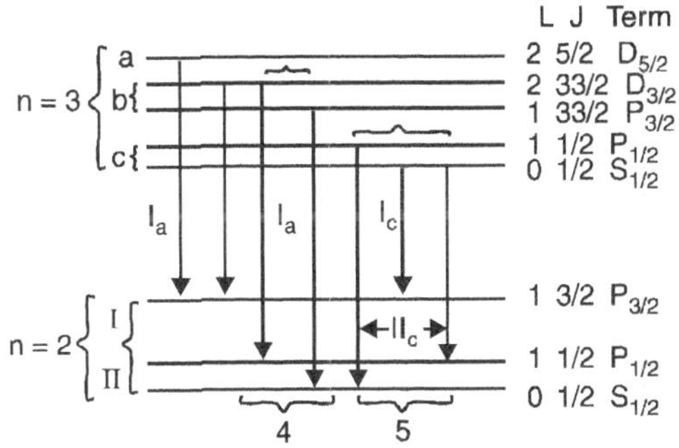

Fig. 5.8

According to the intensity rules, the strongest components are I_a and II_b (1 and 4). The components I_b and II_c (2 and 5) are less strong. I_c (3) is still weaker.

In practice, not all the five components but only doublets can be observed. This is due to the fact that the five components are merged together by the Doppler broadening caused by the thermal motion of the molecules. Very careful experiments conducted by R.C. Williams, in 1938, where the Doppler effect was reduced to a great extent by the use of a liquid air-cooled discharge tube containing deuterium, revealed three components.

Fine structure of the sodium D-line : As we have already seen, ten out of the eleven electrons of the normal sodium atom are interlocked in closed shells, so that the eleventh free electron alone is responsible for the spectrum. The spectral lines arise from the various orbits which this single electron may occupy beyond the two closed shells and the state of the atom as a whole is determined by the state of that electron. For the same reason, all the states of the sodium atom except the ground one are doublets.

With the simple theory of Bohr it has been shown that the D line belongs to the principal series which are due to transitions from the P state to the S state. Now for the *upper P state*, L = 1, J = 3/2 or 1/2 so that two terms are possible : $^2P_{3/2}$ and $^2P_{1/2}$. For the *lower* S *state*, $L = 0$, $L = 1/2$, so that only term is possible : $^2S_{1/2}$.

The transitions that can take place between the two terms of the P state and the single term of the S state are two :

$$^2P_{1/2} \to {}^2S_{1/2} \text{ and } {}^2P_{3/2} \to {}^2S_{1/2}.$$

Now applying the selection rules $\Delta L = \pm 1$ and $\Delta J = 0$ or ± 1 (excluding $O \to 0$), both the transitions are allowed, which explains the doublet line structure of the sodium *D* line. The D_1 component (5896 Å) is due to the transition $P_{1/2} \to S_{1/2}$ and the D_2 component (5890 to $P_{3/2} \to$ Å) $S_{1/2}$ (Fig. 5.9).

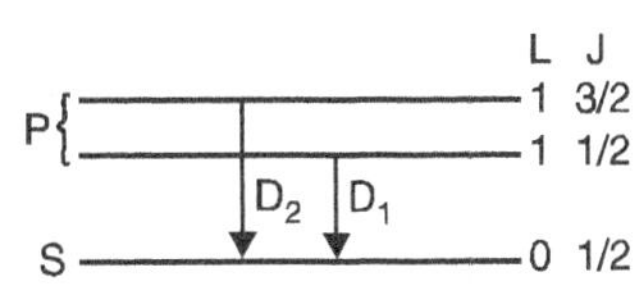

Fig. 5.9

According to the general intensity rules, D_2 will be more intense than D_1, since for D_2, $\Delta L = -1$, $\Delta J = -1$, while for D_1, $\Delta L = -1$, but $\Delta J = 0$.

Applying the total intensity $(2J + 1)$ rule it can be shown that the ratio of the intensities of D_2 to D_1 is 2 : 1, which has been fully confirmed by observation.

Note. Fine *structure of lines in the other series of sodium.* As the *sharp* series arises due to transitions between the S and P states, the lines in it will be *doublets*, as in the principal series.

The *diffuse* series arises due to transitions between the *D* and *P* states and hence the lines in it should be *triplets*, since both the states have two sub-terms each : $D_{5/2}$ and $D_{3/2}$, $P_{3/2}$ and $P_{1/2}$. Four transitions are therefore possible : $D_{5/2} \to P_{3/2}$, $D_{5/2} \to P_{1/2}$, $D_{3/2} \to P_{3/2}$ and $D_{3/2} \to P_{1/2}$. Of these given, $D_{5/2} \to P_{1/2}$ is forbidden, since $\Delta J = 2$. The lines $D_{5/2} \to P_{3/2}$ and $D_{3/2} \to P_{3/2}$ will be intense, since $\Delta L = -1$ and $\Delta J = -1$, while $D_{3/2} \to P_{3/2}$ will be weak, since $\Delta L = -1$ and $\Delta J = 0$. Experimentally, with ordinary resolving power instruments the weak line is not seen. When it was first discovered it was called a "satellite" of the apparent doublet formed by the two brighter lines.

Thus the chief spectrum of all alkalis should consist of doublets and triplets and theory accounts for them very well.

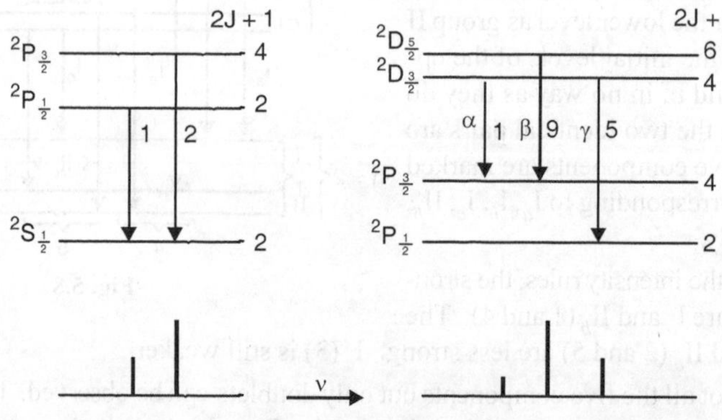

Fig. 5.10. Transitions between doublet levels and (below) relative frequencies of the lines. If *S* lies above *P*, the order of the lines is reversed; similarly if *P* lies above *D*.

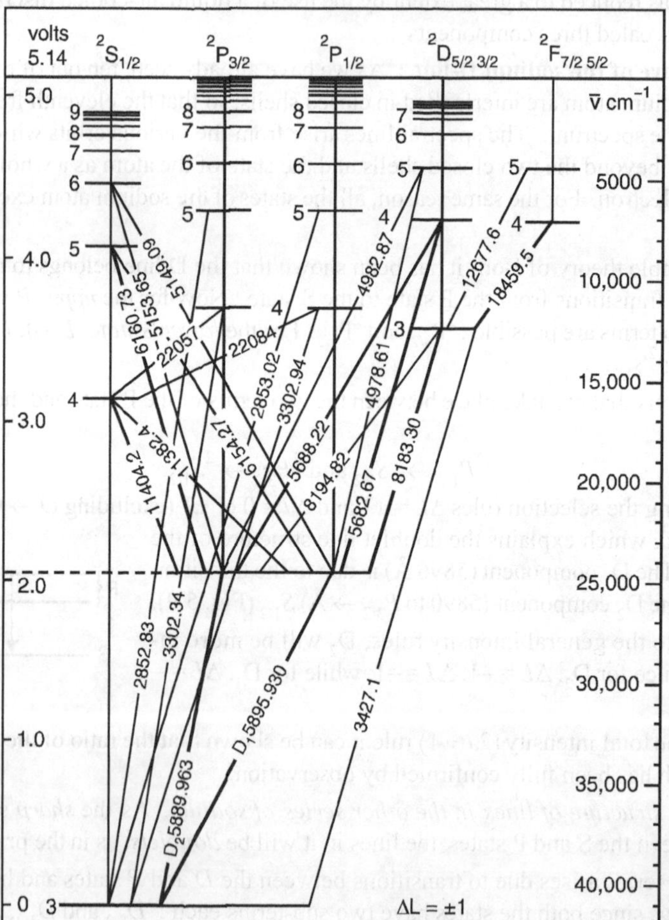

Fig. 5.11. Energy-level diagram of sodium. The numbers along the lines are the wavelengths, in angstroms, emitted during the indicated transitions.

5.8 Shielding

In order to understand the discrepancies at lower values of l, we find that the probability of finding an electron in hydrogen atom varies with the distance from the nucleus.

Smaller the value of l for a given n, the closer the electron gets to the nucleus. The wave functions for sodium (*i.e.* probability) are not identical to that of the hydrogen atom, but the general behaviour is the same.

Therefore, the outer electron in the sodium atom penetrates the core of the inner electrons most often when it is in the S state, less often in the P state and so on. If the outer electron is less shielded from the nucleus, the average force acting on it is greater and therefore its total energy is small.

Hence we see that for small l in sodium the states are displaced downward as compared to hydrogen and there are appreciable differences in energy levels for same n but different l.

5.9 Spectra of Two-Electron Atoms

In the case of atoms of alkaline earth elements, like Beryllium, Magnesium, Calcium, Strontium, Barium and Radium, they have two valance electrons outside the closed shell. Now, the total angular momentum of the atom is the sum of the angular momentum of the two elections. Since the quantum number n is the same for both and $l = 0$ in the normal state, S of the electrons are in the opposite directions.

$$\therefore \qquad \vec{S} = \vec{s_1} + \vec{s_2} = \frac{1}{2} - \frac{1}{2} = 0$$

For the normal state therefore,

$$\vec{J} = \vec{L} \oplus \vec{S} = 0 \quad \therefore \text{ Normal state is a singlet state } i.e., \ ^1S_0.$$

The excited states of the electrons *i.e.*, P states are due to $\vec{L} = \vec{l_1} \oplus \vec{l_2}$ and is equal to unity.

D states $\vec{L} = 2$ etc.

Since $\qquad \vec{S} = \frac{1}{2} + \frac{1}{2} = 1$ or $\vec{S} = \frac{1}{2} - \frac{1}{2} = 0.$

We see that for $\vec{S} = 0$, $\vec{J} = \vec{L}$ $\quad \therefore$ singlet states

$$^1P_1 \text{ for } \vec{L} = 1$$

$$^1D_2 \text{ for } \vec{L} = 2 \text{ etc.}$$

For $\vec{S} = 1$, J values are $L + 1$, L and $L - 1$.

These are triplet states *i.e.*, $^3P_2, {}^3P_1, {}^3P_0$ for $\vec{L} = 1$, $^3D_3, {}^3D_2, {}^3D_1$, etc for $\vec{L} = 2$.
Hence, in a two electron system there are singlet and triplet states.

For $s = 1$, $\vec{L} = 0$, $\vec{J}$ has one value unity.

$\therefore$ These states are denoted as $^3S_1$ and are referred to as triplet states.

In Fig. energy level of calcium is shown. This is for one electron in the 4S state ($n = 4$, $l = 0$) and the other electron changes its state. The transitions giving rise to the special lines are indicated in Fig. 5.12. Note that the singlet terms combine with singlet terms and triplet with triplet. Intercombination is "forbidden", but a few have been found but their intensity is low.

Now, the selection rule is $\quad \Delta l = \pm 1$

and $\qquad \Delta \vec{L} = \pm 2.$

For $\vec{J}$, $\qquad \Delta \vec{J} = 0 \text{ or } \pm 1$

Note $J = 0$ to $J = 0$ is not allowed.

The transitions between singlet states give rise to singlet lines.

The transitions between triplet states give rise to fine structure.

The principal series lines are due to transitions

$$5\,{}^3S_1 \;—\; n\,{}^3P_{2,1,0} \qquad \text{three components.}$$

Sharp series $\quad 4\,{}^3P_{2,1,0} \;—\; n\,{}^3S_1 \quad$ 3 components

Diffuse $\quad 4\,{}^3P_{2,1,0} \;—\; n\,{}^3D_{3,2,1}$

Fundamental $\quad 3\,{}^3D_{3,2,1} \;—\; n\,{}^3F_{4,3,2}$ $\Big\}$ have six components.

Note. $4\,{}^3P_2$ and $4\,{}^3P_0$ states give rise to metastable states and the atom can go only to the normal state from these metastable states.

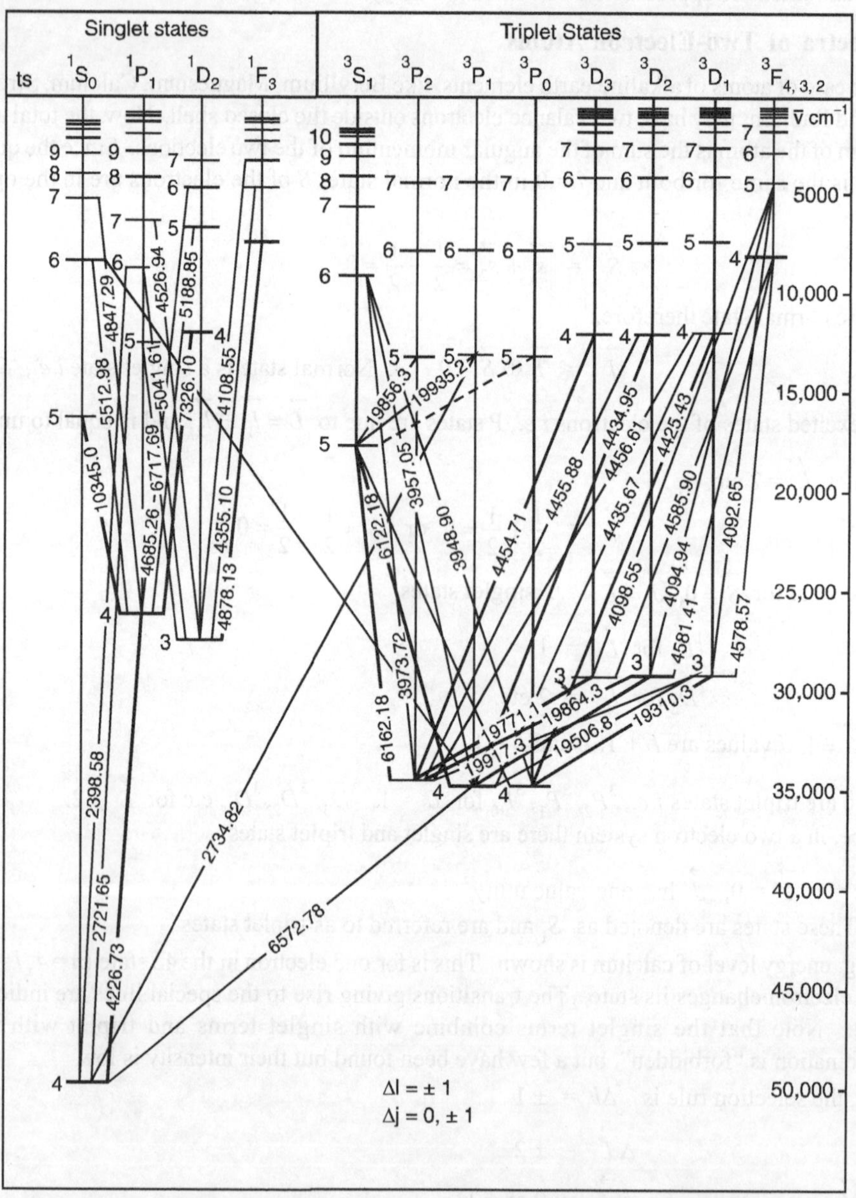

Fig. 5.12. Energy-level diagram of calcium. Wavelengths are in a angstroms.

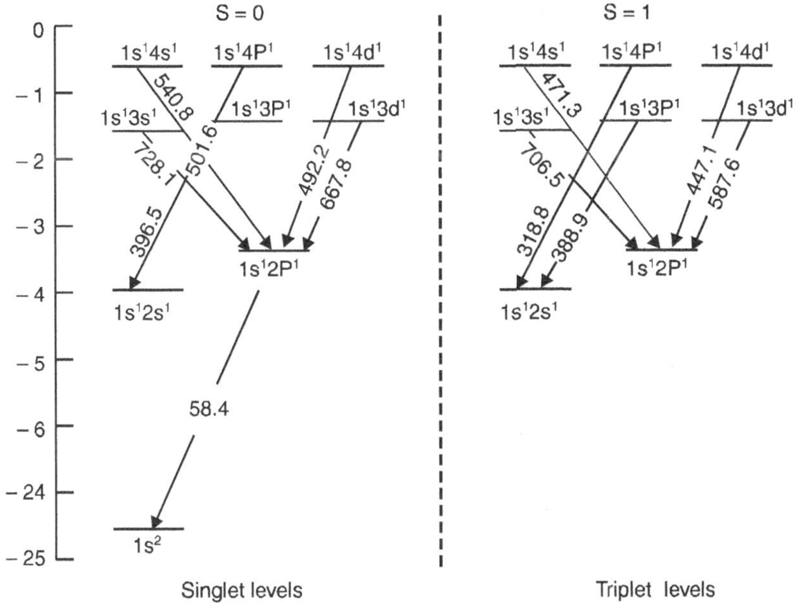

Fig. 5.13. Singlet and triplet energy levels of helium. The triplet levels are split by spin-orbiting coupling, but that splitting is not shown in this figure.

Example 1. Give the diagram for $\vec{l_1} \oplus \vec{l_2}$ giving $\vec{L}$ for $l_1 = 1$ and $l_2 = 1$.

Sol. $$\vec{l_1} \oplus \vec{l_2} = \vec{L}$$

$\therefore$ $$\vec{l_1} = 1 \quad \vec{l_2} = 1$$

gives $\vec{L}$, 0, 1, or 2.

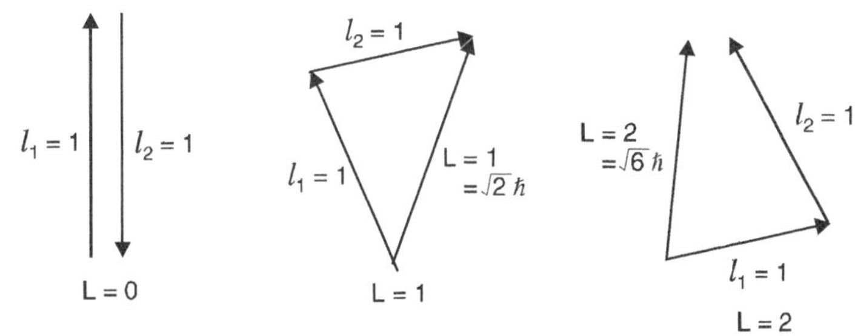

Fig. 5.14.

2. List the term designations $({}^{S}L_J)$ of all possible terms which can result from the following values of L and S :

Example. (*a*) $L = 2,\ S = 1/2$ (*b*) $L = 0,\ S = 2$

Sol. $$\vec{J} = \vec{L} \oplus \vec{S} \quad L > S \text{ multiplicity} = 2S + 1 = 2\ 1/2 + 1 = 2$$

Now $L = 2$ corresponds to a D state.

$$L = 1 \quad S = 1/2 \quad J = 3/2 \quad P \text{ state}$$

$\therefore$ Written as ${}^2P_{3/2}$ and $L = 1, S = 1/2, J = 1/2, {}^2P_{1/2}$

for $L = 2, S = 1/2, J = 5/2$ D state

$\because$ $$J = L \pm S \quad L = 2, \quad S = 1/2 \quad J = 3/2$$

$\therefore$ ${}^2D_{5/2}, {}^2D_{3/2}$.

Example 3.

Find the relative intensities of α, β and γ lines shown in Fig. 5.10 for the case $^2D \rightarrow {}^2P$ state.

Sol. Using the sum rule for initial state gives $\frac{\beta}{\alpha+\gamma} = \frac{6}{4}$ and final state.

$$\alpha + \frac{\beta}{\gamma} = \frac{4}{2}$$

$$\therefore \quad \alpha : \beta : \gamma = 1 : 9 : 5.$$

4. Calculate the angle between $\vec{l}$ and $\vec{s}$ for $^2D_{3/2}$ state.

Sol.
$$\cos\theta = \frac{l^{*2} \times s^{*2} - J^{*2}}{2\sqrt{l(l+1)}\sqrt{s(s+1)}}$$

For $^2D_{3/2}$
$$\cos\theta = \frac{6 \times 3/4 - 35/4}{2\sqrt{6 \times 3/4}}$$

$$= \frac{1}{\sqrt{4.5}}$$

$$= -0.138$$

$$\therefore \quad \theta = 118.1°.$$

EXERCISE V

1. Using vector model determine the possible values of the total angular momentum of an *f*-electron.
2. Determine the values of $\vec{J}$ for the following $\vec{L}$ and $\vec{S}$ vector :

$$L = 1 \quad S = 1$$
$$L = 2 \quad S = 1$$
$$L = 1 \quad S = 3/2$$
$$L = 2 \quad S = 3/2$$

3. Show how Pauli's exclusion principle determines the number of electrons in different shells of an atom.
4. Which of the following are possible atomic states ?
$3\,^3P_{1/2}$, $3\,^2P_{7/2}$, $4\,^2F_{5/2}$, $4\,^2F_{3/2}$, $4\,^1S_1$, $3\,^1P_1$. Explain your answers.
5. Sketch the vector model for $l = 3$, showing all possible values of L_Z.
6. Explain D_1 and D_2 doublet of sodium spectra on the basis of the electron spin.
7. Explain the fine structure of hydrogen line of the hydrogen spectra on the basis of the vector atom model.
8. Explain $R - S\,(L - S)$ and $j - j$ coupling of orbital and spin angular momentum.
9. Give an account of the various quantum numbers used to specify completely the state of an electron in an atom.
10. Apply Pauli's exclusion principle to account for the periodic classification of elements.
11. Explain multiplicity of states and the notation.
12. Explain the spectra of two-electron atoms indicating how singlet and triplet fine structures are obtained.
13. Calculate the possible orientation of the spin vector S with respect to the direction of the magnetic field. (**Ans.** 54°55′, 125°45′)
14. Estimate the relative intensities of lines of compound doublets arising from transitions $^2D \leftarrow {}^2P$ and $^2F \leftarrow {}^2G$.

6
X-RAYS

6.1 Introduction

With a discharge tube exhausted to the stage (10^{-3} mm of Hg), when the Crooke's dark space fills the whole tube, an interesting phenomenon was discovered by Rontgen in 1895. He covered the discharge tube with a thin black cardboard to facilitate better observation of fluorescence and the faint glow discharge. It was noticed that a screen coated with barium platinocyanide present in the laboratory at a distance from the discharge tube also fluoresced brilliantly. Rontgen called these invisible penetrating radiations X-rays. They are also called Rontgen rays. In a series of experiments, he demonstrated the chief properties of the rays. He attributed the origin of these radiations to be due to the bombardment of cathode rays on the walls of the discharge tube. The discovery of X-rays has opened a new field of study particularly in the fields of medicines, surgery, industry etc. and many forms of X-rays tubes have been designed to produce X-rays for different needs. X-rays are produced when fast moving electrons are stopped by an obstacle in their path and metals of high atomic-weight as obstacles are most effective for the purpose. The property of X-rays to produce ionisation in a gaseous medium is used in ionisation chambers.

6.2 Production of X-rays

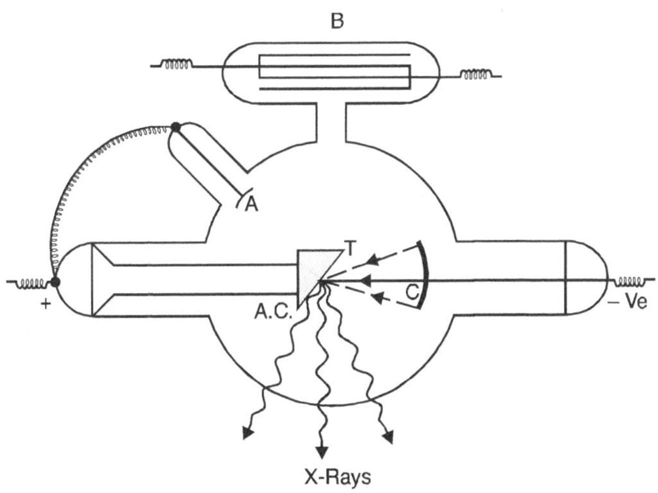

Fig. 6.1

(1) Gas-filled X-ray tube : Fig. 6.1 represents an early form of X-rays tube similar to that of a Rontgen tube. G is a glass bulb exhausted to a pressure of 10^{-3} mm of Hg. C is a concave cathode made of aluminium, which focusses the beam of cathode rays on a target T. Tungsten is used as the target because of its high melting point. This is surrounded by a massive block of copper which helps in the quick dissipation of heat produced by the cathode rays when they are stopped by the target. The target and the copper block together is called the anti-cathode (A.C.), A is a separate anode which makes the discharge steadier and B is occlusion tube containing platinized asbestos in which gases have been occluded. When the X-ray tube stops working due to very low pressure in it, the discharge is passed through the occulsion tube, the gases enter the discharge tube and it again starts working.

The working of the X-ray tube depends on the pressure inside the bulb. If the pressure is comparatively high, the discharge takes place even for a low voltage and the X-rays produced will be very feeble and of low penetrating power (soft X-rays). On the other hand, if the pressure inside is low, a high *voltage* is needed to maintain the discharge and X-rays of higher penetrating (hard X-rays) are produced.

The gas filled tube has serious disadvantages. The gas pressure falls due to the absorption of the ions on the walls of the tube and the intensity of X-rays decreases after some time. Moreover, in such tubes, it is difficult to control the X-ray intensity due to the *voltage* and *pressure* requirements for the discharge.

(2) Coolidge Tube : This tube is commonly used and it is a vacuum tube. It consists of an evacuated glass bulb containing a cathode C and the target T (Fig. 6.2). The cathode is a tungsten filament and can be heated by the current supplied by a low tension DC battery or step down transformer. The tungsten filament emits electrons *i.e.*, thermions which are accelerated to the target by maintaining a difference of potential between the cathode and the target.

By surrounding the cathode with shields of molybdenum maintained at a negative potential, with respect to the cathode, the emitted electrons are focussed on to the target. The target in an X-ray tube is usually tungsten, which has a high melting point. The main advantage of the Coolidge tube is that the current to heat the filament can be controlled and therefore the emission of electrons can be controlled. These tubes have been designed to operate at voltages from 100 volts to 10^6 volts. To control the heat produced in the tube, the target is mounted on a hollow copper tube through which water is circulated. The fins for air colling are also used in some tubes.

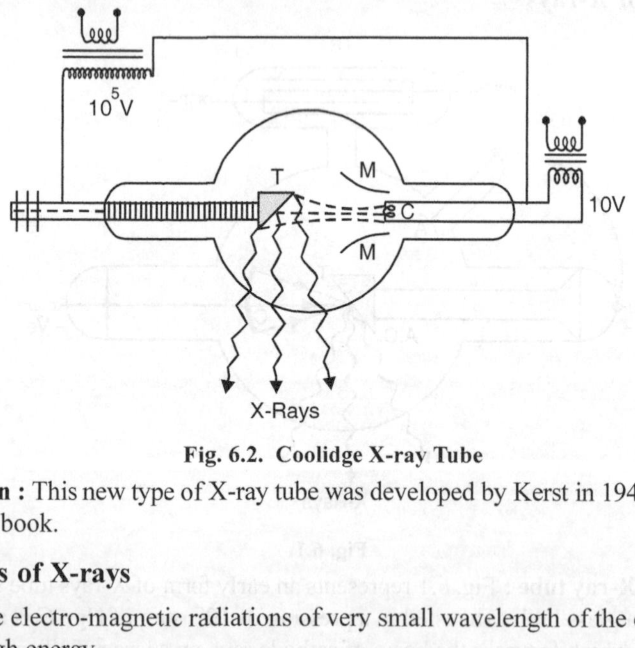

Fig. 6.2. Coolidge X-ray Tube

(3) Betatron : This new type of X-ray tube was developed by Kerst in 1941 and is discussed in detail later in the book.

6.3 Properties of X-rays

1. They are electro-magnetic radiations of very small wavelength of the order of 10^{-8} cm and are photons of high energy.
2. They travel with the velocity of light *i.e.*, 3×10^8 m/s.
3. They affect a photographic plate.

4. They are not deflected by magnetic or electric fields and therefore they do not possess any charge.

5. Similar to light, X-rays, due to their energy, liberate photoelectrons from some metals, when incident on them.

6. They ionize the gas through which they pass.

7. They produce fluorescence in barium-platinocyanide, zinc sulphide and cadmium tungstate.

8. They are highly penetrating and can, pass through many solids which are opaque to visible light e.g., wood, flesh, paper, cardboard, thin sheets of metals, ebonite etc. The penetrating power depends upon the energy of the photons of the X-rays. But X-rays cannot penetrate through heavy metals and bones. Therefore, X-rays cast the shadow of these objects when placed in their path. This property has been used in detecting the fracture of bones and flaws in precious castings.

9. They show interference, diffraction and polarisation, similar to light radiations.

10. When a beam of primary X-rays is incident on a plate of some chosen metal, a part of the radiation goes through the plate and the rest is transformed into heat or into radiations of some other forms. The secondary radiations coming out of the plate contain (1) scattered X-rays, (2) characteristic X-rays, (3) scattered β-rays, (4) characteristic β-rays (Fig. 6.3).

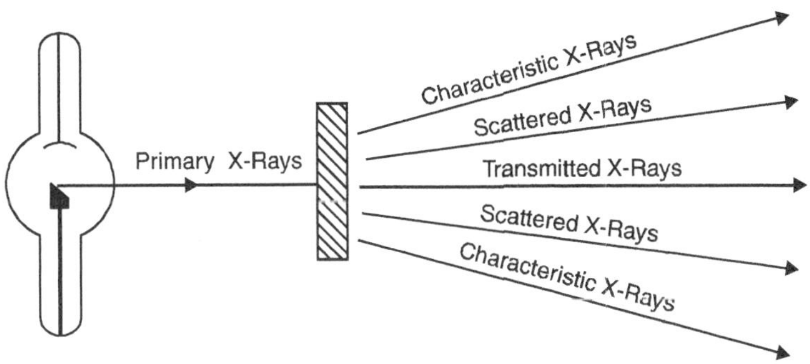

Fig. 6.3

Scattered X-rays have very nearly the same absorption coefficient in a given material as the primary X-rays. The scattered X-rays are also heterogeneous because the X-rays from the X-ray tube are heterogeneous. They have been simply scattered in different directions by the metal plate.

Characteristic X-rays are homogeneous and are typical of the material of the plate. The characteristic radiations of the element increase in hardness with the increasing atomic weight. Barkla studied the absorption coefficient of the characteristic X-rays from various elements and found that the elements have, in general, two types of radiations, *i.e.*, K and L radiations and K-radiations is harder and may have an absorption coefficient smaller than L-radiations.

11. They have destructive effect on living tissues and long exposure of the skin to X-rays is harmful. In some cases X-rays produced reddening of the skin and kill white corpuscles of the blood. X-rays are used for destroying and burning the ulcers and tumours in the body.

6.4 Continuous and Characteristic X-ray Spectra

When resolved by means of a crystal spectrometer, the X-rays emitted by the target of an X-ray tube is found to consist of two distinct types of spectra :

1. A continuous spectrum
2. A sharp line spectrum superposed on the continuous spectrum.

Typical X-ray spectra, are shown in Fig. 6.4 in which the intensity of a given wavelength is plotted against the wavelength in Angstrom units. The curve for 25 kV shows two sharp lines characteristic of the element of the target material. These lines are K_α and K_β. The two distinctively different types of spectra from the same target have different origins.

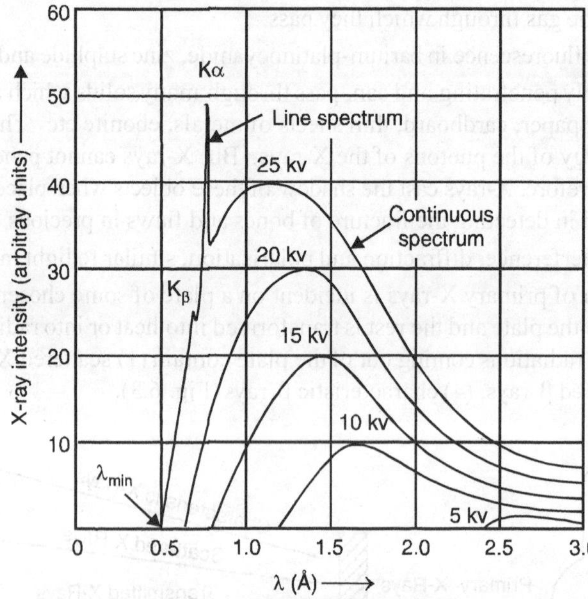

Fig. 6.4

(1) Continuous Spectrum : Most of the electrons in the coolidge tube undergo glancing collisions with the atoms of the target material and thus lose their energy. This results in the increase in temperature of the target.

Some of the electrons are stopped suddenly upon hitting a target. These electrons are rapidly decelerated. The continuous spectrum, results from the radiation emitted by the electrons which are accelerated in the coulomb field of force of the nuclei of the target atoms. The term *bremsstrahlung* or "breaking radiation" is used to describe the radiation emitted by a charged particle that is accelerated in the coulomb field of force of the nucleus. Since all sorts of collisions from direct head-on collision to glancing ones (0° to 180° scattering angles) are possible we get a continuous distribution of energy giving rise to the continuous spectra.

The continuous X-ray spectra presents several interesting features. It will be noted from Fig. 6.4 that there is a definite short-wavelength limit λ_{min} to the Continuous X-ray spectrum independent in of the material of the target.

Duane and Hunt's law : When the voltage across the tube is increased, the wavelength limit λ_{min} is shifted toward smaller values. Duane and Hunt showed that the short wavelength limit of the continuous spectrum varies inversely as the voltage across the tube. If an electron has been accelerated through a difference of potential of V-volts.

$$\nu_{max} = \frac{c}{\lambda_{min}} \text{ and } eV = h\nu_{max} = \frac{hc}{\lambda_{min}}$$

Here eV represents the energy of the electron on striking the target. This is an inverse photoelectric effect ; since the maximum K.E. of an electron striking the target is

$$\frac{1}{2}mv_{max}^2 = eV = h\nu_{max}$$

Substituting values for $h = 6.63 \times 10^{34}$ J.s., $c = 3 \times 10^3$ m/sec and $e = 1.64 \times 10^{-19}$ C, we get,

$$\lambda_{min}\,(A.U.) = \frac{1.24\times 10^{-6}\,m}{V} \qquad ...(6.1)$$

This is Duane and Hunt's law.

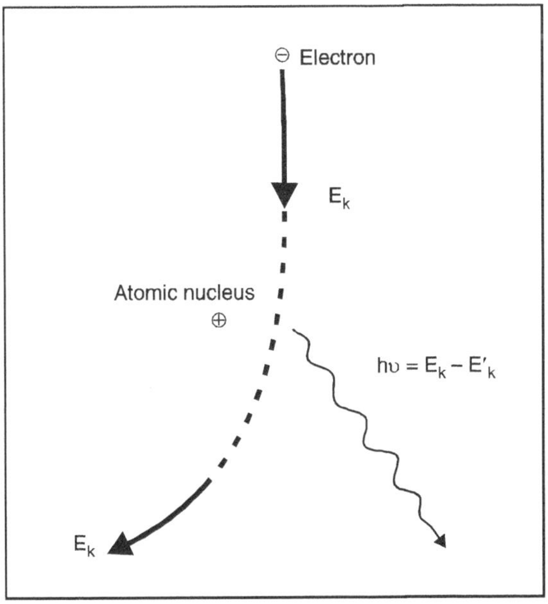

Fig. 6.5 Radiative "collision" of electron and atomic nucleus.

(2) Characteristic Spectra : In collisions between high energy electrons generated in the X-ray tube and the atoms of the target material (termed as anti-cathode), electrons bound to the atoms are knocked out of the inner shells. X-ray producing electrons may also have enough energy to produce ions by removing inner electrons from the atom, even from the innermost K-shell. The vacanices thus created are filled by electron transition from outer shells. Transitions of this kind are accompanied by emission of radiation of energy equal to the energy difference between the two energy levels.

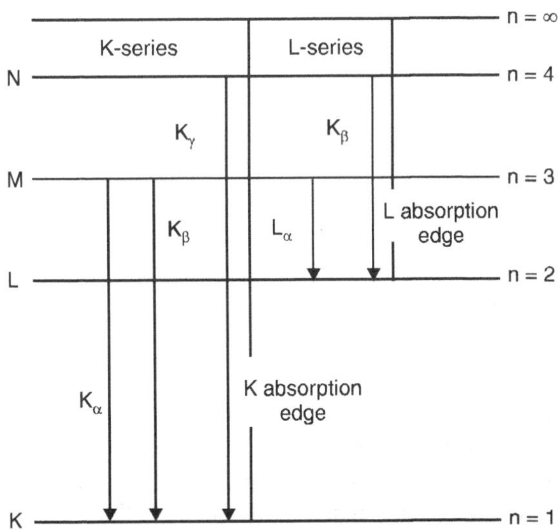

Fig. 6.6

When a K-shell vacancy is created, an electron from the *L*-shell can fill its vacancy [since these are all filled unlike in the case of the hydrogen atom]. This gives rise to X-radiation called K_α line Fig. 6.6. If the K vacancy is filled by an electron from the M-shell, there will be more energetic radiation, K_β. If the vacancy is filled by an electron from the N-shell, K_γ radiation results. In these K-series X-radiation the K_α is the most probable and is most intense. K_β and K_γ are less probable and decrease in intensity.

If the incoming electron has less energy and removes an L-shell electron, the vacancy can be filled by electrons from M, N shells giving the radiations L_α and L_β respectively. Other transitions are not important as far as X-rays are concerned since they give rise to visible light. Again, the L_α line has a greater probability than L_β line and is more intense than L_β.

If an outer N-shell electron is knocked out, the optical spectrum of the target material will be obtained. The M-shell electron gives radiation in the Ultra-violet.

Since K-series radiation are quite energetic as compared to L-series and require greater energy for the incoming electron, they are referred to as "hard X-rays." The L-series on the other hand are called "soft X-rays".

The short wavelength limit of X-ray series terminating at any shell is called the absorption edge wavelength of that shell Fig. 6.6. *i.e.* Transition from $n = \infty$ to the given shell.

Note that when an electron removes a K-shell electron, the vacancy is filled by an electron from the next *L*-shell. In such a case the atom will undergo a transition from the K to the L state, the excess of energy being emitted as a quantum of energy. The K line is due to an atomic process of this kind. The quantum $h\upsilon_\alpha$ is given by Bohr's frequency condition $h\nu_\alpha = E_K - E_L$, where E_K and E_L are the energies of the atom in the K and L shells respectively.

$$\therefore \quad \frac{hc}{\lambda_{K_\alpha}} = E_K - E_L \qquad ...(6.2)$$

Similarly

$$\frac{hc}{\lambda_{K_\beta}} = E_K - E_M \qquad ...(6.3)$$

$$\frac{hc}{\lambda_{K_\gamma}} = E_K - E_N \qquad ...(6.4)$$

The L-series are obtained

$$\frac{hc}{\lambda_{L_\alpha}} = E_L - E_M \qquad ...(6.5)$$

$$\frac{hc}{\lambda_{L_\beta}} = E_L - E_N \qquad ...(6.6)$$

The corresponding absorption edges are

$$h\nu_{K_{\text{abs.edge}}} = E_K \qquad [\because n = \infty \;\; E = 0] \qquad ...(6.7)$$

$$h\nu_{L_{\text{abs.edge}}} = E_L \qquad ...(6.8)$$

6.5 Origin of X-rays and Moseley's Law

The target used in an X-ray tube is made of tungsten (atomic number = 74). In an atom there are two electrons in the K-shell, eight electrons in the L-shell, 18 in the M-shell and so on.

The innermost electrons in the K-shell are nearer the nucleus greater energy is required to knock an electron out of the atom. The energy required to knock out the electrom from K-shell is greater than from the L-shell. When high energy cathode rays bombard the target, an X-ray tube, an electron from the K-shell is knocked out of the atom (Fig. 6.7). The atom is thereby converted into an ion and this state of the ion may be called K quantum state of the ion. Subsequently, electrons from the outer orbits try to jump to the inner orbits to fill up gaps and doing so give out energy in the form of X-rays. The electron from the L-shell jumps into the K-shell and an electron from the M-shell jumps into the L-shells. In actual practice, many atoms of the target simultaneously bombarded by the cathode rays and many lines are emitted simultaneously.

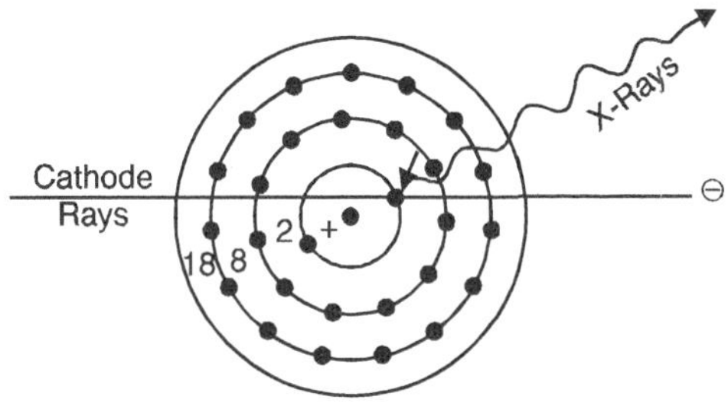

Fig. 6.7

Moseley (1913) studied the X-rays spectra of various elements. Thirty eight different elements were used as target in the X-ray tube. A potassium ferrocyanide crystal was mounted on the spectrometer table and the spectrum was recorded on a photographic plate. The spectral lines observed were of (1) short wavelength-K series and (2) long wavelength L series. K_α line is most intense in the K series. Moreover, it was observed that the wavelength of the K_α lines decreases with increase in the atomic number of the element used as the target.

If a graph is plotted between the square root of the frequency and the atomic number of the element emitting the line, it is a straight line (Fig. 6.8).

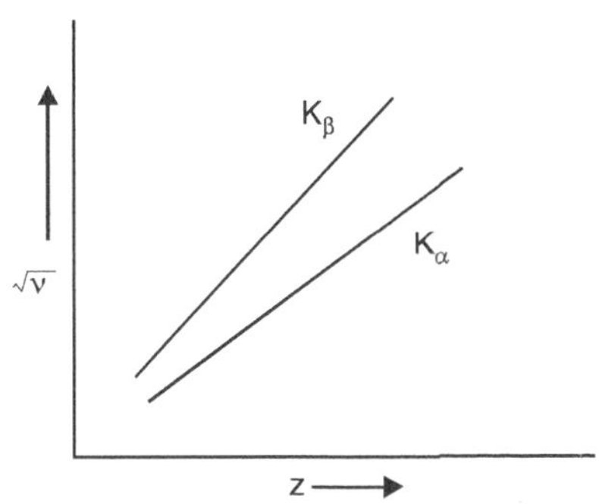

Fig. 6.8

Thus, $\sqrt{v} \propto Z,$

where v is the frequency of the radiation and Z is the atomic number of the element. But according to *Moseley law*, it has been found that $\sqrt{v} \propto (Z-1)$ because the electron which goes from the L-shell to the K-shell moves in an electric field which is essentially that of the positive nuclear charge of the electron still remaining in the K-shell. The electric field is therefore due to the positive charge of magnitude $(Z-1)\,e$. The effect of the electrons in the orbits far away from the nucleus can be neglected because their effect is too small.

According to Bohr's theory, the frequency of a spectral line is given by

$$\nu = \frac{me^4}{8\,t_0^2\,h^3} Z^2 \left[\frac{1}{n_f^2} - \frac{1}{n_i^2}\right]$$

For K_α line $n_f = 1$, $n_i = 2$ ∴

$$\nu = \frac{me^4}{8 \epsilon_0^2 h^3}\frac{3}{4} Z^2$$

or $$\nu = 3/4 \text{ Rc. } Z^2$$

where R = Rydberg's constant

Now, the frequencies v of the K_α lines in Moseley's work on X-rays are given almost exactly if we put the equation $\sqrt{\upsilon} = a(Z - b)$ in the form:

$$\nu = 3/4 \text{ Rc } (Z - b)^2$$

where $b = 1$ *i.e.*, $\nu = \text{Rc } (z - 1)^2 \left(1/1^2 - 1/2^2\right)$

or $$\nu = a^2 (Z - 1)^2;$$

$$(\text{and } a^2 = 3/4 \text{ Rc})$$

Importance of Moseley's work : Moseley found that the square root of the frequencies of vibration is approximately proportional to the atomic weight of the element, but argon, cobalt and tellurium were found in exceptional positions. He suggested that the atomic number rather than the atomic weight, is the characteristic of an element. The discrepancy in the case of argon, cobalt and tellurium disappears when the square root of frequency is plotted against atomic number. Moseley concluded that the number of unit charges on the nucleus is the same as the number of its place in the periodic table and both can be represented by a quantity called the atomic number. Thus, Moseley's work suggested that the position of the element in the periodic table is more directly related to the atomic number rather than the atomic weight and the chemical properties are related to the nuclear, *i.e.*, the atomic number and not to the atomic weight as suggested earlier by Mendeleeff.

6.6 Absorption of X-rays

When a homogeneous beam of X-rays passes through a solid, liquid or gas, a portion of the incident beam is absorbed. Suppose the intensity of the incident beam = I_0 and dI is the intensity absorbed in traversing a thickness dx. The intensity absorbed dI depends upon chance and is proportional to the intensity I and the thickness dx.

∴ $$dI \propto I dx$$

$$dI = -\mu I \, dx$$

where μ is a constant of proportionality and the negative sign indicates that the intensity decreases.

$$\frac{dI}{I} = -\mu \, dx$$

Integrating, $$\log_e I = -\mu x + k \quad \text{...(6.9)}$$

When $$x = 0, \ I = I_0$$

∴ $$\log_e I_0 = k$$

Substituting the value of k in equation (6.9),

$$\log_e I = -\mu x + \log_e I_0$$

$$\log_e \frac{I}{I_0} = -\mu x \text{ or } \frac{I}{I_0} = e^{-\mu x}$$

∴ $$I = I_0 e^{-\mu x} \quad \text{...(6.10)}$$

Therefore, the intensity of X-rays decreases exponentially, and this shows that the beam of X-rays can never be completely absorbed.

The proportionality constant μ is called the *linear absorption coefficient.* Since, μx is a mere number, μ has a dimension reciprocal to length *i.e.*, per cm or per metre. The mass absorption coefficient $= \frac{\mu}{\rho}$, where ρ is the density of the absorbing material.

6.7 X-Ray spectra

Spectral terms : We have seen that each X-ray energy level represents a state of an atom which has one electron missing from a closed shell. Pauli pointed out that in a configuration, in which an electron is missing from a completed shell, the spectral term is the same as if that one electron occupied the shell. This means that the term scheme for X-ray spectra corresponds to the one-electron system, such as hydrogen, alkali metals, etc. Moseley's researches on the characteristic X-rays also suggest the same conclusion, *viz.*, that the terms involved in X-ray spectra are of the hydrogen type.

The notations used are, however, different. The X-ray energy states are designated by the capital letters K, L, M, etc., specifying the shells to which they refer. According to the vector model, not only each of these states is characterised by a total quantum number n (being 1 for the K state, 2 for L state, 3 for M state, etc.), but also all the states except K, are further split into sub-groups, L into two, M into three, N into four, etc.

For, using the fact that an atom emitting X-rays is assimilated to the one-electron system, it follows that in the mechanism of X-ray emission, L = 1, S = s, J = j, so that J = $j = l \pm ½$ and that *every distinct state of an atom except the lowest is a doublet.* Hence the K state ($n = 1, l = 0$) corresponding to the lowest ground state is a *singlet.* The L state ($n = 2, l = 0, 1$) have two sub-groups. The first sub-group ($n = 2, l = 0$) gives one term while the second sub-group ($n = 2, l = 1$) *two terms* corresponding to the two different values of s, *viz.*, 3/2 and 1/2. The L state thus having three possible terms is a *triplet.* The M state ($n = 3, l = 0, 1, 2$) with three sub-groups gives rise to *quintet*, one corresponding to $l = 0$, and $j = 1/2$ (first sub-group), two others to $l = 1$, and $j = 5/2, 3/2$ (third sub-group). The N state ($n = 4, l = 0, 1, 2, 3$) with four sub-groups is a *septet* composed of a singlet for $T = 0$ and three doublets for $l = 1, 2, 3$. The term multiplicity does not, however, go on increasing. Pauli's principle putting a limit, the next *O state is a quintet and P state a triplet.* **There are in all 24 different terms which are quite sufficient to represent completely the X-ray spectra of all the elements.** They are denoted by the Roman numerals attached to the capital letters representing the states, *e.g.*,

$$K, L_I\ L_{II}\ L_{III}, M_I\ M_{II}\ M_{III}\ M_{IV}\ M_V, \text{ etc.}$$

They indicate all the possible combinations of l and j for given values of n.

Fine Structure : The X-ray lines, in general, will arise due to transitions among the 24 different X-ray terms. But by the selection rule transitions can only occur subject to the conditions $\Delta L = \pm 1$ and $\Delta J = 0$ or ± 1, so that the number of possible transitions between two states is thereby considerably reduced and in consequence the number of fine structure components of the X-ray lines will be limited.

If the transition ends in the K state the corresponding radiation is called a K line ; if it ends in the L state, a L line and so on. Fine structure of these general classes are distinguished by small Greek letters in accordance with the following recognised convention : If the transition takes place between the different terms of the L state and the single term of the K state, the lines are called $K_{\alpha 1}$, $K_{\alpha 2}$, etc., if from the different terms of the M state to the K state, $K_{\beta 1}$, $K_{\beta 2}$, etc. Thus a general classification is made as K, L, M lines etc. according to the final state of transition, while the particular classification of the individual fine structure components of each line is made according to the terms of the initial state of transition.

These points are illustrated in Fig. 6.9. The term scheme of the four states K, L, M,.N alone is drawn. The values of n, l and j for each term are indicated. Some of the fine structure of the K and L lines are marked. The continuous lines represent the permitted transitions giving rise to the fine

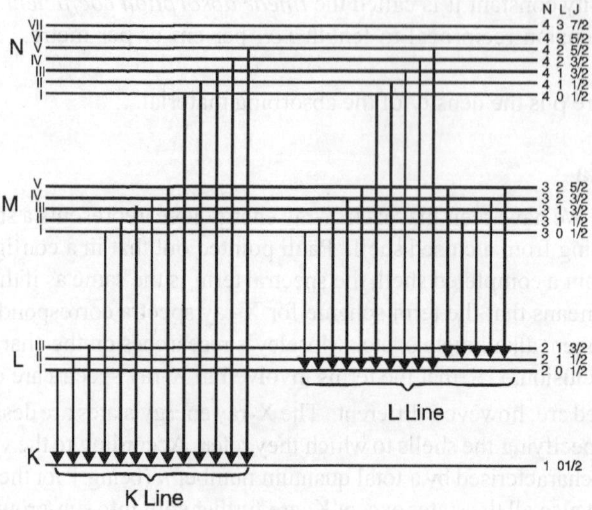

Fig. 6.9. Diagram of X-ray fine structure lines.

structure components, while the dotted lines indicate transitions forbidden by the selection rules. As one proceeds to elements of higher atomic number, the number of terms involved and consequently the number of the fine structure lines will increase. This clear and simple interpretation of the fine structure of X-ray lines in a single scheme was obtained in 1927, thanks chiefly to the work of four physicists, Smekal, Coster and Bohr. Thus, the fine structure of spectral lines, both in the optical and X-ray regions, finds an adequate explanation on basis of spectral terms determined by the vector model and of certain selection rules. It should be emphasized that the fine structure of spectral lines can be satisfactorily accounted for only by the introduction of electron spin.

Example 6.1. *An X-ray tube operates on 50 kV. Calculate the shortest wavelength of the X-rays produced.* (*Punjab*)

Solution.
$$V = 50 \text{ kV}$$

or
$$V = 50 \times 10^3 \text{ volts}$$

$$\text{eV} = h\nu$$

or
$$\text{eV} = \frac{hc}{\lambda}$$

or
$$\lambda = \frac{hc}{\text{eV}} = \frac{6.6 \times 10^{-34} \times 3 \times 10^8}{1.6 \times 10^{-19} \times 50 \times 10^3}$$

$$\lambda = 2.475 \times 10^{-11} \text{ m}$$

$$\lambda = \mathbf{0.2475\ \text{Å}}$$

Example 6.2. *If the minimum wavelength recorded in an X-ray continuous spectrum of 50 kV tube is* 24.7×10^{-12} *m, calculate the Planck's constant.* (*Rajasthan*)

Solution. Here,
$$V = 50{,}000 \text{ volts}$$

$$e = 1.6 \times 10^{-19} \text{ coloumb}$$

$$h = ?$$

$$\lambda = 24.7 \times 10^{-12} \text{ m}$$

$$c = 3 \times 10^8 \text{ m/s}$$

$$\text{eV} = h\nu$$

$$eV = \frac{hc}{\lambda}$$

$$h = \frac{eV\lambda}{c} = \frac{1.6 \times 10^{-19} \times 50,000 \times 24.7 \times 10^{-12}}{3 \times 10^{8}}$$

$$h = \mathbf{6.6 \times 10^{-34}\ joule\text{-}second}$$

Example 6.3. *An X-ray tube operated at 50 kV emits a continuous X-ray spectrum. Calculate the short wavelength limit. Given h = 6.624 × 10^{-34} joule second, e = 1.6 × 10^{-19} coulomb and c = 3 × 10^{8} m/s.* *(Bombay)*

Solution. Here

$$V = 50,000 \text{ volts}$$

$$e = 1.6 \times 10^{-19} \text{ coulomb}$$

$$h = 6.624 \times 10^{-34} \text{ joule-second}$$

$$c = 3 \times 10^{8} \text{ m/s}$$

$$\lambda = ?$$

$$eV = h\nu$$

or

$$eV = \frac{hc}{\lambda}$$

$$\therefore \quad \lambda = \frac{hc}{eV} = \frac{6.624 \times 10^{-34} \times 3 \times 10^{8}}{1.6 \times 10^{-19} \times 50,000}$$

$$\therefore \quad \lambda = \mathbf{0.2484 \times 10^{-10}\ m}$$

or

$$\lambda = \mathbf{0.2484\ Å}$$

Example 6.4. *An X-ray tube is operated at 20 kV, determine the wavelength limit of X-rays.* [*IAS*]

Solution.

$$V = 20 \text{ kV}$$

$$= 2 \times 10^{4} \text{ V}$$

$$e = 1.6 \times 10^{-19} \text{ C}$$

$$h = 6.6 \times 10^{-34} \text{ J-s}$$

$$eV = h\nu = \frac{hc}{\lambda}$$

or

$$\lambda = \frac{hc}{eV}$$

$$\therefore \quad \lambda = \frac{6.624 \times 10^{-34} \times 3 \times 10^{8}}{1.6 \times 10^{-19} \times 2 \times 10^{4}}$$

$$\lambda = 6.21 \times 10^{-11} \text{ m}$$

$$\lambda = 0.621 \text{ Å}$$

Example 6.5. *What is the minimum voltage applied to an X-ray tube to produce X-ray of 1 Å ?*

Given :

$$e = 1.6 \times 10^{-19} \text{ C}$$

$$h = 6.625 \times 10^{-34} \text{ J-s}$$

(Delhi 1990)

Solution. Here

$$eV = h\nu$$

$$\text{eV} = \frac{hc}{\lambda}$$

$$V = \frac{hc}{e\lambda}$$

$$h = 6.625 \times 10^{-34} \text{ J-s}$$

$$c = 3 \times 10^8 \text{ m/s}$$

$$e = 1.6 \times 10^{-19} \text{ C}$$

$$\lambda = 1 \text{ Å} = 10^{-10} \text{ m}$$

$$\therefore \quad V = \frac{6.624 \times 10^{-34} \times 3 \times 10^8}{1.6 \times 10^{-19} \times 10^{-10}}$$

$$V = \mathbf{1.242 \times 10^4 \text{ volts}}$$

$$V = \mathbf{12.42 \text{ kilo volts}}$$

Example 6.6. *What is the minimum wavelength emitted by an X-ray tube with an accelerating voltage of 4.5×10^4 V?* (*Osmania, 1992*)

Solution. Here,

$$\text{eV} = h\nu = \frac{hc}{\lambda_{\text{min}}}$$

$$\lambda_{\text{min}} = \frac{hc}{\text{eV}}$$

$$h = 6.624 \times 10^{-24} \text{ J-s}$$

$$c = 3 \times 10^8 \text{ m/s}$$

$$e = 1.6 \times 10^{-19} \text{ C}$$

$$V = 4.5 \times 10^4$$

$$\therefore \quad \lambda_{\text{min}} = \frac{6.624 \times 10^{-34} \times 3 \times 10^8}{1.6 \times 10^{-19} \times 4.5 \times 10^4}$$

$$= 0.276 \times 10^{-10} \text{ m}$$

$$= \mathbf{0.276 \text{ Å}}$$

Example 6.7. *Calculate the critical voltage required to stimulate the emission of the characteristic lines of the K-series in tungsten, if the K absorption edge is 0.178 Å.*

Solution.

$$\text{eV} = h_{\nu_k} = \frac{hc}{\lambda_k}$$

$$= \frac{6.63 \times 10^{-34} \times 3 \times 10^8}{(0.178 \times 10^{-10} \text{ m})(1.6 \times 10^{-19})}$$

$$= 69.8 \times 10^3 \text{ eV}$$

Thus

$$V = \frac{6.98 \times 10^3 \text{eV}}{e}$$

$$= \mathbf{69.8 \times 10^3 \text{ volts}}$$

The critical voltage is therefore 69.8 kV.

Example 6.8. *Determine theoretically the grating space in rock salt and hence the wavelength of X-rays. Given $\rho = 2.163$, θ for Ist order = 6° and Molecular wt. = 58.45.*

Solution:

Let molecular weight of rock salt (NaCl) = M and density be ρ.

Then the volume occupied by one gram mole = M/ρ. Now, the no. of moles in 1gm mole = N(Avogadro's number).

$\therefore$ No. of molecules per c.c. = $N/(M/\rho) = N\rho/M$

These are two atoms in a molecules of NaCl.

$\therefore$ No. of atoms per c.c. = $2N\rho/M$.

Let d be the grating space for 100 planes.

$\therefore$ The no. of atoms in 1cm length of an edge of the unit cell = $1/d$. Hence the no. of atoms in a cm cube of rock salt is $(1/d)^3$.

$$\therefore \qquad \frac{2N\rho}{M} = \left(\frac{1}{d}\right)^3$$

$$\text{or} \qquad d = (M/2N\rho)^{1/3}.$$

For Na = 23, Cl = 35.45 $\therefore$ Molecular wt of NaCl = 58.45. Now, $\rho = 2.163$ for NaCl.

and $N = 6.025 \times 10^{23}$. On substituting we get $d = 2.83 \times 10^{-8}$cm.

Using the Bragg Eq. for $\theta = 6°$. We have,

$$\lambda = 2d \sin\theta$$

$$\lambda = (2 \times 2.83 \times 10^{-8} \sin 6°) \text{ cm}$$

$$= \mathbf{0.59 \text{ nm.}}$$

Example 6.9. *Calculate the wavelength of the K_α line, from a copper target assuming Moseley's law.*

R for $H_2 = 1.097 \times 10^7 \text{ m}^{-1}$, screening constant = 1.

Solution:

According to Moseley's law, the frequency of the K_α for copper is given as

$$\nu = Rc(z-1)^2 \left[1/1^2 - 1/2^2\right]$$

z for Cu = 29 and $c = 3 \times 10^8$ m/s.

$$\nu = 1.097 \times 10^7 \times 3 \times 10^8 \times (28)^2 \times 3/4$$

$$= 1.935 \times 10^{18}$$

$$\therefore \qquad \lambda \text{ for } K_\alpha = c/\nu = \frac{3 \times 10^8}{1.935 \times 10^{18}}$$

$$= 1.55 \times 10^{-10}\text{m.}$$

$$= \mathbf{1.55 Å.}$$

EXERCISE VI

1. Describe the continuous and characteristic X-ray spectra of an element.
2. State Duane and Hunt's law. What is its significance ?
3. State Moseley's law for X-rays and explain its importance.
4. Describe the Coolidge tube for X-ray production. Enlist some properties of X-rays.
5. Describe the fine structure of X-ray spectra.

6. If the critical voltage for the emission of K-line from copper is 9000 volts and the wavelength of the K-absorption edge is 1.377 Å, calculate the ratio h/e.[**Ans.** 4.31×10^{-15} v-sec]
7. What is the wavelength of X-rays emitted when 100 KeV electrons strike a target ? What is their frequency ?
8. A KCl crystal has a density of 1.98×10^3 kg/m^3. The molecular weight of KCl is 74.55. Find the distance between the adjacent atoms. **Ans.** 3.14A
9. If the spacing of certain planes in calcite is 3.14Å, find the angles at which Ist and 3rd order diffraction maxima are found for X-rays of $\lambda = 0.8$Å. **Ans.** 7.3° and 22.5°

7

MOLECULAR PHYSICS

7.1 Molecular Spectra

The spectra emitted by atoms as a result of energy changes in their electronic system, which we have considered so far, are known *as atomic spectra.* There is another class of spectra called the *band* or molecular *spectra*, emitted by molecules, *i.e.*, when the emitting element is in the molecular state. The name 'band spectra' was given to these at a time when spectroscopes of low resolving power alone were available so that the component lines of the spectra were indistinguishable and appeared merged together into strips of shaded intensity, giving the impression of a continuous spectrum but divided into several bands. The fact that spectra of this type are actually due to molecules is readily established by the disappearance of the bands when the emitting substance is heated to temperatures at which the molecule dissociates into atoms.

With high resolving power instruments band spectra disclose a threefold structure : (*i*) Each band is composed to a *large number of lines* which are arranged with great regularity often stretching in both directions from a gap in the band and crowding together at the band-head, (*ii*) several bands are found to follow one another in a regular sequence, constituting a *group of bands*. (*iii*) a close and regular arrangement of different groups of bands forming a *band system*. Several such band systems are comprised in a complete molecular spectrum ; sometimes adjacent bands overlap and obscure the regularity of arrangement of lines, which makes the analysis of band spectra very difficult.

Many substances are known to produce band spectra in different spectral regions, *viz.*, in the *far infra-red* (about 150 Å to 30 μ, 1 μ = 10,000 Å). The *near infra-red* (about 5 μ to 1 μ) and *visible and ultra-violet region* 700 Å to 1000 Å. These three systems of bands according to the accepted ideas of the mechanism of their production, are called *pure rotation bands, rotation-vibration bands* and *electronic bands* respectively.

7.2 Experimental Study

The *rotation bands* in the far infra-red were first noticed in the absorption spectra of the hydrogen halides, by Czerny, who found in the case of HCl a series of seven absorption maxima in the region 120 μ to 44 μ and was able to show that the observed absorption lines could be represented by the relation :

$$\bar{\nu} = 20.84\, m - 0.001814\, m^3$$

where $m = 4, 6, 7, 8, 9, 10, 11.$

He obtained similar relations in the case of HBr and HI. The above relation shows that the group of lines that constitutes the rotation bands are separated by an approximately constant frequency difference, since m assumes successive integral values, the missing integers corresponding to lines not observed. The second term in m^3 indicates that the frequency interval between the successive lines is not quite constant, but decreases with higher wave numbers.

Water and ammonia also show complicated absorption spectra in this region. Till of late, it has not been easy to observe the pure rotation bands in many cases owing to experimental difficulties. But with the recent introduction of the microwave spectrometer, it has now become comparatively easy to observe and analyse these bands and interesting results have been obtained.

ROTATIONAL SPECTRA

Theoretical Explanation

7.3 Rigid Diatomic Molecule

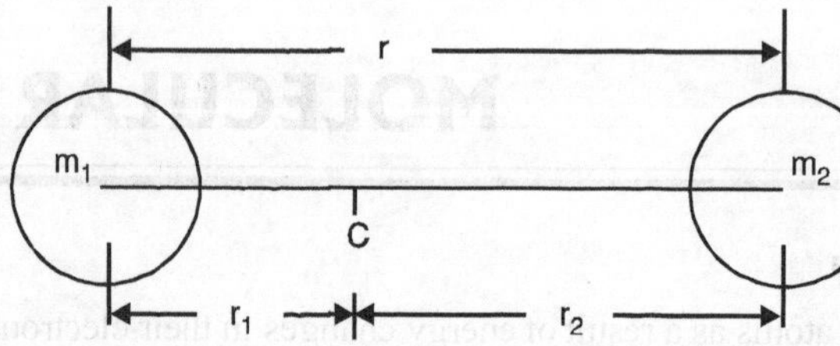

Fig. 7.1

A diatomic molecule consists of two atoms of masses m_1 and m_2 and seperated by a bond length r. Then $r = r_1 + r_2$.

C is the centre of mass of the system.

Taking moments about C.M.

$$m_1 r_1 = m_2 r_2 \qquad \text{...(7.1)}$$

The moment of inertia of the molecules about an axis through C.M and perpendicular to the line joining the atoms is

$$I = m_1 r_1^2 + m_2 r_2^2 \qquad \text{Subst. from (7.1)}$$

$$= m_2 r_2 r_1 + m_1 r_1 r_2$$

$$= r_1 r_2 (m_1 + m_2) \qquad \text{...(7.2)}$$

$$\because \quad r_1 = r - r_2$$

$$\therefore \quad m_1 r_1 = m_2 r_2 = m_2 (r - r_1)$$

$$\therefore \quad m_1 r_1 + m_2 r_1 = m_2 r$$

$$\therefore \quad r_1 = \frac{m_2 r}{(m_1 + m_2)} \quad \text{Similarly} \quad r_2 = \frac{m_1 r}{(m_1 + m_2)}$$

Subst. in Eq. (7.2), we have

$$I = \frac{m_1 m_2}{m_1 + m_2} r^2 = \mu r^2 \qquad \text{...(7.3)}$$

where $\mu = \dfrac{m_1 m_2}{m_1 + m_2}$ is the **reduced mass** of the system (molecules).

Using Schrödinger's equation the allowed energy levels for the diatomic molecule are :

$$E_J = \frac{\hbar^2}{2I} J(J+1) \text{ where J} = 0, 1, 2, 3, \ldots.$$

J is called the rotational quantum number ; $\hbar = h/2\pi$

Now angular momentum $L = I\omega$ and energy $E = \dfrac{1}{2} I\omega^2 = \dfrac{L^2}{2I}$, $\upsilon = \dfrac{\Delta E}{h}$ = the frequency

$$\therefore \quad \overline{\upsilon}_J = \frac{E_J}{hc} = \frac{\hbar^2}{2Ic} J(J+1) \qquad J = 0, 1, 2, \ldots.$$

or

$$\overline{\upsilon}_J = B J (J+1) \qquad J = 0, 1, 2, 3.$$

where B = rotational constant given by

$$B = \frac{\hbar^2}{2Ic} = \text{rotational const.}$$

$$I = \text{Moment of inertia about an axis.}$$

For $J = 0$ $\qquad \bar{\upsilon}_J = 0 \quad$ *i.e.* molecule is not rotating.

For $J = 1$ $\qquad \bar{\upsilon}_1 = 2B$

Frequency of photon

$$\Delta\upsilon = \upsilon_{J \to J+1} = \frac{\Delta E}{h} = \frac{E_{J+1} - E_J}{h} = \frac{\hbar}{2\pi I}(J+1) \qquad \text{...(7.4)}$$

$$\because \quad E_J = \frac{1}{2}I\omega^2 = \frac{L^2}{2I} \text{ and } L = \sqrt{J(J+1)}\,\hbar \qquad \text{...(7.5)}$$

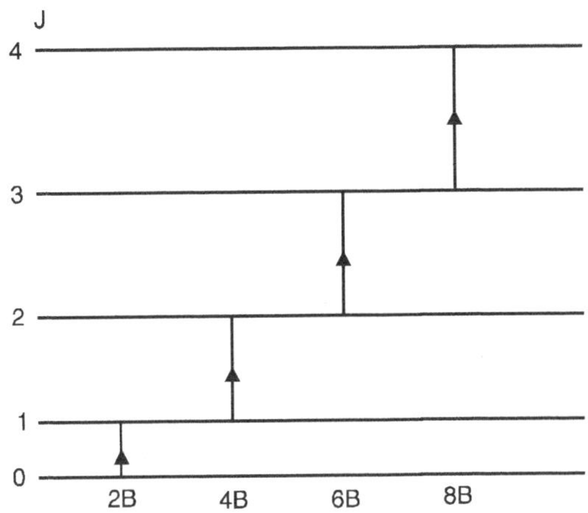

Fig. 7.2

Differences between the levels gives rise to the spectra. Let *a* molecule absorb enough incident radiation to raise it from $J = 0$ (gnd state rotational level) to $J = 1$ state.

$\therefore$ Energy absorbed will be

$$\bar{\upsilon}_{J=0} - \bar{\upsilon}_{J=1} = 2B - 0 = 2B\ m^{-1}$$

i.e.

$$\bar{\upsilon}_{J=0 \to J=1} = 2B\ m^{-1}$$

Similarly for $J = 1$ to 2

$$\bar{\upsilon}_{J=1 \to J=2} = \bar{\upsilon}_2 - \bar{\upsilon}_1 = 6B - 2B = 4B\ m^{-1}$$

In general the energy require to raise from J to $J + 1$ state is

$$\begin{aligned} \bar{\upsilon}_{J \to J+1} &= B(J+1)(J+2) - BJ(J+1) \\ &= B[J^2 + 3J + 2 - (J^2 + J)] \\ &= 2B(J+1)\ m^{-1}. \end{aligned}$$

Note that a transition can also occur from $J = 0 \to J = 2$ but this is spectroscopically forbidden.

$\therefore$ The selection rule for a rigid diatomic molecule is $\Delta J = \pm 1$ [both absorption and emission].

7.4 Intensities of Rotational lines

Transitions can take place from $J = 0$ to $J = 1$, $J = 1$ to $J = 2$, etc. Since $\Delta J = \pm 1$.

Intensities of these lines depends on the relative probabilities of these transitions.

The probability depends on the no. of molecules in the different states.

Using the Boltzmann's formula

$$N_J = N_0 \exp\left(\frac{-E_J}{KT}\right) \qquad ...(7.6)$$

and

$$E_J = BcJ(J+1) \qquad ...(7.7)$$

where K = Boltzmann's constant = 1.38×10^{-23} J/K.

For $T = 300$ k, $B = 200\ \text{m}^{-1}$ and $J = 1$

$$N_1 = N_0 \exp\left\{\frac{-2\times10^2 \times 6.63\times10^{-34} \times 3\times10^8 \times 2}{1.38\times10^{-23} \times 300}\right\}$$

$$N_1 = N_0 \exp(-0.019) = 0.98\, N_0$$

∴ The number of molecules in the two states are almost equal at room temperature *i.e.* for $J = 1$ and $J = 0$ state.

∴ Intensities differ for different values of J.

∴ J and $J + 1$ intensities differ by

$$J = \sqrt{\frac{KT}{2Bhc}} - \frac{1}{2}.$$

7.5 Vibrational Spectra

When a molecule is suitably excited it undergoes simple harmonic oscillations ; where the bond length between the two atoms gets compressed and extended. This is similar to a spring obeying Hooke's law.

$$\therefore \qquad f = -K(r - r_{eq})$$

where f = restoring force, K = force constant

r_{eq} = minimum equilibrium distance or bond length (internuclear)

and r = internuclear distance.

Now, the energy curve has the form

$$E = \frac{1}{2}K(r - r_{eq})^2$$

which is parabolic in nature *i.e.* simple harmonic oscillator model.

∴ On the basis of classical mechanics the oscillatory frequency is given by

$$\nu_0 = \frac{1}{2\pi}\sqrt{\frac{K}{\mu}}\ \text{Hz} \quad \text{where } \mu = \text{reduced mass of the system.}$$

Using Schrödinger's equation for a simple harmonic oscillator the energies turn out to be

$$E_v = \left(v + \frac{1}{2}\right)h\nu_0$$

where v is the vibrational quantum number ($v = 0, 1, 2, 3$ .....)

∴ The lowest vibrational energy for which $v = 0$ is

$$E_0 = \frac{1}{2}h\nu_0 \qquad \text{[referred to as zero-point energy} \neq 0]$$

In general, ∴

$$E_v = \left(v + \frac{1}{2}\right)\hbar\sqrt{\frac{K}{\mu}}$$

Using Schrödinger's equation gives a simple selection rule for the harmonic oscillator *i.e.* $\Delta v = \pm 1$. Note that the vibrations involve a change in the dipole moment of the molecules.

$\therefore$ Vibration spectrum is observed only in hetanuclear diatomic molecules.

$\therefore$ By the selection rule we have

$$\Delta E = E_{v+1 \to v} = h\left(v+1+\frac{1}{2}\right)\upsilon_0 - h\left(v+\frac{1}{2}\right)\upsilon_0 \quad \ldots (7.8)$$

$$= h\upsilon_0 \text{ for emision}$$

and $\Delta E = E_{v+1 \to v} = h\upsilon_0$ for absorption.

$\therefore$ The levels are equally spaced as shown in Fig. 7.3.

Hooke's law is not obeyed exactly by real molecules. For large extensions the bond is not elastic. Hence we have an anharmonic oscillator. For an anharmonic oscillator, the expression for the energy was given by Morse. The distribution for the vibration levels gets closer for large *v*.

Morse energy is given by

$$E = D\,[1 - \exp\{\alpha\,(r - r_0)\}]^2 \quad \ldots (7.9)$$

where r = interatomic seperation

r_0 = equilibrium seperation of the two molecules.

D = dissociation energy of the molecule.

d = width of the potential well (constant for a particular molecule).

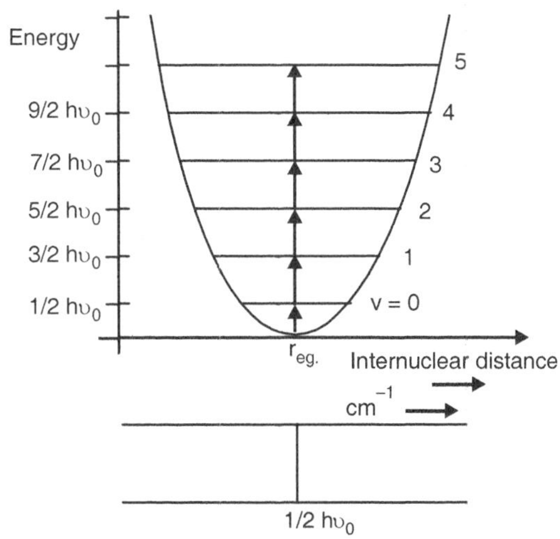

Fig. 7.4

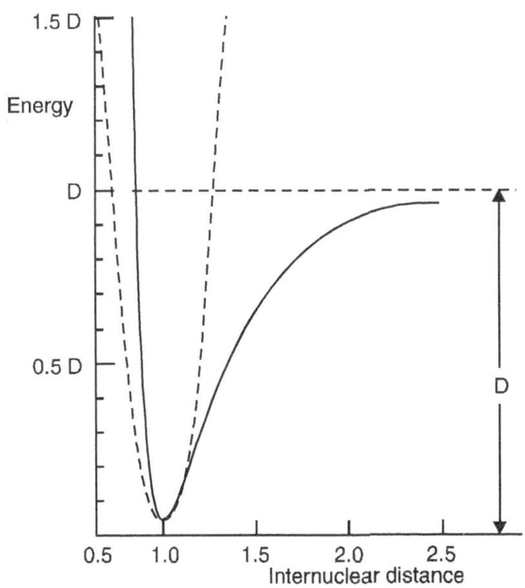

Fig. 7.4

This is an empirical formula and fits the vibrational energy levels of a diatomic molecule very closely.

Morse energy spreads out more than for the parabola at distances far from the minimum (Refer Fig. 7.5).

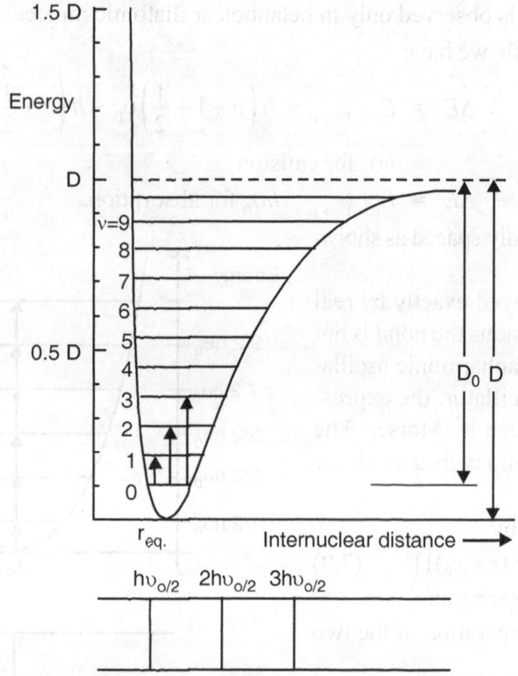

Fig. 7.5

Selection rules for the anharmonic oscillator are found to be

$$\Delta v = \pm 1, \pm 2, \pm 3 \ldots\ldots$$

Here there is a possibility of larger jumps.

The vibrational energy levels are :

$$Ev = \left(v + \frac{1}{2}\right) h\upsilon_0 - \left(v + \frac{1}{2}\right)^2 h\upsilon_0\, x_e \qquad \ldots (7.10)$$

where x_e = anharmonic constant.

Thus for ground state ($v = 0$)

$$Ev_0 = \frac{1}{2} h\upsilon_0 \left(1 - \frac{1}{2} x_e\right)$$

$\therefore$ The zero point energy differs slightly from that of the harmonic oscillator.

Now using Boltzmann's formula

$$\frac{N_{v=1}}{N_{v=0}} = \exp\left\{-\frac{6.63 \times 10^{-34} \times 3 \times 10^{10} \times 10^3}{1.38 \times 10^{-23} \times 300}\right\}$$

Spacing is of the order of 10^3 cm^{-1}

$$= \exp(-4.8) \simeq 0.008$$

$\therefore$ $v = 1$ state has a population of 0.01% of ground state population. Therefore, we ignore transitions from $v = 1$ or more states.

$\therefore$ (1)$v = 0 \rightarrow v = 1$ $\Delta v = +1$ has considerable intensity.

$$\Delta E = E_{v=1} - E_{v=0}$$

$$= \left(1+\frac{1}{2}\right)h\upsilon_0 - x_e\left(1+\frac{1}{2}\right)^2 h\upsilon_0 - \left\{\frac{1}{2}h\upsilon_0 - \left(\frac{1}{2}\right)^2 x_e\, h\upsilon_0\right\}$$

$$= h\upsilon_0 (1 - 2x_e)$$

(2) $v = 0 \rightarrow v = 2$, $\Delta v = +2$ small intensity

$$\Delta E = 2(1 - 3x_e)\, h\upsilon_0$$

(3) $v = 0 \rightarrow v = 3$ $\Delta v = +3$ negligible intensity

$$\Delta E = 3(1 - 4x_e)\, h\upsilon_0$$

The three transitions are shown. $\because$ $x_e \simeq .01$ the lines lie close to $h\upsilon_0$, $2h\upsilon_0$ and $3h\upsilon_0$.

$h\upsilon_0$ is called the fundamental absorption.

$2h\upsilon_0$ and $3h\upsilon_0$ are called the first and second overtones respectively.

7.6 Rotation-Vibration Bands

The *rotation-vibration bands* in the near infra-red have also been observed in the absorption spectra of the hydrogen halides. The general structure of a typical band of this sort (HCl) is shown in Fig. 7.6. The form is seen to be double—the so-called *double band*, in the sense that a curve drawn through the intensity maxima of the individual lines shows two maxima, as indicated by the dotted line. There is a gap between the two bands—the *missing central line*. This feature of a double band with a central gap distinguishes the rotation vibration spectra from the pure rotation spectra which have only a single band with no gap.

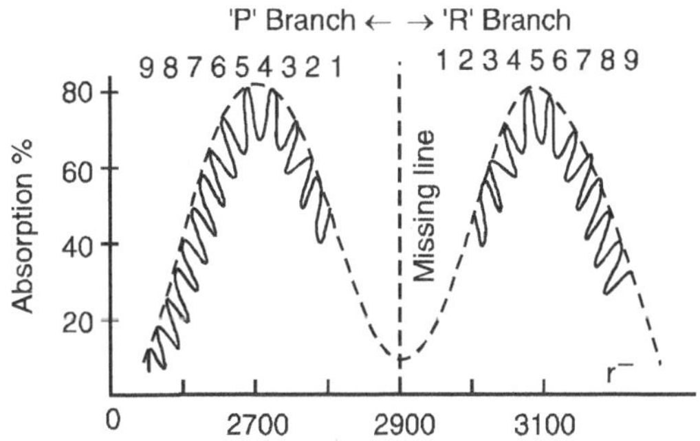

Fig. 7.6

The wave numbers of the individual line maxima are well represented by the relation :

$$\bar{v} = 2886.2 + 20.5379\, m - 0.30318\, m^2 - 0.001814\, m^3$$

The first term indicates the region in which this type of spectra appears (2886 cm^{-1}), *i.e.*, near infra-red ; the second term gives the approximate distance between the lines (20.5 cm^{-1}), which is seen to be nearly the same as the distance between the lines of the pure rotation spectra (20.8 cm^{-1}). The other two terms are correction factors which indicate that the frequency interval between the line is not constant, but decreases with higher members. The lines which are on the higher wave number side of the central missing line constitute the so-called *R* branch, while those on the lower wave number side the *P* branch.

Rotation-vibration bands are found also with substances other than the hydrogen halides, but their structure is often more complex with no gap in the centre of the band. Thus the absorption spectrum of CO_2 shows many rotation-vibration bands from 1.46 μ to 15 μ that of water vapour in.

7.7 Theoretical Explanation of Rotational-Vibrational Bands

To a first approximation we can consider a diatomic molecule to execute rotations and vibrations quite independently. This is known as Born-Oppenheimer approximation. Therefore according to this approximation energies of rotation and vibration are additive.

$$E_{\text{Total}} = E_{\text{rot.}} + E_{\text{vib.}} \text{ (Joules)} \quad ...(7.11)$$

Substituting, for $E_{\text{rot.}}$ and $E_{\text{vib.}}$ we have,

$$= \frac{\hbar^2}{2I} J(J+1) + \left(v+\frac{1}{2}\right)h\nu_0 - \left(v+\frac{1}{2}\right)^2 h\upsilon_0\, x_e \text{ Joules} \quad ...(7.12)$$

where x_e = anharmonic constant.

$$\therefore \quad E = E_v + J(J+1)Bh$$

where $B = h/8\pi^2 I$.

In Fig. 7.7 rotational levels are indicated for $v = 0$ and $v = 1$.

Now selection rules are $\Delta v = \pm 1, \pm 2$ etc. $\Delta J = \pm 1$

If $\Delta v = 0$ is allowed then we get only rotational transitions.

Note $\Delta J = 0$ is not allowed *i.e.* vibrational changes must be accompanied by a simultaneous rotational change. Since Bh is small all lines of vibration changes lie close together and form a *band*. The combined vibration-rotation spectrum consists of equally spaced lines on each side of $h\nu_0$.

Lines of lower frequency *i.e.* $\Delta J = -1$ are called P branch and lines of higher frequency $\Delta J = +1$ are called the R branch.

∴ Lines arising from	$\Delta J =$ −2	−1	0	+1	+2	
called	O	P	Q	R	S	branch

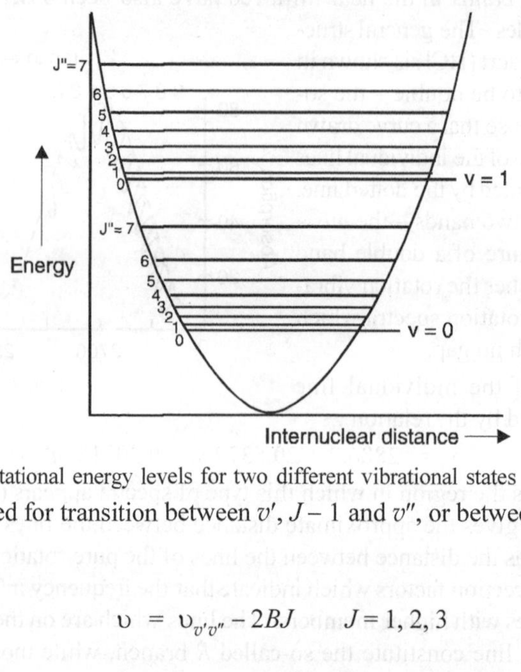

Fig. 7.7. The rotational energy levels for two different vibrational states of a diatomic molecule.

Frequency emitted for transition between v', $J-1$ and v'', or between v', J and v'' $J-1$ respectively are :

$v', J-1, v'', J$

$$\upsilon = \upsilon_{v'v''} - 2BJ \qquad J = 1, 2, 3$$

$v', J; v'', J-1$

$$\upsilon = \upsilon_{v'v''} + 2BJ \qquad J = 1, 2, 3.$$

$$\therefore \quad \nu_{v'v''} = \frac{E_{v'} - E_{v''}}{h} \quad \text{Assuming } v' > v'' \quad E_v' > E_v''.$$

∴ Lines in a given vibration-rotation band emitted by a diatomic molecule should be equally spaced in frequency with a separation = $2B$.

Note $\upsilon = \upsilon_{v'v''}$ cannot occur. ∴ No line corresponding to a pure vibration can be seen.

Nearest frequencies are

$$\upsilon = \nu_{v'v''} + 2B \text{ and}$$

$$\upsilon = \upsilon_{v'v''} - 2B \text{ as shown in Fig. 7.8.}$$

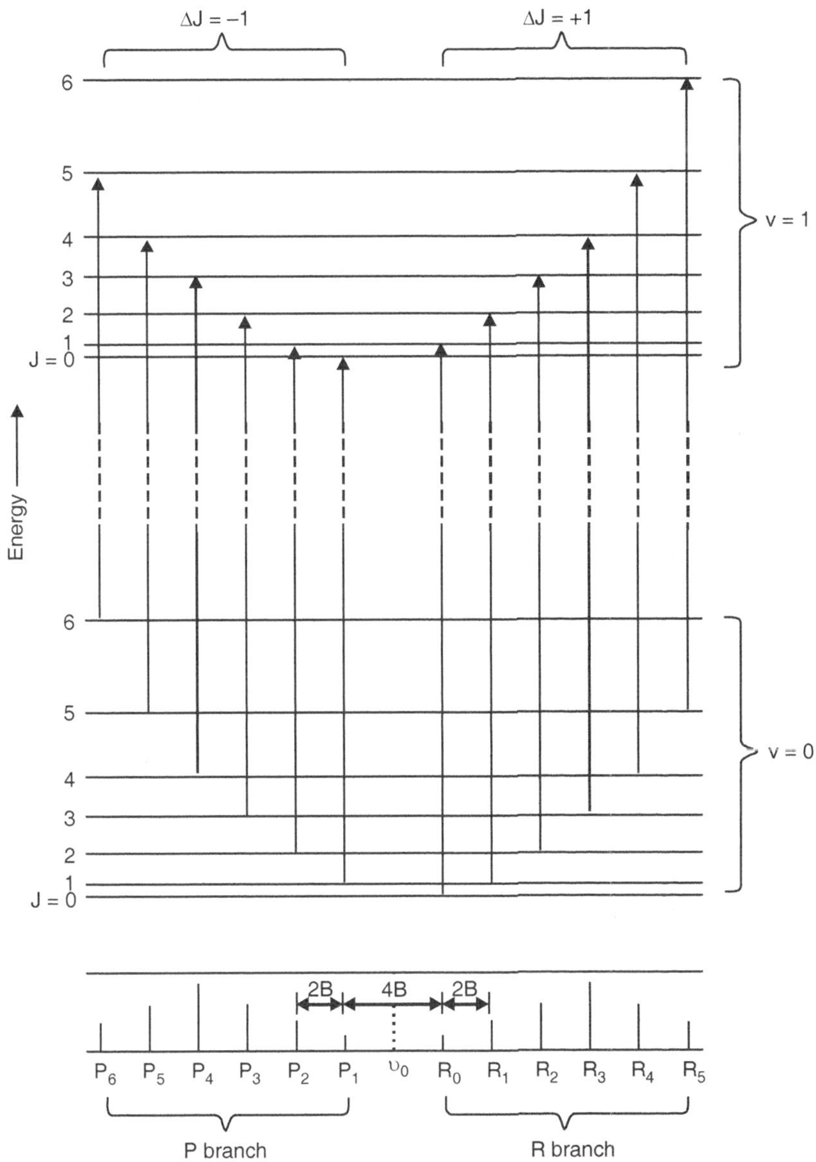

Fig. 7.8. Some transitions between the rotational-vibrational energy levels of a diatomic molecule together with the spectrum arising from them.

7.8 Rotational-Vibrational Electronic Spectra

It is possible for the electronic states of molecules to be excited when there is enough energy. This gives the general molecular-spectra.

This gives rise to energy

$$hv = hv_e + (v + 1/2)hv_0 + J(J+1)\frac{\hbar^2}{2I}. \qquad ...(7.13)$$

The outermost valence electrons are involved in the process. These give rise to transitions in the visible and ultra-violet regions. For each electronic state there are a number of vibrational states and for each vibrational state there are a number of rotational states.

In transitions of different electronic states given by v' and v'' of v different changes of J occur and these form a band. All transitions for a given pair of electronic states for v' and v'' form a band system. Hence molecular spectra consists of many band systems.

For electronic band the selection rules are :

$$\Delta J = 0 \quad \text{or} \pm 1.$$

Lines for which ΔJ have opposite sign to that for Δv constitute the *P* branch.

$\Delta J = 0$ gives rise to the *Q* branch.

ΔJ in the same direction as Δv form the *R*-branch. Normally the *Q* branch in missing. In the Fig. 7.9, these branches are show. They are referred to as *Fortrat diagrams.*

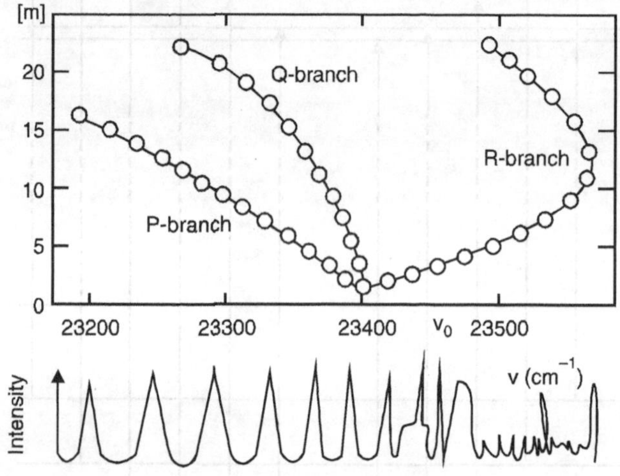

Fig. 7.9. Fortrat diagram (AlH band for λ = 4291 A).

Franck-Condon Principle : Intensity distribution

The intensity distribution of the vibration in an electronic band varies in a complex way. This can be explained by the Frank-Codon principle.

It states that *an electronic transition in a molecule occurs so rapidly, compared to the vibrational motion of the atoms, that during the transition the vibrating molecule does not change its internuclear distance appreciably.*

Fig. 7.10 shows three possibilities : (see next page)

In Fig. (*a*) The upper electronic state has the same internuclear distance as the lower electronic state. As per the Frank-Condon principle a transition occurs vertically as shown. If we both electronically and vibrationally the molecule is initially in the ground state (ε'' and υ'') then the most probable transition is to (ε' $\upsilon' = 0$). This is strongest line (0, 0). There is some chance of the transition ending on $v' = 1, 2$ etc. states. The (1, 0), (2, 0) etc. lines diminish in intensity as shown in Fig. (*a*).

In Fig. (*b*) the executed electronic state has a greater internuclear separation than the ground state. In this case a vertical transition from $v'' = 0$ most likely occurs to $v' = 2$ state. Transitions to lower and higher v' states being less probable the intensities diminish on both sides.

In Fig. (*c*) the upper state separation is drawn as considerably greater than the lower state and the vibrational level to which the transition takes place has a high value of v'. Further, transitions occur to states which have energy in excess of the dissociation energy. Hence in transitions to these levels we get a continum and not lines as shown.

The situation tends to be more complex for emission or absorption spectra from an excited vibrational state, since transitions now take place from both ends of the vibrational limits with equal probability. Hence we get two maxima which coincide only if the equilibrium separation are the same in both states.

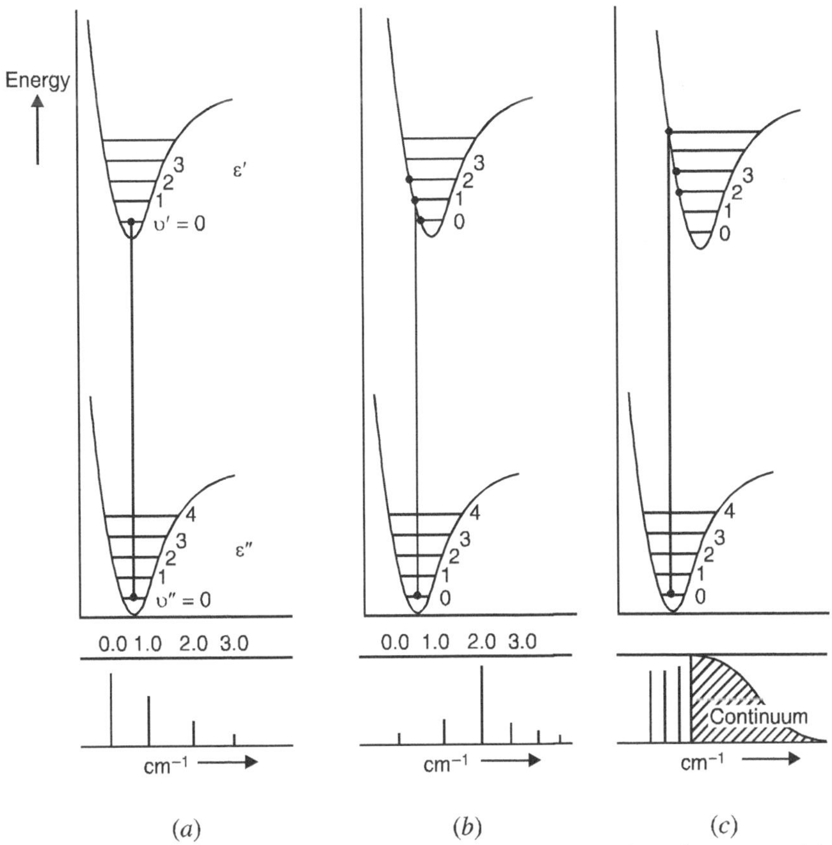

Fig. 7.10. The operation of the Franck-Condon principle for (*a*) internuclear distances equal in upper and lower states, (*b*) upper-states internuclear distance a little greater than that in the lower state, and (*c*) upper-state distance considerably greater.

Dissociation of Molecules : Heat of Dissociation

The amount of work required to be done in order to separate the constituent atoms of a molecule is known as *heat of dissociation* and is expressed in calories.

Using the wave mechanical formula $E_n = h\nu_0(n + 1/2)$ leads to the energy levels of the nuclear motion.

The natural frequency $$\upsilon_0 = \frac{1}{2\pi}\sqrt{\frac{1}{m}\left(\frac{d^2V}{dr^2}\right)_{r_0}}$$

$\left(\frac{d^2V}{dr^2}\right)_{r_0}$ gives the curvature at point $r = r_0$.

$\therefore$ Greater the curve greater is the natural frequency of the nuclear motion and higher are the corresponding energy levels.

For highly excited states with large quantum number n the curve is no longer parabolic and not a harmonic oscillator. With increase in n, the energy levels get crowded more and more closely together upto a *convergence limit.* This limit corresponds to the dissociation of the molecule, it requires energy that corresponds to the depth of the potential well below the zero level.

Beyond this the atoms or ions are split up and fly apart with a velocity depending on the excess of energy. Thus in molecular spectra this is given by a band with a convergence limit followed by a continum.

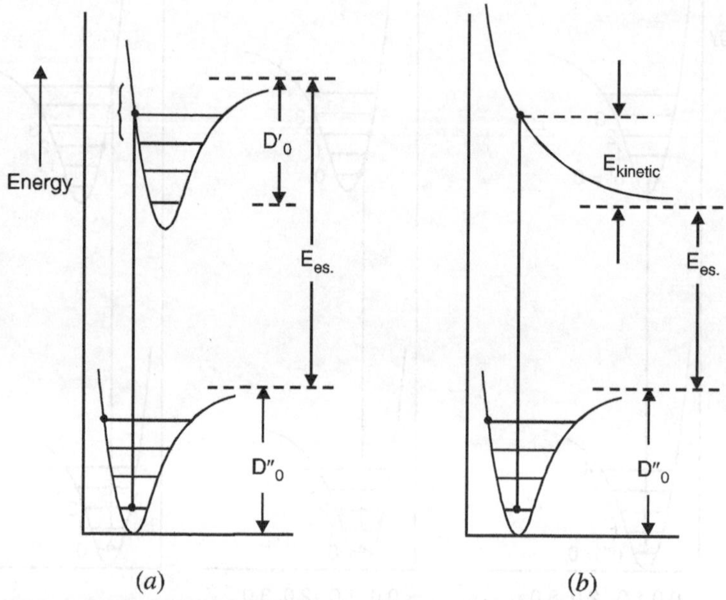

Fig. 7.11. Illustrating dissociation by excitation into (*a*) a stable upper state, and (*b*) a continuous upper state.

Hence we see that the molecules absorbs energy in quantised form only upto a certain limit, *i.e.* equal to energy of dissociation, beyond this the absorption of energy is continuous.

In Fig. 7.11 D_0'' is the Dissociation energy. In Fig. (*b*) we have an illustration of a case where the upper electronic state is *unstable,* there is no minimum in the energy curve and so when the molecule is raised to this state it immediately dissociates.

Example 1. *Calculate the frequency of Oscillation of a hydrogen molecule of its force constant is 4.8 × 10^2 N/m and mass of hydrogen atom = 1.67 × 10^{-27} kg.*

Solution.

$$\upsilon = \frac{1}{2\pi}\sqrt{\frac{K}{\mu}}$$

Here μ is the reduced mass of the system.

$$\frac{1}{\mu} = \frac{1}{m} + \frac{1}{m} = \frac{2}{m}$$

$$\mu = \left(\frac{m}{2}\right)$$

Here

$$K = 4.8 \times 10^2 \text{ N/m}$$

$$\mu = \frac{1.67\times10^{-27}}{2} = 0.835 \times 10^{-27} \text{ kg.}$$

$$\therefore \quad \upsilon = \frac{1}{2\pi}\sqrt{\frac{4.8\times10^2}{0.835\times10^{-27}}}$$

$$\upsilon = 1.2 \times 10^{14} \text{ Hz}$$

Example 2. *The first line in the rotation spectra of CO is 3.842 cm^{-1}. Find the bond length. Reduced mass of CO = 11.384 × 10^{-27} kg.*

Solution.

$$\bar{\upsilon} = 2B = 3.842 \text{ cm}^{-1}$$

$$\therefore \qquad B = 1.921 \text{ cm}^{-1}$$

$$\because \qquad I = \frac{\hbar^2}{2IBC} \quad \therefore \quad I = \frac{(6.63 \times 10^{-34})}{4\pi^2 \times 2 \times 2.997 \times 10^{10} \times 1.921 \text{ cm}^{-1}}$$

$$I_{CO} = 14.569 \times 10^{-47} \text{ kg-m}^2$$

$$r^2 = \frac{I}{\mu} = \frac{14.569 \times 10^{-47}}{11.384 \times 10^{-27} \text{ m}^2}$$

$$= 1.279 \times 10^{-20} \text{ m}^2$$

$$r = 0.113 \text{ m} = 1.31 \text{ Å}$$

Example 3. *Find the ratio of the intensities with J = 10 and J = 0 for the HCl molecules.*

Solution. Energy is given by

$$E_J = J(J+1) \times 1.3 \times 10^{-3} \text{ eV}$$

Compared to room temperature kT of 0.025 eV. The statistical weight of each J state is $2J + 1$. J, $J - 1$, ... D, ... $-J$.

$\therefore$ Ratio of number of HCl molecules with $J = 10$ to the number with $J = 0$ at room temperature,

$$\frac{n_{10}}{n_0} \alpha (2J+1) e^{-(E_{10} - E_0)/KT} = \frac{21}{1} e^{-143/25} = 21 e^{-5.7}$$

$$= 0.07$$

$\therefore$ The states $J = 10$ occur 0.07 times as often as states with $J = 0$. Hence the intensity ratio. since line intensities are directly proportional to the populations.

Example 4. *The bond length R of carbon monoxide (CO) is 0.113 nm and the reduced mass is 1.14×10^{-26} kg. Find the energy and angular velocity of the CO molecule in its lowest state.*

Solution.

$$I = \mu R^2 = (1.14 \times 10^{-26} \text{ kg})(1.13 \times 10^{-10} \text{ m})$$

$$= 1.46 \times 10^{-46} \text{ kg-m}^2$$

Energy for $J = 1$ is

$$E_J = \frac{J(J+1)\hbar^2}{2I} = \frac{\hbar^2}{I} = \frac{(1.054 \times 10^{-34} \text{ J.s})^2}{1.46 \times 10^{-46} \text{ kg-m}^2}$$

$$= 7.61 \times 10^{-23} \text{ J}$$

$$\omega = \sqrt{\frac{2E}{I}} = \sqrt{\frac{2 \times 7.61 \times 10^{-23}}{1.46 \times 10^{-46}}}$$

$$= 3.23 \times 10^{11} \text{ rad/sec.}$$

EXERCISES VII

1. Give the theory of the rigid distance molecule and account for the rotational spectra.
2. How are the intensities obtained for the different rotational lines ?
3. Give the theory of the vibrational spectra. Find the spacing between the energy levels.
4. Show that for an anharmonic type oscillator the diatomic molecules vibrational spectra are unequally spaced.
5. Give the theoretical account of the rotational vibrational bands and show that Q-branch is missing.
6. What is heat of dissociation of a molecule ?
7. Describe the rotational-vibrational electronic spectra of molecules.
8. What are Fortrat diagrams ? State and explain the Franck-Condon principle for the Intensity distribution.

8

RAMAN EFFECT AND SPECTROSCOPIC TECHNIQUES

8.1 Raman Effect

Raman effect may be said to be the *optical analogue* of Compton effect, since it also belongs to the category of *incoherent scattering*, which finds an adequate interpretation only on the basis of the quantum theory principles. Hence the inclusion of Raman effect among the applications of Bohr's theory may be justified. Moreover, as Raman effect is closely connected with molecular spectra the study of the former naturally follows the latter.

Discovery : In 1928, Sir C.V. Raman was studying the scattering of light by liquids with the intention of reproducing such natural phenomena as the *blue of the sea and of the sky*. Already, in 1871, Lord Rayleigh had given a correct explanation of the blue colour of the sky. He attributed the phenomenon to the scattering of light by the individual air molecules or groups of them found in abundance in the earth's atmosphere. He showed also that the *coefficient of scattering of light in any medium is inversely proportional to the fourth power of the wavelength of light*. From this it follows that the intensity of the scattered light must increase rapidly as the wavelength decreases. Hence when sunlight passes through the atmosphere, the intensity of the blue colour in the scattered light will be very high compared with most of the other components of white light, since its wavelength is smaller than those of most of the other visible colours. This is the reason why the sky appears blue. It is to be noted that this type of scattering, known as *Rayleigh scattering*, is also called *coherent scattering*, since, though there is a change in the intensity of the scattered light, there is no change in the spectral character, *i.e.*, in the wavelength.

In his experiments, Raman found that when a beam of monochromatic light was passed through organic liquids such as benzene, toluene, etc., the scattered light contained other frequencies in addition to that of the incident light. Although such an effect has been predicted on theoretical grounds as early as 1923 by Smekal, Raman was the first to observe it experimentally. The arrangement used by him was simple in design. A round-bottomed glass flask was filled with dust-free benzene and the liquid was strongly illumined by the 4358 line from a mercury arc, suitably filtered and concentrated by a lens. The scattered light was examined by means of a spectroscope placed transversely, *i.e.*, in a direction at right angles to that of the incident radiation. In the spectrum of the scattered light, number of new lines were observed on both sides of the main line. Those on the low frequency side were numerous and more intense than those on the high frequency side ; the spacing between the lines was symmetrical about the main line ; most of them were strongly polarised. The observed spectrum is now generally referred to as *Raman spectrum* and the lines ; as *Raman lines* ; more specifically those on the low frequency side as *Stoke's lines* and those on the high side as *anti-Stroke's lines*.

Raman, having thus observed a set of a new discrete frequencies in the scattering process, contrary to the expectation of classical theory, was able to establish that they constituted a new phenomenon distinct from both the simple Rayleigh or coherent scattering and the more complex Stoke's of fluorescent scattering. He argued that in Rayleigh scattering no frequency change was caused, but only some of the already existing frequencies were selected, whereas in the phenomenon observed new lines of different frequencies appeared even when a single frequency (monochromatic light) was scattered.

Nor could the observed effect be considered similar to fluorescence, although there was a certain amount of superficial resemblance between the two, in so far as with a given exciting line new lines made their appearance in both. For, in the first place, the frequencies in the fluorescent spectrum were always less than the incident frequency, while the Raman lines had frequencies both higher (anti-Stocke's lines) and lower (Stoke's lines) than that of the incident line. Secondly, the frequencies of the fluorescent lines were really independent of that of the exciting line, provided the latter was able to produce fluorescence, but in the case of the Raman lines, their frequencies were directly related to that of the incident light, since if the latter was varied the former changed at the same rate. Thirdly, while the frequencies of the fluorescent lines were fixed by the nature of the scatterer, it was the frequency shifts of the Raman lines (also known as *Raman frequencies*) that were determined by the scatterer rather than the frequencies themselves. Fourthly, the lines observed by Raman were strongly polarised unlike the lines in the fluorescent spectrum. Finally, the most important fact that confirmed Raman's discovery as a new effect was that the Raman frequencies were either actual infra-red frequencies in the absorption spectrum of the scatterer or differences in such frequencies which was not the case in fluorescent scattering. This fact proved further that the observed effect was molecular phenomenon. Hence the modified frequencies observed in the scattering process was a new type of secondary radiation, to which the name *Raman Effect* was given and for which Raman was honoured by the scientific world in 1930 by conferring on him the Nobel Prize.

Almost simultaneously with Raman, Landsberg and Mendelstam in Russia, studying the light scattered by certain crystals discovered the effect in the case of solids. It was soon found that many liquids, vapours, gases and transparent solids exhibited the effect which thus proved to be a general phenomenon.

Experimental study : Raman effect has been extensively investigated by a large number of workers. The *general principle* used in these researches is to illumine the substance under test with an intense monochromatic source of light and photograph the scattered radiation by means of a spectrograph arranged in a transverse direction. But the technical details vary according to the nature of the substance studied, *i.e.*, solid, liquid or gas, the chief aim being to obtain the best and quick results.

Apparatus : The original simple arrangement of Raman was not quite efficient and required very long exposures, about 100 hours and more, to obtain good records of the Raman spectrum. Hence improvements were made as regards the container of the substance, the source of radiation, filters, spectrograph, etc.

The apparatus shown in Fig. 8.1 is the one, first designed by Wood and now ordinarily used in the study of Raman effect in *liquids*.

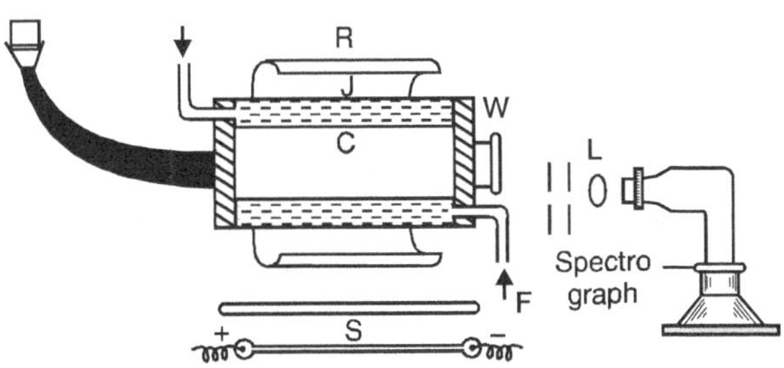

Fig. 8.1. Apparatus for the study of Raman effect.

The container C of the liquid to be investigated, called the *Raman tube*, consists of a glass tube of about 1 or 2 cm in diameter, and 10 to 15 cm long, one end of which is drawn out into the shape of a horn and blackened outside to provide a suitable background, the other end being closed with an optically plane glass plate constituting the window W through which the scattered light emerges. The container is surrounded by a water jacket in which cold water is circulated to percent overheating of the liquid in the Raman tube due to the proximity of the exciting arc.

An ideal *source S* would be light from a helium discharge tube filtered by nickel oxide glass, giving a strictly monochromatic line of wavelength 3888 Å. But on account of the many technical difficulties involved in the construction and manipulation of this source, it is not widely used. The source ordinarily employed is the mercury arc, the next best available, from which it is possible to get single wavelengths by the use of suitable filters. Thus, for instance, to obtain the 4358 Å line, slightly acidulated quinine sulphate solution contained in a *novial* glass vessel is used as a filter, which cuts off all the other lines except the 4358 Å. To get the 4046 Å line, a solution of iodine in carbon tetrachloride contained in a novial glass cell is found to be a very satisfactory filter. The filter solution may be arranged either to surround the Raman tube or in front of the arc. The mercury arc is placed as close to the Raman tube as possible, which results in a large intensity of the incident light. A semi-cylindrical aluminium reflector R, enhances the intensity of illumination still further.

The chief features of a *spectrograph,* suited for the study of Raman spectra, are : (1) large light gathering power, (2) special prisms of high resolving power, and (3) a short focus camera. A lens *L* in front of the window *W* directs the scattered radiation upon the slit of the spectrograph, which is carefully aligned along the axis of the Raman tube and screened from the direct rays of the arc. The more intense Raman lines of a liquid such as CCl_2 can be photographed in about an hour with a small spectrograph, but the recording of the complete spectrum may require up to ten to fifteen hours, depending largely on the intensity of the incident light, the speed of the spectrograph and the intrinsic brilliance of the Raman lines. It may be noted that instruments of high resolving power such as gratings are not used with advantage, on account of their poor luminosity which necessitates long exposures.

Certain modifications in the apparatus are necessary for the excitation of Raman effect in solids and gases. In the case of *solids* which are available as *large and transparent blocks*, like gypsum, quartz etc., a container is not required and the light from the arc can be directly focussed on the substance with a large condensing lens. With solids which are in the form of *loose crystal powders*, Raman effect can be obtained as was first shown by Bear and Menzies, by reflecting light from the crystal surfaces. But special precaution has to be taken to avoid the masking effect due to large amount of direct light coming out of the container by repeated reflection at the crystal faces and entering the spectrograph. To achieve this, two techniques have been used : *one* the use of complementary filters as developed by Ananthakrishnan at the Raman laboratory and the *other* a special type of spectrograph with two parts each part having a prism and two lenses with a common slit in between the two, devised by Billroth, Kohlrausch and Reitz in Germany. Both give very good results with crystal powders and the latter even with very small quantities of the substance.

In the case of gases, the intensity of the light scattered is normally very weak, but the difficulty has been overcome by intense illumination of the gas under high pressure and the use of a spectrograph of large light-gathering power. Wood employed a very long tube of HCl gas and obtained its Raman spectrum at atmospheric pressure using a specially made mercury arc which was placed in contact with the gas tube, hollow cylindrical reflectors enclosing both of them. The illumination produced in this way was very intense as the light from the arc was reflected back and forth between the walls of the reflectors. Rosetti was the first to develop the technique of exciting Raman effect in gases under high pressure, which shortened considerably the time of exposure. He used a thick-walled quartz tube 20 cm long and 2.2 cm internal diameter, which could stand pressures of 10 to 15 atmospheres. With the 2537 Å line of the mercury arc he was able to obtain the Raman spectra of several gases under pressure. Bhagavantam has constructed a Raman tube for gases which can stand pressures up to 50 atmospheres. It is made of transparent silica and enclosed in an outer steel tube for protection. He has been able to obtain with his arrangement good spectrographs of Raman spectra of gases in about 40 or 50 hours time of exposure using the 3650 Å, 4046 Å and 4358 Å mercury lines.

8.2 Results of Raman Effect

Raman effect has been observed and studied in a large number of liquids, organic and inorganic, many vapours and gases and some solids both crystalline and amorphous. An idea of the type of Raman spectra produced by solids, liquids and gases may be formed by inspecting the photographs with benzene, carbon tetrachloride (liquids), acetylene (gas), calcite and diamond (solids) [Fig. 8.2].

It is seen that a number of new lines and bands, exhibiting a variety of characters of intensity and fine structure are recorded on either side of the exciting radiation. Each line in the incident spectrum, is of sufficient intensity, gives rise to its own set of lines or bands. We shall now outline the main results obtained from the study of such photographic records.

Raman effect in liquids : About a hundred liquids, so far examined show the effect in an unmistakable manner. With benzene, the frequency shifts of the Raman lines correspond to an infra-red wavelength of 3.27 μ, in which region benzene exhibits a strong band in its absorption spectrum. A clear examination, however, of the infra-red absorption spectrum of benzene shows that none of the Raman lines are represented in it. An interesting feature with CCl_4 is the triad of Stoke's and anti-Stoke's, lines equally spaced on either side of the exciting 4358 Å line. With well purified water, two broad bands are found instead of sharp lines at about $\lambda = 3$ μ. Solutions of salts in water give Raman spectra characteristic of the salts and water.

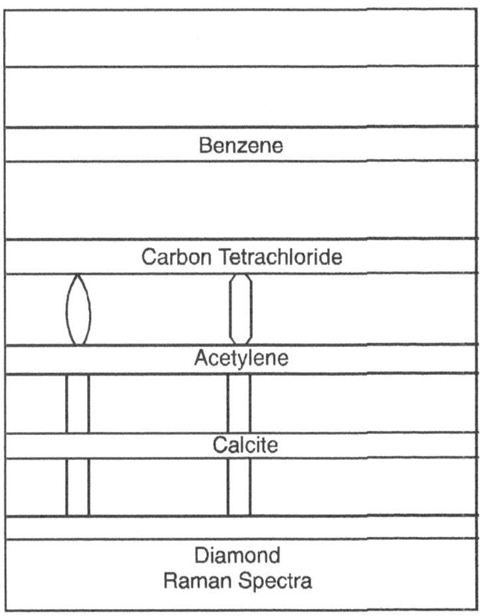

Fig. 8.2. Raman spectra of different substances.

Raman Effect in Gases : Several gases such as HCl, CO, CO_2, NO, N_2O, CS_2, hydrogen, oxygen, nitrogen, ammonia, etc. have been studied. The frequency shift of the Raman line of HCl gas corresponds to 3.46 μ, which is almost exactly the wavelength of the central missing line of the infra-red absorption band of HCl. CO gives a Raman line whose frequency shift is equal to the frequency of its infra-red bands, while CO_2 gives a line whose shift is equal to the frequency difference of two of its infra-red bands. Oxygen, hydrogen and nitrogen give a pattern of equally spaced lines. The intensities of the individual lines alternate in the case of nitrogen and hydrogen, while with oxygen the alternate lines are absent. These peculiarities in intensity of alternate lines have led to very significant conclusions are regards molecular and nuclear structure.

Raman effect in solids : The Raman lines obtained with crystals are sharp, becoming diffuse with rise of temperature. In calcite ($CaCO_3$) two lines which are nearest the exciting line have been identified with the oscillations of the crystal lattice. Diamond exhibits a strong and sharp line of a comparatively large frequency shift, which has to be described to a lattice oscillation.

Intensity of Raman lines : The intensity of a Raman line when expressed as a function of the parent line is usually a few hundredths in liquids and few thousands in gases. In spite of the practical difficulties in measuring the intensities of Raman lines results of fundamental importance have been obtained. The variations in intensity as we pass from line to line and from substance to substance are of great significance in the study of molecular structure and chemical constitution. The Stoke's lines are always more intense than the corresponding anti-Stoke's lines ; these, however, grow more intense as the temperature is raised. All the Raman lines move inward towards the parent line with increase of temperature.

Polarisation of Raman lines. Just as the Raman lines vary greatly in their intensities, so also their states of polarisation. The fact that different lines are differently polarised is probably connected with their relative intensity.

The experimental arrangement used for determining the polarisation of Raman lines is essentially the same as that described earlier, but with the following modifications. The light from the source is concentrated by means of a condenser into the substance contained in the Raman tube. A suitably orientated double image prism whose function is to separate the vertical and horizontal components in the scattered light is placed in front of the slit of the spectrograph, so that two images, one above the other are formed on the slit, which are simultaneously photographed.

The state of polarisation of a Raman line is measured by a quantity known as the *depolarisation factor* which is simply, the ratio of the intensities of the horizontal and vertical components when the incident light is vertically polarised. The ratio is readily obtained from the traces photographed as described above, by one of the usual methods employed for comparing the intensities of two beams of the same wavelength. In order to get fairly accurate values of the depolarisation factor, the following precautions should be taken : (*i*) Crystalline quartz should not be used for condenser, spectrographs or windows, since its optical activity complicates the phenomenon of polarisation; (*ii*) the window of the Raman tube through which the scattered light emerges should be stain free and plane; and (*iii*) errors arising from oblique refraction at the prism surfaces, want of transversality in the incident beam and slit width should be eliminated.

Cabannes found that the Raman lines in crystals, such as quartz, are differently polarised, the intensity and depolarisation of the lines depending upon the orientation of the crystal. Menzies has investigated the polarisation of the Raman lines in liquids, such as CCl_4, in directions perpendicular and obliquely forward to the incident beam and has shown that many of the observed facts could be accounted for by considering the initial and final directions of vibrations in the molecule involved to be parallel in the case of polarised lines, perpendicular in that of unpolarised lines and at an oblique angle for partially polarised lines. The following are some of the important result obtained :

(*a*) The depolarisation factor varies from 0 to 0.86 for the vibrational Raman lines, while it has a constant value of 0.86 for the rotational lines.

(*b*) With circularly polarised incident light, part of the Raman line is circularly polarised in the reverse direction and circularly polarised in the same direction as that of the incident light. Highly depolarised rotational lines exhibit reverse circular polarisation.

(*c*) Sharp and strong lines are ordinarily characterised by low depolarisation factors, while diffused and weak ones by high depolarisation factors.

(*d*) Corresponding line in molecules having similar structures have nearly the same depolarisation factors. This is to be expected as the polarisation of a Raman line is decided by the symmetry of the oscillation.

Nature of Raman effect : From the many and varied experimental data, it is clear that Raman effect is a molecular phenomenon. In the case of free molecules, as in gases, three different kinds of Raman effect can therefore be distinguished, *viz.*, rotational, vibrational and electronic. A mixed "rotation-vibration" effect can also take place under certain circumstances. Solids can exhibit another type, in which the crystal lattice as a whole takes the place of the molecule. Most of the observed Raman lines and bands with moderate or large frequency shifts are due to the vibrational effect, since pure rotational Raman lines lie very close to the parent line and are often masked by it except in the case of certain light gases. The vibrational Raman lines are due to the various normal modes of vibration of the molecule or the crystal lattice. The lines arising from oscillating crystal lattice are characteristic only of the solid state and are not present in liquids or gases. Raman lines due to the electronic effect are rarely observed, except in the single case of NO obtained by Rosetti.

The rotational and vibrational Raman spectra resemble closely the far infra-red and near infra-red absorption spectra of molecules respectively. Further, in many cases the Raman frequencies are equal either to the actual infra-red frequencies of the absorption spectrum of the scatterer or to the differences of such frequencies. But there seems to be little or no correlation between the relative intensities of Raman lines and the intensities of the corresponding infra-red bands. In addition, not all Raman lines have their corresponding infra-red bands, nor all infra-red bands their corresponding Raman lines. Hence it appears that although the ultimate result as far as the molecule is concerned, is the same in both the Raman and band spectra, yet the mechanism of production and the intervening laws are quite different in the two cases.

Theoretical explanation. A simple and satisfactory explanation based on the quantum theory was put forward by Prof. Smekal, in 1923, which accounts for the essential features of the Raman effect. According to the quantum theory, a beam of monochromatic light of frequency υ_0 has its energy distributed in quanta, each of energy $h\upsilon_0$. When such a light quantum or photon hits a molecule of the scatterer, three things might happen. The molecule might merely deviated the photon without absorbing its energy, which would result in the appearance of the *unmodified* line in the scattered beam. The molecule might, on the other hand, absorb part of the energy of the incident photon, giving rise to the *modified Stoke's line*, whose frequency would evidently be less than that of the incident radiation. It may also happen that the molecule itself, being in an excited state, imparts some of its intrinsic energy to the incident photon and this would produce the *anti-stoke's line* or frequency greater than that of the incident radiation. The mechanism of Raman scattering can therefore be analytically expressed as follows :

Treating the phenomenon as a kind of collision between the photon and the molecule and applying the principle of conservation of energy we may write

$$E_p + \frac{1}{2}mv^2 + h\nu_0 = E_q + \frac{1}{2}mvV^2 + h\nu' \quad ...(8.1)$$

where E_p and E_q are the intrinsic energies of the molecule before and after collision respectively, m the mass of the molecule, v and v' its velocities before and after impact, ν_0 and ν' the frequencies of the incident and scattered photon.

As the collision does not cause any appreciable change of temperature, we may assume that the kinetic energy of the molecule remains practically unaltered in the process and hence simplifying the above relation as

$$E_q + h\nu_0 = E_q + h\nu'$$

$$\therefore \quad h(\nu' - \nu_0) = E_p - E_q$$

or

$$\nu' = \nu_0 + \frac{E_p - E_q}{h} \quad ...(8.2)$$

In this expression, if $E_p = E_q$, $\nu' = \nu_0$ which accounts for the unmodified line; if $E_p < E_q$, $\nu' < \nu_0$, which refers to the Stoke's line; and finally if $E_p > E_q$, $\nu' > \nu_0$, which corresponds to the anti-Stoke's line.

Considering the molecule involved in the process, since the same property of possessing energy in quanta appears in molecules and atoms, we may apply quantum principles to the change in the intrinsic energy of the molecule and write

$$E_p - E_q = nh\nu_m \qquad ...(8.3)$$

where $n = 1, 2, 3$, etc. and υ_m is the characteristic frequency of the molecule.

Hence equation (8.2) reduces, in the simplest case of $n = 1$, to :

$$\nu' = \nu_0 + \nu_m \qquad ...(8.4)$$

The frequency of scattered radiation *i.e.*, of Stoke's line is

$$\nu_s = \nu_0 - \nu_m \qquad ...(8.5)$$

The frequency of scattered radiation *i.e.*, of Anti-Stoke's line is

$$\nu_{as} = \nu_0 + \nu_m \qquad ...(8.6)$$

This expression shows that the difference in frequency ($\nu_0 - \nu'$) between the incident and scattered lines in the Raman effect corresponds to the frequency ν_m characteristic of the molecule, which thus accounts for the experimentally observed fact that the Raman liens are symmetrically situated on either side of the parent line at intervals corresponding to the characteristic frequency of the molecule and hence to the infra-red absorption lines of the scattered.

8.3 Practical Importance of Raman Effect

Over and above the great theoretical interest attached to it as further confirmation of the quantum theory, the Raman effect is of immense practical importance on account of its many useful applications in Physics and Chemistry. The universal nature of the phenomenon, the relative simplicity of experimental technique, the ease with which the effect can be controlled, making it appear as a part of the visible spectrum, which can be chosen at will (its position depending only on the choice of the incident frequency) and the complementary character of the spectrum obtained with reference to infra-red absorption (the former containing lines which are not found in the latter) make the Raman effect a very convenient and powerful tool of research in problems concerning the intimate structure of matter, chiefly as regards the constitution of molecules in gaseous, liquid and solid states. As an illustration, we shall now briefly state how the Raman effect is utilised in the study of molecular structure.

8.4 Raman Effect and Molecular Constitution

Raman spectra are, in general, determined by those factors which affect most the nature of vibration *viz.* the number of atoms in the molecule, the masses of the atoms and the strength of the chemical bonds between the atoms.

Taking a *diatomic molecule*, its natural frequency of vibration is given by $\nu_0 = (1/2\pi)\sqrt{F/\mu}$, where F is restoring force per unit displacement and μ the resultant mass. This means that (*a*) there is only one vibration frequency in diatomic molecules, (*b*) a molecule containing light atoms should have a higher frequency than the one containing heavier atoms, and (*c*) a molecule in which the force binding the atoms is large should have higher characteristic frequency than the one in which the force is weak. This force depends on the nature and strength of the inter-atomic bonds. Thus a diatomic molecule with a double bond should have a higher frequency than the one containing only a single bond. The Raman lines are also expected to appear with high intensity when the bond is of the covalent type (homopolar or electrically non-polar molecules) while the reverse in the case when the bond is of the electrovalent type (polar or heteropolar molecules). The reason for this is to be found in the fact that the Raman lines essentially depend upon the symmetry of the molecules and the extent to which the polarisability is affected by the oscillations. In covalent molecules the binding electrons remain common to the nuclei, so that the polarisability of the molecule is considerably modified by nuclear oscillations and this vibration in poloriability gives use to Raman lines. The electrovalent molecules the binding electrons definitely change over from one nucleus to the other in the formation of the molecule so that the polarisability of the molecule is little affected by nuclear oscillation, which means no Raman lines will occur.

Considering *poly-atomic molecules* which are constituted by more than two atoms, the Raman spectra will naturally be much more complex. First of all more than one characteristic frequency is to be expected *e.g.*, triatomic molecules will, in general, have three such frequencies. Secondly, the arrangement of atoms, such as symmetrical or unsymmetrical, linear or non-linear symmetry, etc. is an additional factor by which the intensity of the Raman lines is essentially determined. Hence, from the number and intensity of the observed lines in the Raman effect, in conjunction with infra-red data, it is possible to draw important conclusions about molecular structure. Theory leads to the following important rule, known as *the rule of mutual exclusion.* For molecules with a centre of symmetry transitions that are allowed in the infra-red are forbidden in the Raman spectrum and *vice versa.* The rule implies that for molecules without a centre of symmetry there are transitions that can occur both in the infra-red and the Raman effect. It does not , however, imply that all transitions that are forbidden in one must occur in the other. For, some transitions may be forbidden in both.

The following examples may serve to illustrate these remarks :

Diatomic molecules : Some of the diatomic molecules studied are H_2, D_2, N_2, O_2, HCl, HBr and HI. The first four are homonuclear molecules *i.e.*, constituted with identical atoms, while the last three heteronuclear molecules, *i.e.*, made up of non-identical atoms. In all these cases there is only one vibration frequency and its value obtained from Raman spectra is given below. The restoring force per unit displacement F in each case, calculated with the help of the relation $\nu_0 = (1/2\pi)\sqrt{F/\mu}$, is also given.

	H_2	D_2	N_2	O_2	HCl	HBr	HI
ν_0 (cm^{-1})	4156	2993	2331	1556	2880	2558	2233
$F \times 10^{-5}$	5.1	5.3	22.4	11.4	4.8	3.8	2.9

It is readily seen from such a tabulation that (1) the heavier the molecule the lower is the vibration frequency, and (2) the values of F, which is measure of the binding strength, may be roughly classified in the ratio of 3 : 2 : 1, thereby indicating the existence of three different types of bonds triple, double and single, between the atoms in the molecule.

Triatomic molecules : Among the several triatomic molecules analysed, we shall consider here only three, *viz.*, CO_2, N_2O and H_2O as typical cases illustrating some special points of interest in molecular structure.

Carbon dioxide (CO_2) is one of the most frequently and thoroughly studied molecules. It has two very strong bands in the infra-red absorption spectrum at 668 and 2349 cm^{-1}, while only one strong band in its Raman spectrum at 1389 cm^{-1}. As none of these occurs both in the Raman and infra-red spectrum, it follows from the rule of mutual exclusion that the molecule of CO_2 must have a centre of symmetry. For triatomic molecules this implies that the molecule is linear and symmetric. Hence the molecular structure of CO_2 is O-C-O. This symmetric structure is confirmed by the rotational Raman spectrum of CO_2, which consists only of alternate (odd) levels, like that of O_2. The three bands stated above, two in the infra-red spectrum and one in the Raman spectrum represent the three fundamental frequencies of CO_2. Carbon disulfide (CS_2) also belongs to this class.

Nitrous oxide (N_2O) : This molecule has the same number of electrons as CO_2 and one might therefore expect it to have a linear symmetric structure. But analysis of the infra-red and Raman spectra of N_2O shows clearly that this molecule although linear, is not symmetrical. The tree fundamental frequencies of N_2O are 2224, 1285 and 589 cm^{-1}. All the three appear in the infra-red absorption and two of them, *viz.*, 2224 and 1285, appear in the Raman spectrum; 589 has, however, not been recorded, due to weak intensity. At any rate, the two Raman lines appear also in the infra-red, which argues to the absence of the centre of symmetry in the molecule. For, it was a centre of symmetry, only one fundamental should appear in the Raman spectrum and this fundamental should not appear in the infra-red, according to the rule of mutual exclusion. Hence the molecule has the unsymmetrical structure N-N-O. This conclusion is confirmed by the rotational Raman spectrum of N_2O, which consists of both (odd and even) sets of lines without any alteration in intensity. Other molecules of this type are HCN, ClCN, BRCN, ICN, etc.

Water (H_2O) : The observation of a strong pure rotational spectrum and its structure, as well as that of the rotational-vibrational spectrum lead unambiguously to the conclusion that H_2O, though symmetrical, is not linear but bent (Fig. 8.3). According to theory, all triatomic molecules of bent symmetrical structure should give rise to three Raman lines all of which are represented in the infra-red spectrum. There are two very strong bands at 1595 and 3756 cm^{-1} in the infra-red absorption spectrum, which are very probably two of the three fundamental frequencies. There is also evidence to show that the bend in the system is 120°. In the Raman spectrum, two frequencies have been recorded at 1665 and 3605 cm^{-1} which correspond roughly with the infra-red frequencies. There is some difficulty in recording the Raman lines of H_2O, which are very weak, due evidently to the fact that the movements of the light hydrogen atoms do not cause appreciable variation in the polarisability tensor. Water gives some other extra bands which have to be ascribed to polymerised molecules -$(H_2O)_2$, $(H_2O)_3$, D_2O, S_2O and H_2S have similar bent symmetrical structures.

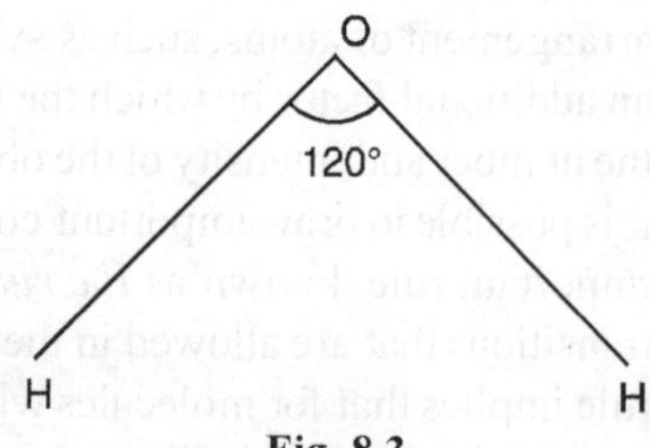

Fig. 8.3

Thus the study of Raman spectra of different substances enables one to classify them according to their molecular structures.

8.5 Spectroscopic Techniques

Spectrum arising out of the transitions between the molecular energy levels can be divided into three types :

1. *Pure Rotational Spectra* in the range 150 μ to 30 μ : The emitted radiation is in the far infra-red or microwave region.
2. *Rotational-Vibrational Spectrum* in the range of 5 μ to 1 μ. The radiation emitted is in the near infra-red region.
3. *Electronic Spectrum* 1000 to 7000 A° lies in the visible and ultraviolet regions.

Infra-red Spectroscopy

1. Source : Nernst glower and Globar.

Nernst filament is a mixture of Zirconium and Yitrium oxides which are formed into small hollow rod about 2 mm in diameter and 80 mm in length. It is heated to 1,500°C to 2000°C. It gives radiation of 1.4 μ. It requires pre-heating.

Globar is a rod of carborundum (Silicon-Carbide). It is thicker and longer than Nernst. It does not require pre-heating and is a conductor at room temperature. It works in the wave-length region longer than 15 μ.

2. Sample : Gas samples are studied in glass cells that are about 5 or 10 cms long closed at their ends with rock salt windows. The cell is evacuated and the sample is filled through a stop-cock or needle valve. The material of the window is transparent over the infra-red region. The usual materials are NaCl, KCl and cesium iodide.

For liquid samples the thickness is chosen so that the transmittance lies between 15 to 70%. Hence, a very thin layer is used for most liquids. Solid samples are dissolved and examined as dilute solutions.

For 2.5 - 15μ a mixture of carbon tetrachloride and carbon-disulphide is useful.

Table 8.1 Window Materials for Infra-red

Material	*Range*
Glass	Visible - 2.6 μ
Quartz	Visible - 4.5 μ
NaCl Rock Salt	200 μ - 17 μ
Syloine KCl	15 μ - 20 μ
KBr	20 μ - 26 μ
Cesium iodide	1 μ to 40 μ

3. Detectors : There are two main types :

(*a*) Sensing the heating effect of Radiation

(*b*) Photoconducting.

In both these types the greater the effect at a given frequency on conductivity, the greater the transmittance (less absorbtion) of the sample at that frequency. Thermopile, Bolometers and Golay cells measure radiant energy by virtue of its heating effect.

Thermopiles are a series of thermocouples linked together and is based on the thermocouple principle *i.e.*, when one of the two metal junctions is heated by the incident radiation an *emf* is developed which is proportional to the amount of radiation falling on it.

Bolometers give an electrical signal due to a change in the resistance of a metal with temperature.

An example of temperature method is the *Golay cell* which is pneumatic in operation. Radiation is made to fall on a very small cell containing air, and temperature changes are measured in terms of pressure changes within the cell which can be recorded directly as "transmittance".

(*b*) *Photoconductors* are a particular class of conductors in which the energy required comes from the incident radiation and some substances (lead sulphide) are sensitive to IR (Infra-red) so that they make excellent detectors.

Photographic plates have to be specially sensitised for IR to be used as detectors as ordinary plates are not sensitive to IR.

Radiometer is a D'Arsonual type of moving coil galvanometer. A single thermocouple is suspended by a quartz-fibre. When radiation falls on the meter the couple produced turns the coil.

Mirrors : Mirrors are used for collinating and focussing of IR radiation.

4. Dispersing Medium : Choice of prism material depends on the region of spectra under investigation. Some materials have been listed in table.

Higher dispersion is obtained from a reflection grating used in place of a prism in the Littrow mounting.

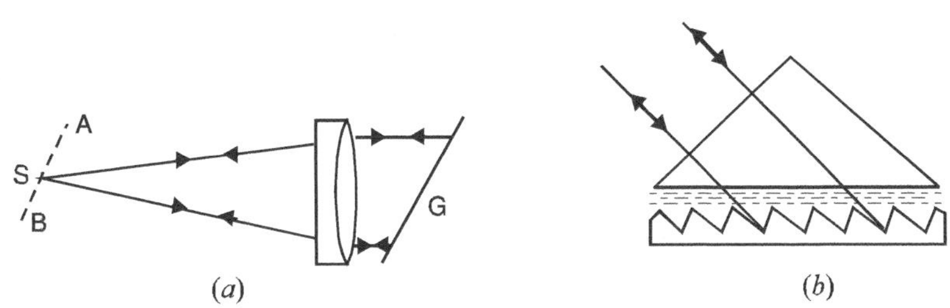

Fig. 8.4(a) & (b). Littrow mountings for plane gratings.

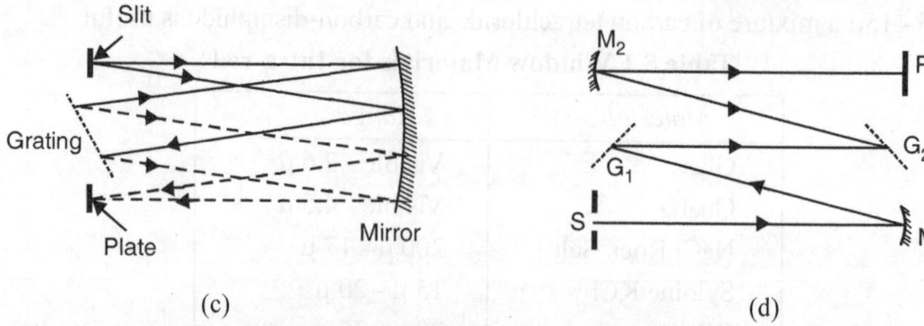

Fig. 8.4. Spectrum is scanned by rotating the mirror.

Wadsworth mounting : The Wadsworth arrangement is the one important mounting which is not based on the Rowland circle; the optical system is shown in Fig. 8.5. Light from the slit S is collimated by the mirror M (originally Wadsworth used a lens) and falls on the concave grating G. The plate P is placed on the grating normal, and to investigate different spectral regions the angle of incidence must be varied to bring the middle of the range on to the normal. The surface on which the spectral lines are in focus has a shape which depends upon the angle of incidence employed. This means that the plate holder must be such that the curvature of the plate can be varied. The Wadsworth arrangement is confined within little more than half the area occupied by the Eagle mounting.

The important advantage of Wadsworth is that astimagtism is very much reduced. In addition the grating is used with a relatively large angular aperture and this, plus the absence of astimagtism, renders the lines more intense. For spectrochemical analysis the Wadsworth mounting is more convenient than Rowland and Paschen mountings.

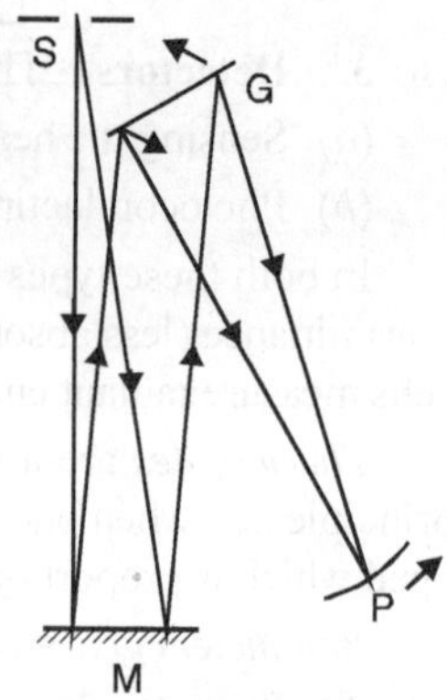

Fig. 8.5. Wadsworth mounting.

Double Beam Instruments

Two equivalent beams are taken form one and the same source. One beam is focussed on the sample, while the other serves as a reference beam. The reference beam passes through a variable callibrated attenuator. Both the beams are presented to a monochromator detector system on alternate half cycles by means of a rotating mirror. (Refer Fig. 8.6).

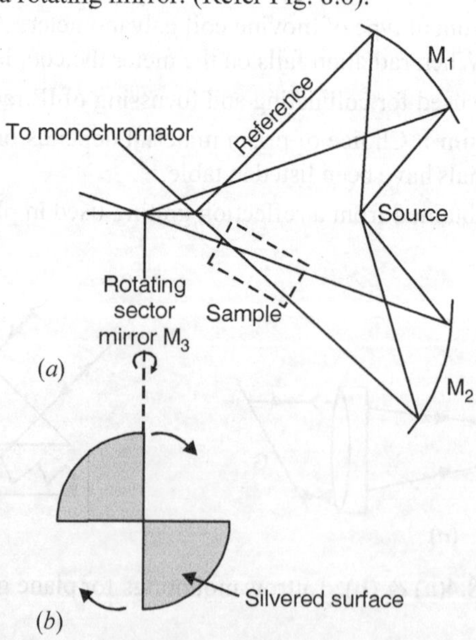

Fig. 8.6. (*a*) Schematic diagram of a double-beam infra-red spectrometer. (*b*) A plan view of the rotating sector mirror M_3.

As the mirror rotates it alternately reflects the reference beam or allows the sample beam into the monochromator. As the monochromotar scans the wavelength range, the detector and the amplifier system notes and discriminates between the two signals. The difference is used by a servo amplifier to actuate a motor to drive the attenuator into the reference beam till there is no difference between the two beams. Servo amplifier acts in a direction to correct for any difference and thus maintains a balance. The attenuator position for each wavelength gives the absorption of the sample. The electrical signal is given to a recorder.

By balancing sample and reference beams, the absorption of the atmosphere, CO_2 and H_2O do not appear in the IR spectrum since both beams are reduced in energy to the same extent. This is the main advantage. It is easy to amplify a.c. signal than the d.c. signal from a single beam. One is able to present appropriate quantity of pure solvent into the reference beam eliminating the solvent spectrum from the final trace.

UV (Ultra Violet Spectroscopy)

Copper and Mercury arcs are suitable as sources for the uv radiation.

The spectrum is produced by a prism or a grating. The prism material should not absorb the UV radiation.

Table 8.2

Materials	*Range*
Glass	3300 A°
uv galss	2000 A°
Quartz	850
Fluoride	1200 A°

Gratings are useful for region below 1100A°.

Detecting is done by the photographic method which uses special plates such as Q-plates which are sensitive to the *u-v* region.

Since air absorbs all radiations below 1800 A°, vacuum spectrographs are used.

Ultraviolet Spectrographs

Quartz region near and short ultra-violet regions.

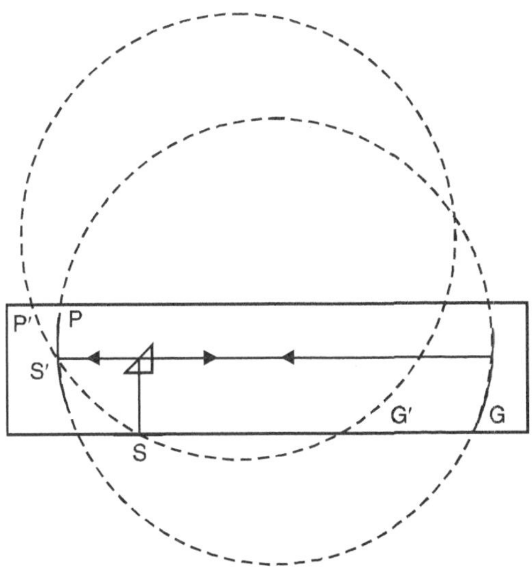

Fig. 8.7. Eagle mounting.

Quartz spectrograph is used since quartz polarises and produces double images due to birefrigence, fused quartz is used in constructing the prisms and lenses.

Since quartz does not transmit *uv* beyond 1850 so vacuum spectrographs with fluorite prisms and lenses are used.

In the Lyman region or extreme ultra-violet, concave grating vacuum spectrographs are used. Eagle mountings are generally used.

The relative compactness of the eagle mounting causes it to be widely used in vacuum spectrographs since a much smaller volume has to be evacuated (Air absorbs wavelengths below 1800 A°). Temperature control is also easier with this smaller apparatus. This is important since a variation in temperature causes a significant variation of the grating interval.

Example 8.1. *In observing the Raman spectrum of a sample, using 2537 Å as the exciting line, one gets stoke line at 2683 Å deduce the Raman shift in cm^{-1} units. Compute the wavelength in Å for corresponding Stoke's and anti-Stoke's line if the exciting line is 5461Å.* [*I.A.S. 1987*]

Solution. (*i*) Here $\upsilon_s = \upsilon_a - \upsilon_m$

$$\upsilon_m = \nu_0 - \nu_s$$

$$= \frac{c}{\lambda_0} - \frac{l}{\lambda_s}$$

$$\frac{c}{\lambda_m} = c\left[\frac{1}{\lambda_0} - \frac{1}{\lambda_s}\right]$$

$$\frac{1}{\lambda_m} = \frac{\lambda_s - \lambda_0}{\lambda_0 \lambda_s}$$

Here $\lambda_0 = 2537 \text{ Å} = 2537 \times 10^{-10} \text{ m}$

$$\lambda_s = 2683 \text{ Å} = 2683 \times 10^{-10} \text{ m}$$

$$\frac{1}{\lambda_m} = \frac{2683 \times 10^{-10} - 2537 \times 10^{-10}}{2537 \times 10^{-10} \times 2683 \times 10^{-10}}$$

$$\frac{1}{\lambda_m} = 2.145 \times 10^5 \text{ per meter}$$

Raman Shift,

$$\upsilon_m = \frac{1}{\lambda_m} = 2.145 \times 10^5 \text{ per metre}$$

$$= 2.145 \times 10^3 \text{ per cm}$$

(*i*) Here $\lambda_0 = 5461 \text{ Å} = 5461 \times 10^{-10} \text{ metre}$

$$\frac{1}{\lambda_m} = 2.145 \times 10^5 \text{ per metre}$$

$$\lambda_s = ?$$

$$\upsilon_s = \upsilon_0 - \upsilon_m$$

$$\frac{c}{\lambda_s} = \frac{c}{\lambda_0} - \frac{c}{\lambda_m}$$

$$\frac{1}{\lambda_s} = \frac{1}{\lambda_0} - \frac{1}{\lambda_m}$$

$$\frac{1}{\lambda_s} = \left(\frac{1}{5461 \times 10^{-10}}\right) - 2.145 \times 10^5$$

$$\frac{1}{\lambda_s} = 18.31 \times 10^5 - 2.145 \times 10^5$$
$$= 16.165 \times 10^5$$
$$\therefore \quad \lambda_s = \frac{1}{16.165 \times 10^5}$$
$$= \mathbf{6186 \times 10^{-10}\ m}$$
$$= 6186\ \text{Å}$$

(*ii*) Similarly, for anti-Stoke's line

$$\nu_{as} = \upsilon_0 + \upsilon_m$$
$$\frac{c}{\lambda_{as}} = \frac{c}{\lambda_0} + \frac{c}{\lambda_m}$$
$$\frac{1}{\lambda_{as}} = \frac{1}{\lambda_0} + \frac{1}{\lambda_m}$$
$$\frac{1}{\lambda_{as}} = \left(\frac{1}{5461 \times 10^{-10}}\right) + 2.145 \times 10^5$$
$$\frac{1}{\lambda_{as}} = 18.31 \times 10^5 + 2.145 \times 10^5$$
$$\frac{1}{\lambda_{as}} = 20.455 \times 10^5$$
$$\lambda_{as} = \frac{1}{20.455 \times 10^5}$$
$$= \mathbf{4889 \times 10^{-10}\ m}$$
$$= \mathbf{4889\ \text{Å}}$$

Example 8.2. *In an experiment with benzene the wavelengths of a pair of Stoke's and anti-Stokes's lines were measured to be 4554 Å and 4178 Å respectively. Calculate the wavelength of the corresponding infra-red absorption line.* [*Calcutta, 1992*]

Solution. Here
$$\upsilon = \upsilon_0 \pm \upsilon_m$$
$$\therefore \quad \upsilon_1 = \upsilon_0 + \upsilon_m$$
and
$$\upsilon_2 = \upsilon_0 - \upsilon_m$$
$$\therefore \quad \upsilon_1 - \upsilon_2 = 2\upsilon_m$$
$$\frac{c}{\lambda_1} - \frac{c}{\lambda_2} = \frac{2c}{\lambda_m}$$
$$\lambda_m = \left[\frac{2\lambda_1\lambda_2}{\lambda_2 - \lambda_1}\right]$$

Here
$$\lambda_2 = 4554\ \text{Å}$$
$$\lambda_1 = 4178\ \text{Å}$$
$$\lambda_m = \frac{2 \times 4554 \times 4178}{4554 - 4178}$$
$$\lambda_m = \mathbf{1.012 \times 10^5\ \text{Å}}$$

EXERCISE VIII

1. Explain Raman effect.
2. How is Raman effect theoretically explained ?
3. Give briefly the experimental arrangement to study Raman effect.
4. What are Stoke's and anti-Stoke's lines in Raman scattering ?
5. With regard to infra-red spectroscopy, describe : (*i*) source, (*ii*) sample, (*iii*) Detectors, (*iv*) Dispersing medium.
6. Describe Wadsworth and Eagle mounting.
7. Give a brief account of ultra-violet spectroscopy.
8. Describe a double-beam infra-red spectrometer.
9. Draw a diagram of Littrow mounting used in I.R. Spectroscopy.

9

INTERACTION OF CHARGED PARTICLES AND NEUTRONS WITH MATTER

We first look at the main features of the interaction of all types of radiation with matter : *i.e.* charged particles, photons and neutrons. The interaction of any type of radiation always involves charged particles, at some stage in the process. For γ-rays and neutrons *i.e.* uncharged radiation, there is a transfer of some of the energy to charged particles before there is any effect on the absorbing medium.

9.1. Charged Particles

A charged particle loses its energy due to the action of its electric field on atoms it encounters as it passes through matter. Most of the energy is lost to electrons. Due to the small mass of the electron it carries away most of the incoming energy more than the heavy nucleus. Unless the energy is very high, the ions have very little effect on the energy loss. For a head-on collision with an ion of mass M and energy E, an electron of mass m initially at rest emerges with a speed approximately twice that of the incident ion and its energy

$$\Delta E = E\left(\frac{4m}{M}\right)$$

This energy loss is small compared to E and there are many collisions that take place before the ion comes to rest. The heavy particle is deflected very little by the interactions with electrons and moves almost in a straight line path. Due to the energy loss there is a definite *range* depending on mass, charge, energy and the nature of the stopping medium. The recoiling electrons give rise to secondary electrons which lose their energy.

The rate at which a particle loses its energy per unit length of its path is known as the specific energy loss and its average value is the *stopping power* of the medium.

The stopping power of the medium is given by Bethe—Block formula (relativistic).

$$-\frac{dE}{dx} = \left(\frac{Ze^2}{4\pi \in_0}\right)^2 \frac{4\pi Z\rho N}{Amv^2}\left[\ln\left(\frac{2mv^2}{I}\right) - \ln(1-\beta^2) - \beta^2\right] \qquad ...(9.1)$$

where β = v/c ion velocity Ze = electronic change of mass m

N = Avogadro number. OA = Atomic mass number

Z = Atomic number, ρ = density of stopping material

Ze is the base nuclear charge

β^2 term is neglected if speeds are small compared to the velocity of light c.

I = mean energy required to ionise an atom of the medium. Usually it is taken to be 11ZeV as it is difficult to evaluate.

$\because$ E decreases with distance $\therefore$ dE/dx is negative.

$\frac{Z\rho N}{A}$ = density of electrons in the material.

We can obtain the energy loss per unit mass per unit area by dividing by the density ρ. The quantity $\frac{dE}{\rho dx}$ is independent of the density of the stopping medium.

The range of a particle is also given in units of mass per unit area = range × density.

It is clear from equation 9.1 that the stopping power does not depend on the mass of the particle but is only a function of its velocity and charge.

∴ Stopping power of α-particles is four times for a proton of the same velocity, except towards the end of the range.

9.2. Energy Dependence

For any charged ion, the energy dependence is dominated by the $1/v^2$ term in the Eq. 9.1 and there is a weaker dependence on v which comes from the logarithmic term.

Empirically, a simple power law for the energy can be stated as :

$$\frac{dE}{dx} = \frac{\text{Const.}}{E^K} \quad ...(9.2)$$

$K = 0.8$ is a useful approximation when the energy loss is required in calculations.

Range of an ion is given by $R = \int dx = \int_E^0 \frac{dE}{\left(\frac{dE}{dx}\right)} \propto E^{1+K}$ using the approximation of Eq. 9.2 for the stopping power. Now $R \propto \frac{1}{\rho}$, where ρ is the density of the medium.

$$\frac{dE}{dx} \propto \rho$$

∴ Range can be expressed as $R' = \rho R$ as mass per unit area and does not depend on ρ^2.

For low values of energy the curve for dE/dx vs E reaches a maximum and then decreases and tends to zero. This is because the ion speed is so low that there is maximum energy transfer equal to I and log term goes to zero. Electron capture and loss by the ion begins to be significant and it cannot be assumed that its charge Z is due to the bare nucleus.

9.3. Bragg Curve

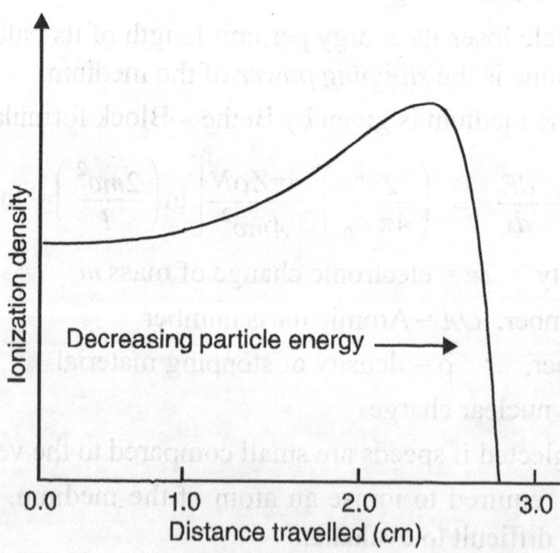

Fig. 9.1. Variation of ionization density vs distance travelled by an α-particle in air (the Bragg curve).

Eq. 9.2 gives the rate of energy loss and it is found to increase as the particle energy decreases and so the number of ions produced in the medium per unit distance (ionisation density) increases as the particle slows down. When most of the energy is lost, $-dE/dx$ reaches a maximum. The variation of the ionisation density with distance gives the Bragg curve and shows how the particles distribute their energy in matter. As the particle loses its energy the ionisation density reaches a peak called Bragg peak close to the end of its path. After this it falls down abruptly to zero, and hence a range can be defined. Equation 9.1 can be written as :

$$dE/dx \propto Z^2 f(v) \qquad ...(9.3)$$

Since

$$R = \int_E^0 \frac{dE}{dx} \propto \frac{m}{Z^2} F(v) \qquad ...(9.4)$$

as

$$dE = mv\, dv$$

Equations 9.3 and 9.4 are useful in estimating the relative values.

Geiger found that the range of natural α particles in air is given by

$$R \simeq 0.32\, E^{1/2} \qquad ...(9.5)$$

where E is the energy in MeV.

The approximation is good for 4 MeV $< E <$ 10 MeV.

9.4 Dependence of the Stopping medium

Eq. 9.1 can be written as

$$\frac{1}{\rho}\frac{dE}{dx} \propto \frac{Z}{A} \ln\left(\frac{2mv^2}{I}\right)$$

$\therefore$ For mass numbers A and density relation we have the **Bragg-Kleeman** rule as

$$\frac{R_1}{R_2} \simeq \frac{\rho_2}{\rho_1}\frac{\sqrt{A_1}}{\sqrt{A_2}} \quad \text{or} \quad \frac{R_1'}{R_2'} \simeq \frac{\sqrt{A_1}}{\sqrt{A_2}} \qquad ...(9.6)$$

where R is in units of distance ; R' distance and mass per unit area.

This gives the relative ranges of an ion in material of different mass numbers.

Since range is of a statistical nature there is a spread in the observed values. This is called **Straggling.** For α-particles of 5 MeV there is a standard deviation of range distribution of 1%. For heavy particles straggling is not a large effect.

Hence we see that identical charged particles, all having the same initial velocity, do not have the same range. The range is distributed over a small interval about the mean range. This phenomenon is called range **Straggling**.

9.5 Electrons

Loss of energy by electrons to atomic electrons via the electric interaction is the same as for heavier charged particles. But their speeds are greater and they are lighter. Hence $-\frac{dE}{dx}$ is smaller. Therefore, the range is larger for the same energy.

The loss of energy is greater in a single collision as compared to the heavy ion. It may produce secondary electrons these radiate energy by **bremsstrahlung** on changing their direction.

$\therefore$ Electrons loss in energy is due to :

(1) collision with electrons

(2) radiation energy loss by bremsstrahlung.

Radiation loss is strongly dependent on Z of the stopping medium and rises rapidly with energy and is the dominant contribution.

The critical energy at which the two terms are equal is given by $E_c \simeq 1600\, mc^2/Z$. E_c = critical energy with mc^2 = 0.511 MeV formula by Bethe and Heitler. For example Z = 82 for lead gives $E_c \sim 10$ Mev.

The range is difficult to compute due to statistical variation in energy loss and due to electron scattering. Range is defined as the distance at which the intensity falls off to 50%.

As seen in Fig. 9.2 the intensity of α-particles falls off very fast as compared to electrons where R_C is the range due to scattering and statistical variation.

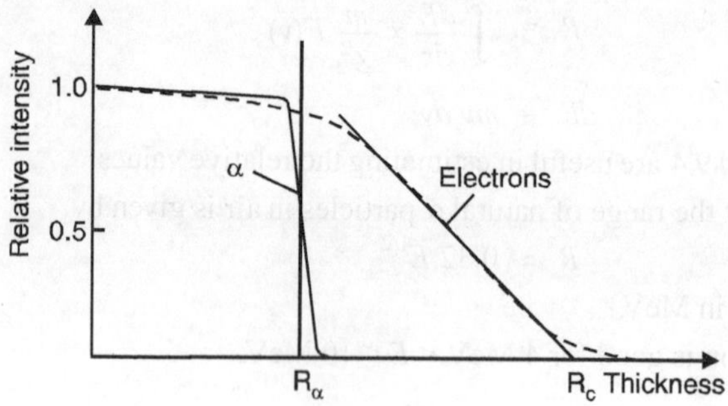

Fig. 9.2

9.6 Gamma Rays

There are mainly three processes by which γ-rays or X-rays interact with matter. They are : photoelectric effect, Compton scattering and pair production.

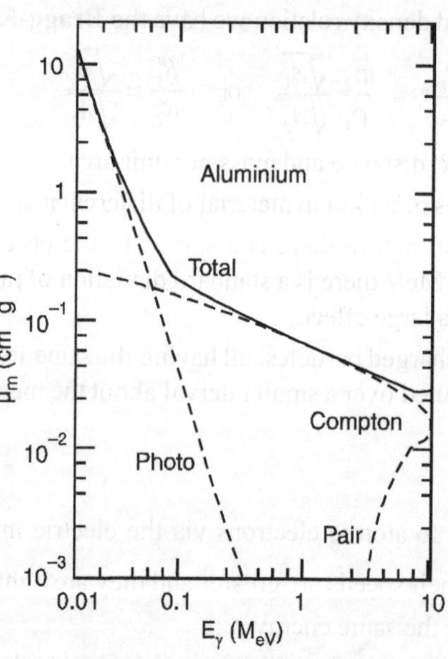

Fig. 9.3

Fig. 9.3 shows the relative importance as a function of photon energy.

9.7 Photoelectric Effect

When a photon collides with an atom it strikes against the orbital electrons and transfers all of its energy to the electron and removes it from the atom.

This process is termed as the photoelectric effect and the equation by Einstein is :

$$h\upsilon = E_{Kin} + E_B \qquad ...(9.7)$$

where E_B is the binding energy of the ejected electron.

This effect is the predominant one for energies below 0.1 MeV for γ-rays. The contribution of the two K-electrons is much greater than that of all the rest of the electrons put together. The atom which loses the electron may get de-excited by giving out less tightly bound electrons. Electrons given out by this process are called Auger electrons. An electron from a higher shell may also fill the vacancy with a characteristic X-ray photon being emitted. This X-ray may interact and may also be absorbed in the medium.

Conservation of energy and momentum does not allow the photoelectric effect from occurring on a free electron. The process must take place near an atom, which can take away the excess momentum. Due to its large mass it acquires very little of the original energy. Note that smaller the electron's binding energy with respect to the γ-ray energy, the more it appears like a free electron and the smaller is the probability of the photoelectric effect (P.E.). Hence there is a strong dependence on energy and atomic number. The P.E. cross-section is given by the formula

$$\sigma_{\text{P.E.}} \propto \frac{Z^5}{E_\gamma^{3.5}} \text{ for } \gamma \text{ energies of a few hundred KeV} \quad ...(9.8)$$

Z^5 dependence means that for a given photon energy the process is much more important in heavy absorbers than in light ones.

In the case of light atoms Auger effect is predominant.

9.8. Compton Scattering

In this interaction, there is a collision of a photon of energy E with a free electron. The photon emerges with a lower energy E' and the electron recoils with kinetic energy T depending on the scattering angle.

By conservation of energy of electron

$$T = E - E' = E_T - mc^2 \quad ...(9.9)$$

where E_T = Total energy of the recoil electron and mc^2 rest mass of the recoil electron.

$$\because \quad p = E/c \quad \text{and } p' = E'/c$$

momentum of photon before and after the collision and p momentum of electron.

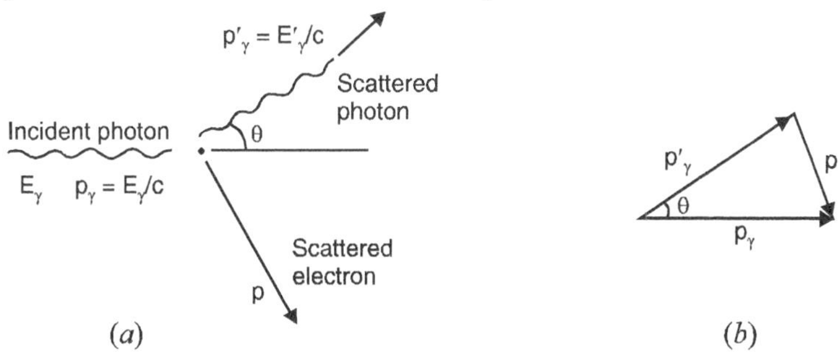

Fig.. 9.4

Adding the momentum of the electron vectorial Fig. 9.4 (b). By the cosine rule, and substituting for p and p' we get,

$$(pc)^2 = E^2 + E'^2 - 2EE' \cos\theta = E_T^2 - m^2c^4 \quad ...(9.10)$$

$$\because \quad E_T^2 = p^2c^2 + m^2c^4 \text{ for the electron.}$$

Eliminating E_T from (9.9) and (9.10), we get for the scattering energy of the photon

$$E' = \frac{E}{1 + \left(\frac{E}{mc^2}\right)(1 - \cos\theta)}$$

$\theta = 0$ gives $E = E_{max}$ to $\theta = 180°$ E' which is a minimum. E min approaches $mc^2/2$.

The electron energy goes from zero to a maximum for $\theta = 180°$.

Cross-section is in the range of 0.5 MeV to 10 MeV. It is less dependent on E and Z. It depends directly on the density of electrons in the medium multiplied by Z/A.

Since Z/A is approximately constant, Compton contribution to mass alternation coefficient is independent of Z.

$$\sigma_c = \text{Compton cross-section.}$$

9.9. Pair Production

The entire photon energy is converted into the creation of an electron-position pair in the presence of an atom.

$$T_- + T_+ = E_\gamma - 2mc^2 \quad ...(9.11)$$

∴ Pair production requires the presence of a heavy body for conservation of energy and momentum. There is some dependence of its probability on Z. Threshold energy is $2mc^2 = 1.022$ MeV.

∴ Pair production does not become important until E exceeds several MeV.

Only part of the initial energy gets converted to kinetic energy of the charged particles which gets transferred to the absorbing medium.

The positron slows down and is attracted to an ordinary electron. Anhillation gives rise to two γ-rays of energy 0.511 MeV each. These are emitted in opposite directions to conserve momentum and interact in the absorbing medium by P.E. or Compton scattering.

9.10 Attenuation

For a well collimated beam all three processes are possible. Photons which undergo P.E. or P.P disappear and most of those that are Compton scattered are deflected away.

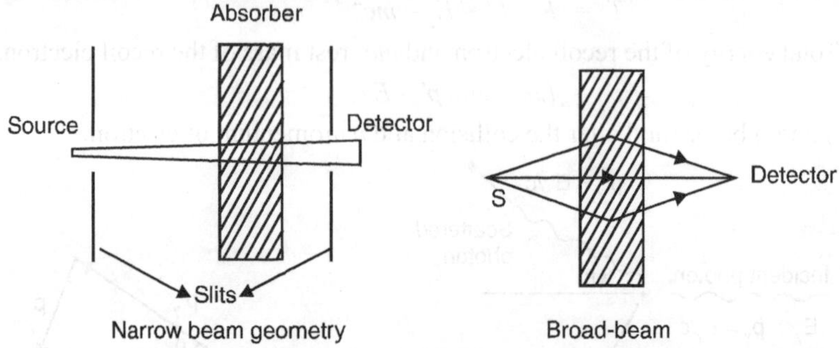

Fig. 9.5

For a well collimated beam it is easy to calculate the dependence of the photon intensity on the absorber thickness.

For a beam of intensity I of photons incident per sec on a section of thickness dx of a slab of material with atomic density N and if the total interaction cross-section per atom of material is σ, the rate of removal of photons from the beam is given by

$$dI = -N\sigma I\, dx \quad ...(9.12)$$

drop in intensity

where $\sigma = \sigma_{P.E.} + Z\sigma_c + \sigma_{P.P.}$ ...(9.13)

σ_c = Compton scattering c-s for a single electron

Integrating Eq. 9.12 $I = I_0 \exp(-N\sigma x) = I_0 \exp(-\mu x)$

where I_0 = incident intensity, I = intensity on emerging

x = depth in units of length

μ = linear attenuation coefficient (= $N\sigma$) dimension of (length)$^{-1}$

mass attenuation coefficient μ_m is defined as linear att. coeff. ÷ ρ (density) of the medium

$$\mu_m = \mu/\rho \qquad ...(9.14)$$

For absorber half thickness equals ln $2/\mu$

i.e. the thickness which reduces the intensity to half its initial value.

μ (or μ_m) is independent of x only for monoenergetic photons.

∴ It is difficult to stop γ-rays compared to charged particles. Note if the beam is uncollimated there is a contribution by Compton scattered γ-rays as well. For thick absorbers it can be more than ten times larger due to the scattered γ-rays.

9.11 Neutrons

Since neutrons are uncharged particles their interaction with matter is mainly by nuclear reactions which depend strongly on the energy and the nucleii on which the neutrons collide. Neutrons could be scattered in which case the recoiling nucleii takes away most of the energy. Also, the neutrons could well be absorbed by any of the processes discussed before.

Neutrons of a few MeV, fusion, contributes mainly to the absorption cross-section. The Binding Energy (B.E.) of a neutron is typically 7-8 MeV. The neutron is captured and forms a compound nucleus which is in an excited state. The energy of the incoming neutron is then later released in the form of reaction products such as γ-rays, charged particles or neutrons. The release of energetic charged particles is important as it helps to detect the neutrons.

Due to scattering the neutron slows down due to successive collision. The neutron is slowed down till it reaches thermal equilibrium with the surroundings. These are called thermal neutrons and has a typical energy of 0.025 eV. (kT : k = Boltzman's constant 1.38×10^{-23} J/K, T – Absolute temp. = 300, 1 eV = 1.6×10^{-19} J). These neutrons when captured give rise to MeV of energy in the form of other radiations. The neutron absorption cross-section is highest for many materials for slow neutrons and hence easy to detect.

Neutrons can be absorbed or scattered through large angles like γ-rays. If the neutron beam is very well collimated then scattering and absorption contribute to loss in intensity I.

$$\therefore \quad I = I_0 \exp(-N\sigma x) \text{ where } \sigma = \sigma_s + \sigma_a \qquad ...(9.15)$$

$$I = I_0 \exp(-\Sigma x) = I_0 \exp(-x/\lambda) \qquad ...(9.16)$$

where Σ = $N\sigma$ macroscopic total Cross-section.

λ = $1/\Sigma$ is the mean attenuation length.

The scattering mean free path $\lambda_s = 1/\Sigma_s$ is the average distance between successive scatterings.

Similarly $\lambda_a = \dfrac{1}{\Sigma_a}$ neutron absorption mean free path

$$\therefore \quad \Sigma = N(\sigma_s + \sigma_a) = \Sigma_s + \Sigma_a \qquad ...(9.17)$$

$$\therefore \quad \text{Total mean free path} \quad \frac{1}{\lambda} = \frac{1}{\lambda_s} + \frac{1}{\lambda_a} \qquad ...(9.18)$$

Boron detectors are based on the reaction

$${}_5B^{10} + {}_0n^1 \longrightarrow {}_3Li^7 + {}_2He^4 + 2.78 \text{ MeV.}$$

For thermal neutrons cross-section for this reaction is 3770 barns.

The 2.78 MeV energy is carried off by the Li nucleus and α-particles producing considerable ionisation in its path.

For natural Boron the Cross-section is reduced, B^{10} is less abundant. Ionisation chambers with Boron coated electrodes or proportional counter filled with Boron trifluoride vapour (BF_3) is used in detection of neutrons.

9.12 Neutron Moderation

Consider an elastic collision between a neutron of mass m, energy E and travelling with a velocity v which strikes a target nucleus (mass M) which is initially at rest. Ref. Fig. 9. 6 for laboratory frame and centre of mass frame (*C-M* frame).

After the collision, let E_1 be the energy of the neutron and v_1 its velocity in the laboratory frame. The velocity of the centre of Mass V_{cm} is defined by

$$mv = (m + M) V_{cm} \text{ as } M \text{ is at rest} \quad ...(9.20)$$

In *C-M* frame, the centre of mass is at rest and velocity of the target nucleus V' is equal to V_{cm}. Now, since kinetic energy is conserved in an elastic collision we have v' and V' are unchanged by the collision.

$$V_{cm} = \frac{Mv}{(m+M)} \quad ...(9.21)$$

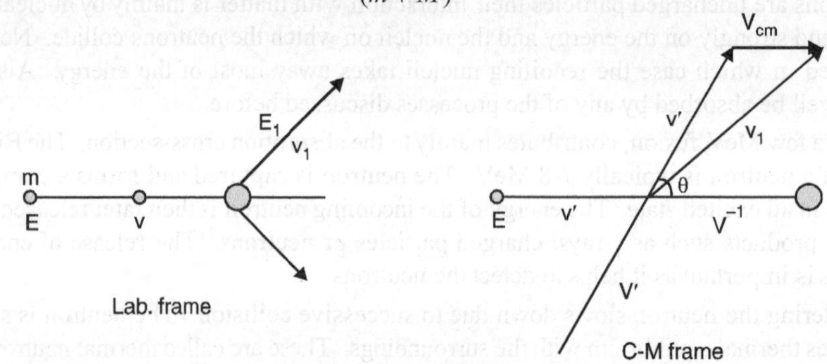

Fig. 9.6

Relation between the velocities in the two frames is

$$V' = V_{cm}$$

$$v = v' + V_{cm}$$

The scattered neutron velocity in the two frames is

$$\vec{v}_1' = \vec{v}' + \vec{V}_{cm} \text{ from Fig. } (b)$$

Squaring Equation : $$v_1^2 = v'^2 + V_{cm}^2 + 2v' V_{cm} \cos\theta \quad ...(9.22)$$

θ being the angle between v' and V'.

Substituting of v' and V_{cm} in Eq. (9.20) and (9.21) and noting that

$$\frac{E_1}{E} = \left(\frac{v_1}{v}\right)^2$$

Eq. (3) gives for the lab energy of the scattered neutron.

$$E_1 = E\left(\frac{M^2 + m^2 + 2Mm\cos\theta}{(M+m)^2}\right) \quad ...(9.23)$$

For $\theta = 180°$ $$E_{1\,(\min)} = E\left(\frac{M-m}{M+m}\right)^2 = \alpha E$$

$$E_{1\,(\min)} = \alpha E$$

where $$\alpha = \left(\frac{M-m}{M+m}\right)^2$$

Substituting for neutron $m = 1$ and $M = A$ in above equation,

$$1-\alpha = 1-\frac{(A-1)^2}{(A+1)^2} = \frac{4A}{(A+1)^2} \qquad ...(9.24)$$

It can be shown that the scattering angles in the two frames are almost identical, since $A >> 1$.

Let $P(E_1)\,dE_1$ be the probability distribution of the scattered neutron having an energy between E_1 and $E_1 + dE_1$.

From Eq. 9.23 there is a one to one correspondence between E_1 and centre of mass scattering angle θ.

For the probability distribution in the angle, we have

$$-p(\theta)\,d\theta = P(E_1)\,dE,$$

where negative sign shows the energy decreases as the angle increases.

Let the scattering be isotropic in *C-M* frame.

$\therefore$ The probability of scattering into solid angle $d\Omega$ at an angle θ is $dP = \frac{d\Omega}{4\pi} = \frac{2\pi \sin\theta\, d\theta}{4\pi}$

$$\therefore \qquad p(\theta)\,d\theta = \frac{2\pi \sin\theta\, d\theta}{4\pi} = \frac{1}{2}\sin\theta\, d\theta$$

Substituting in Eq. (4) for *M* and *m*, we get

$$\frac{dE_1}{d\theta} = -\frac{2AE\sin\theta}{(A+1)^2}$$

Now $$1-\alpha = \frac{4A}{(A+1)^2}$$

$$\therefore \qquad P(E_1)\,dE_1 = \frac{(A+1)^2}{4AE}\,dE_1 = \frac{dE_1}{(1-\alpha)E} \qquad ...(9.25)$$

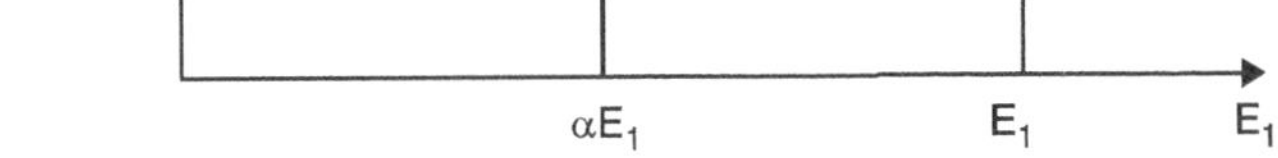

Fig. 9.7.

Energy distribution is uniform between αE and E. Area under the curve is unity.

The quantity $(1-\alpha)\,E$ is the range of all post collision energies.

$\therefore$ The average neutron energy after first collision is

$$\overline{E_1} = \frac{1}{2}(1+\alpha)\,E$$

$$\text{Average energy loss} = \frac{1}{2}(1+\alpha)\,E \qquad \therefore \quad \frac{E_1}{E} = \frac{1-\alpha}{2}$$

$\therefore$ Fractional energy loss per collision is independent of E. After second collision same fractional energy is lost etc.

$\therefore$ After n collisions the average energy is

$$\overline{E_n} = E \times \left(\frac{\overline{E_1}}{E}\right)^n \qquad ...(9.26)$$

A better quantity which characterises the distribution of neutron energy is the logarithmic energy decrement ξ defined as average value of $\left(\frac{E}{E_1}\right)$ after 1st collision.

$$\therefore \qquad \xi = \ln\left(\frac{E}{E_1}\right) = \int_{\alpha E}^{E} \ln\left(\frac{E}{E_1}\right) P(E_1)\, dE_1$$

Substituting for $P(E1)$ from eq. (5) and substituting $x = \frac{E_1}{E}$.

$$\xi = \frac{1}{1-\alpha}\int^{\alpha} \ln x\, dx = 1 + \frac{\alpha}{1-\alpha}\ln\alpha$$

$\therefore$ From equation it is seen that the logarithmic energy decrement.

$$\xi = \ln E - \overline{\ln E_1}$$

With each collision E decreases by ξ, therefore after n the average value of $\ln E_n$ is

$$\overline{\ln E_n} = \ln E - n\xi \qquad \text{...(9.27)}$$

In terms of A

$$\xi = 1 + \frac{(A-1)^2}{2A}\ln\left(\frac{A-1}{A+1}\right) \qquad \text{...(9.28)}$$

For $A > 10$ approximate value of $\xi = \dfrac{2}{A + \frac{2}{3}}$ ...(9.29)

This gives an error of only 3%.

To estimate the number n of collisions required to slow the neutrons from E to the median energy E_n' after n collisions we have

$$\ln E_n' = \overline{\ln E_n} \qquad \text{...(9.30)}$$

Combining (9.27) and (9.30) we get

$$n = \frac{1}{\xi}\ln\frac{E}{E_n'}$$

Table 9.1

	A	α	ξ	n (to thermalise)
^{1}H	1	0	1.00	18
^{2}H	2	0.111	0.725	25
^{4}He	4	0.225	0.425	43
^{9}Be	9	0.640	0.209	86
^{12}C	12	0.716	0.158	115
^{238}U	238	0.983	0.0084	2172

9.13 Nuclear Detectors

Proportional Counter : When particles of low specific ionisation passes through an ionisation chamber, the pulses produced is too small to detect. One way of amplifying the signal obtained from a gas-filled detector is to increase the electric field so that the electrons gain enough energy between collision with gas atoms to cause further ionisation. In such a case the electrons produce further ionsiation, and a rapid amplification of the original cascade occurs in what is called as Townsend avalanche. It should be noted that the size of the output signal is proportional to the number of ions formed by the primary ionisation process. Such a chamber is called a proportional counter.

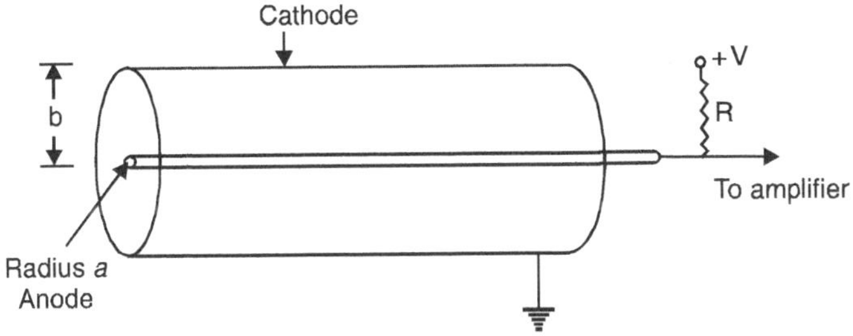

Fig. 9.8

The proportional counter consists of a cylindrical gas filled tube (radius b Fig.) with a very thin central wire (radius a) which is insulated from the tube. The central wire is positive with respect to the tube which serves as the collecting electrode. This is connected to a pulse amplifier.

The electric field varies with the radial distance r and is given by $E = \dfrac{V}{r \ln \dfrac{b}{a}}$ [for $b = 10$ and $a = .01$ mm $V = 1$ kV and $E = 10^7$ Vm^{-1}].

Usually Argon gas and a mixture of some polyatomic gas such as methane is used (90% + 10%) or Helium and isobutane are used (96% + 4%). The window of the proportional counter is either Aluminium or mica. Voltages of 1500 - 5000 volts are required to get proper amplification at atmospheric pressure. In a proportional counter the avalanche is produced at the point where the primary ionisation occurs and does not spread along the whole anode wire, unlike the G.M counter (to be described). Electron collection times are 1 - 10 *ns*, positive ion collection is slow. Proportional counters are widely used for neutron counting. For slow neutrons BF_3 is used as the counter gas. ^3He gas is also used as filling for slow neutrons. Fast neutrons can be detected using H_2 gas as filling.

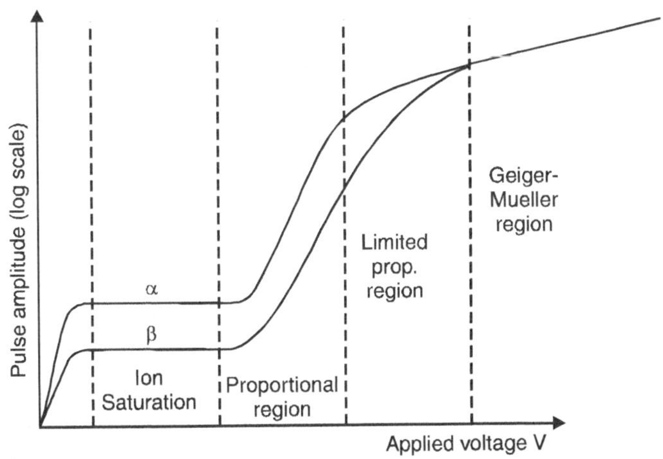

Fig. 9.9

Fig. 9.9 shows the pulse height with applied voltage V. Two curves show the amount of energy by the gas for α and β particles. Initially the output increases with V as recombination of ions and electrons is reduced. In the plateau region there is a full charge collection and the output is independent of V. Beyond this stage charge multiplication occurs and we enter the proportion region where the output pulse increases rapidly with voltage.

9.14 Geiger-Mueller Counter (G.M counter)

It consists of a cylindrical sealed glass tube of radius 2-3 cm enclosing a co-axial metal cylinder serving as the cathode. A very thin metal wire of diameter 0.1 mm (tungsten) is stretched along the axis of the cylinder and serves as the anode. Electrical connections are provided to cathode and anode by metallic leads sealed through the glass tube. It, therefore uses the same geometry as the proportional counter but the applied voltage is increased. The electric field near the wire becomes so high that the electrons can gain sufficient energy between collisions to excite inner electrons of the gas atoms. The (U-V) photons have enough energy to ionise other atoms through the volume of the counter giving more electrons. These give rise to further electrons giving an avalanche in a short time (1 μs) and the whole process grows until a discharge extends along the whole length of the wire. Electrons are collected quickly leaving positive ions behind and the discharge stops when the build-up of positive charge near the wire reduces the electric field below the Geiger action level. The resulting voltage pulse is large and can be counted without amplification.

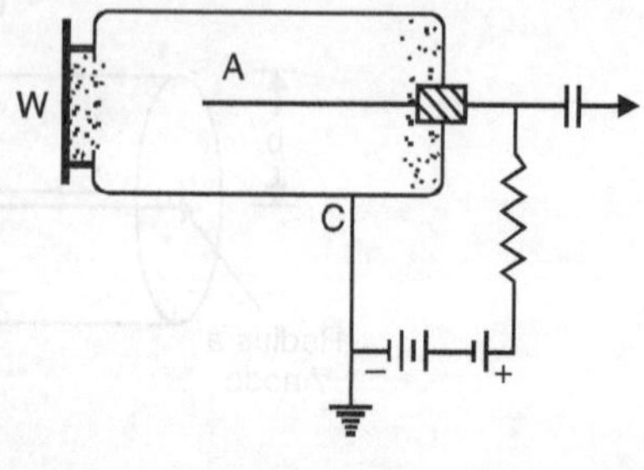

Fig. 9.10

The +ve ions drift towards the cathode (time about 10^{-4} s) after the discharge and are neutralised. In order that secondary electrons which are released from the surface do not have any effect to trigger further discharge a molecular gas is used for quenching. In this case the atomic ions of the primary gas exchange their charges with the molecular gas. Therefore only molecular ions reach the cathode. The excess energy is used in molecular dissociation and not in the release of free electrons. Quenching periods are a few hundred microseconds and thus limits the use of counter for high rate counting. Quenching gases are ethanol or a halogen (90% Argon and 10% ethanol mixture). As the quencting gas is used up, the G-M tube has a finite operating life time.

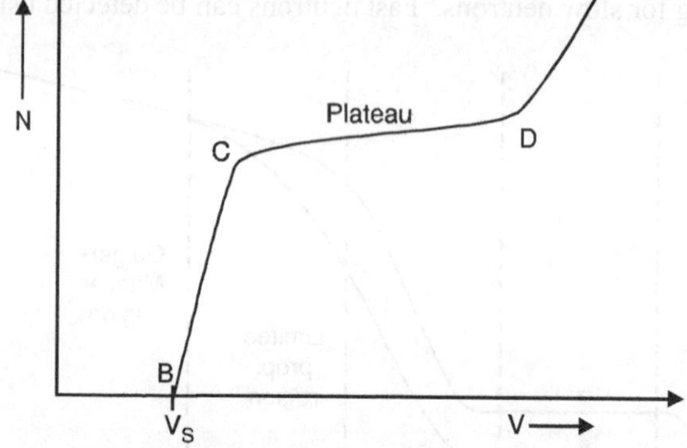

Fig. 9.11. Characteristics.

Characteristic curve of the G-M counter is a function of the voltage *vs* the rate of counting. The nature of the curve is shown in Fig. 9.11. For low voltage, the counter operates in the ionisation chamber region where there is no gas amplification. Voltage pulses are small and the counts are not able to the recorded. Unless the voltage across the tube exceeds a certain minimum *Vs*, known as threshold voltage no counts are recorded. As the voltage is increased beyond *Vs*, the count rate increases gradually as the gas amplification now sets in and output size increases gradually. This is the region of proportional counting where more and more low energy particles are counted till pt *C*. Beyond this region, the counting rate becomes constant. *C* is the commencement of the Geiger region in which the primary events are recorded, independent of the energy. The flat part of the curve is known as the plateau region of the counter.

The G.M counter is operated in the plateau region. There is a small slope for the plateau in practice. The nature of the plateau depends on the filling gas. Usually the plateau extends over 100 to 200 volts.

Dead time of the counter is defined as the time interval between the production of the initial pulse and the initiation of the second Geiger discharge. This is usually ~50 to 100 μs. This occurs due to the slow movement of the heavier positive ions from anode to cathode.

Recovery time is the time interval after which the counter returns to its original state to produce the full sized pulses again.

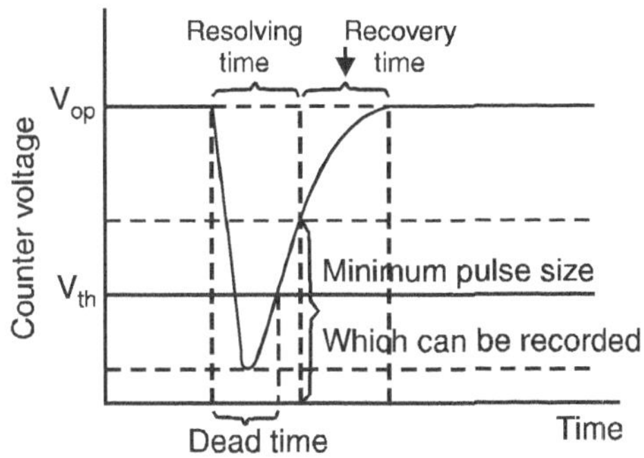

Fig. 9.12.

Resolving Time

Assume that a counting system with a resolving time (τ) responds at a rate n counts per unit time when exposed to N initiating events per unit time. In unit time the total insensitive time is $n\tau$ and the number of counts missed is $Nn\tau$.

Now, the number of counts missed = Error in counting

$$\therefore \quad Nn\tau = N - n.$$

$$\therefore \quad \text{Actual count rate } N = \frac{n}{(1 - n\tau)}$$

Hence the actual count rate can be calculated by knowing τ.

9.15 Scintillation Counter

A scintillation counter is used for the detection and counting of particles emitted by radioactive substances and these counters have replaced the spinthariscope and Geiger counters.

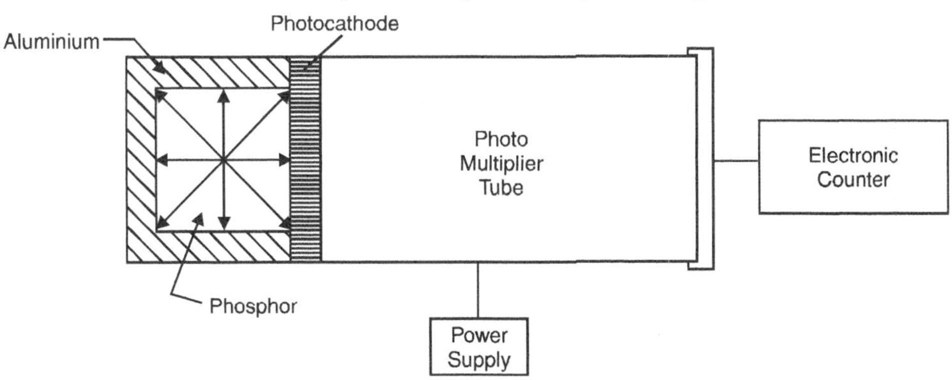

Fig. 9.13

It consists of a photo-multiplier tube in conjunction with suitable fluorescent materials called phosphors (Fig. 9.13). The light coming from the scintillation produced by the incidence of an α-particle on the phosphor strikes the photo-cathode and produces one or more phto-electrons. The ejected electrons are multiplied by the action of the photo-multiplier tube and are detected by a suitable circuit. The phosphor is usually covered with a thin layer of aluminium. The visible radiations continues until the radiations reach the photo-cathode. The most commonly used phosphors are anthracene, napthalene and stilbene. Sodium iodide crystals and potassium iodide crystals activated with thallium are also used as phosphors in scintillation counters. By this arrangement scintillations at the rate of 10^6 per second can be counted.

Organic scintillators consist of molecules which interact with each other rather weakly. Organic scintillators are effective for electrons and fast neutrons but have low γ-rays detection efficiency. Inorganic scintillators improves their efficiency for γ-rays. Plastic scintillators are used as their decay times, of the order of 1 *ns*, are much less than the 200 *ns* of NaI crystals. These are preferred for high-energy experiments.

9.16 Cloud Chamber

C.T.R Wilson designed a cloud chamber which was the fast method for the visualisation of tracks of ionising particles.

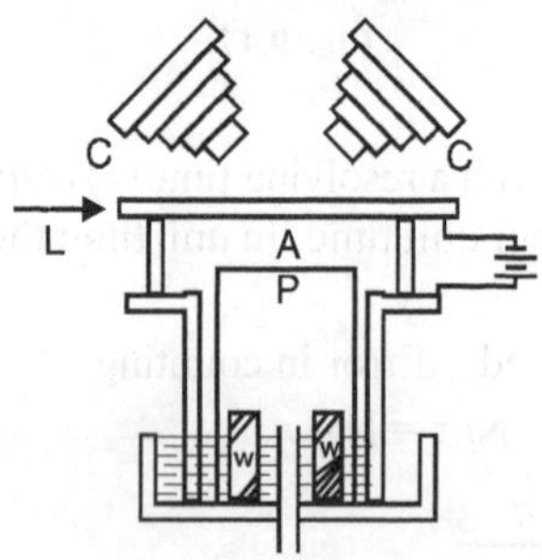

Fig. 9.14

Principle : When there is a sudden expansion of saturated vapour in a chamber, supercooling of the vapour occurs. Charged ions serve as nucleii of condensation of water droplets in a dust free atmosphere of air-water vapour mixture under the condition of supersaturation.

In Fig. 9.14, *A* is an airtight closed chamber containing air or some other gases mixed with the vapour of some volatite liquid (ethyl alcohol and water used). The walls and roof are made of glass. An adiabatic expansion can take place by means of a moveable piston.

On expansion of the air, particles are allowed to enter the chamber through a side window; droplets are formed on the ions produced by the particle leaving behind a track. If the chamber is illuminated by light *L* the track is visible. These can be photographed. The cloud chamber requires a definite recovery time after the expansion, which is its limitation.

Working:

Let V_0, T_0 be the volume and temperature of the gas mixed vapour and P_0 be the saturated vapour pressure. The mass m_0 of vapour is given by

$P_0V_0 = m_0 R T_0$, where R is the molar gas constant.

On Adiabatic Expansion, the new volume and temperature are related as

$T_0 V_0^{\gamma-1} = TV^{\gamma-1}$, γ is the ratio of the two specific heats. If no vapour condenses, then the new pressure P_1 is given as

$$P_1V = m_0RT.$$

If P is the S.V.P at temp T, then it is much less than P_1 and the vapour will be supersaturated and most of the vapour will condense till the pressure is P. If the mass of the vapour left is m, then

$$PV = mRT.$$

The supersaturation ratio δ is defined as the ratio of the vapour densities at stable and unstable conditions.

$$\therefore \qquad \delta = \frac{m_0 / V}{m / V} = m_0 / m \text{ or } \delta = P_1 / P.$$

If the initial pressure is one atoms at a temp. of 20°C, the expansion ratio 1.31 gives δ = 6.

Adjusting the value of δ to 6 and above with a gauge meter, the expansion is allowed to take place just when the particles enter the chamber. The tracks formed are photographed after illuminating the chamber with an intense light beam.

The thickness of the tracks is charateristic of the particle. α-particle tracks can be easily distinguished from those made by electrons. α-particles on an average produce about 50,000 ion pairs per cm compared to β-particles which produce only 50 ion pairs per cm.

∴ α-tracks are short and thick whereas the β-tracks are long and thin.

By applying a magnetic field curved tracks are obtained. These can be photographed and studied. Using the Wilson Cloud Chamber position was discovered.

9.17 Spark Chamber

When a metal plate or wire is charged to a high potential and another conductor is placed nearby, a spark jumps the gap between the conductors and a "bridge" of ions is created across the gap when ionising particles pass.

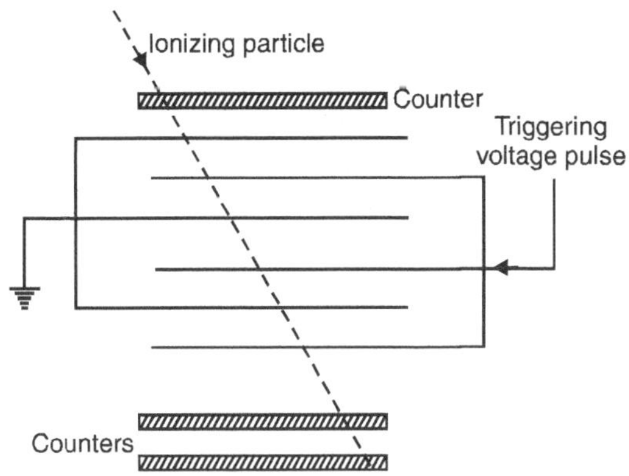

Fig. 9.15

A common form of spark chamber is constructed from a series of closely spaced parallel metal plates with alternate plates charged to a potential by a short pulse trigged on the arrival of ionising particles. Sparks go across the noble gas (mixture of Helium and neon) filling the region between the plates and revealing the path of the particles by the location of the sparks. They can be photographed. Another form uses a two-dimensional array of parallel wires to locate the spark path.

9.18 Nuclear Emulsions

We know that charged particles affect photographic plates. The optical photographic plates are not suitable for nuclear radiations. The nuclear emulsions differ from the optical emulsions in that they have a considerably higher silver halide content and smaller grain size. In nuclear emulsion, the thickness is greater than optical emulsions. The smaller the grain size, the more sensitive the emulsions are to ionising particles. Hence different commercially available emulsions, differing in grain size, can be used to differentiate between different particles.

At the present time, the nuclear emulsion technique of special interest for physicists is based on the recording of the tracks of individual ionizing particles in photographic emulsion, in which the photosensitive material (AgBr) is present in much larger concentrations (about 3 to 4 times as large) than in ordinary photographic plates. The emulsions are also made much thicker (between 25 to 2000 microns compared to 2 or 3 microns in ordinary plates). Further, the silver bromide grains are much smaller in size in nuclear emulsions (~ 0.1 to 0.6 μ diameters) than in ordinary plates.

When the high energy ionizing particle passes through the emulsion, it leaves a trail of ions along its track. It is his ionization which, most probably, affects the AgBr grains and produces a latent image of the track, which becomes visible on development of the plate. Some of the important characteristics of the nuclear emulsion plates are compared with those of the ordinary optical emulsion in Table 7.3.

The sensitivity of the emulsion for the detection of different types of particles depends on the grain size. Smaller the grain diameter, less is the sensitivity. Particles, which are highly ionizing, require less sensitive plates for recording their tracks. Thus, the fission fragments, which are the most ionizing, require the least sensitive emulsions, while high energy electrons, which produce very few ions along their tracks, require the most sensitive emulsions.

Table

Property	Optical Emulsion	Nuclear Emulsion
AgBr : Gelatin (mass)	47 : 53	80 : 20
AgBr : Gelatin (volume)	15 : 85	45 : 55
Grain size (micron)	1 – 3.5	0.1 – 0.6
Thickness (microns)	2 – 3	25 – 2000
Sensitivity to light	Very high	Poor.

Advantages of nuclear emulsion detectors.

1. These detectors have a large stopping power for very energetic particles and provide a continuously sensitive medium for recording of low, medium and high energy particles.

2. They can measure a large no. of parameters for most of the particles. These include estimation of charge, mass, energy, life time etc.

3. They have excellent spatial resolution ~ 0.5μm and high angular resolution ~ 10^{-4} radians which are required for accurate measurements in high energy interactions which involve multiple particle production.

4. They are relatively cheap. They are siutable for high altitude cosmic ray experiments.

5. These have been used in cosmic ray studies leading to the discovery of π and K-mesons.

Example 9.1. *Find the maximum energy transferred by a 3 MeV α-particle to an electron which is initially at rest. α-particle is assumed to be massive compared to the electron.*

Solution. In a head-on collision there is maximum transfer of energy. If the initial speed of approach of α-particle is v

$\because \quad M >> m_e$

Speed of separation of electrons is $2v$.

$$\text{Maximum electron energy} = \frac{1}{2} m\,(2v)^2$$

$$= 4E_\alpha \frac{m}{M} = \frac{4 \times 3\ MeV}{4 \times 1836} = 1.633 \text{ keV}.$$

Example 9.2. *The specific rate of energy loss per unit mass per unit area of a 5 MeV proton in silicon is 59 keV mg^{-1} cm^2 and its range R′ is 50 mg cm^{-2}. Calculate values of* $\frac{-dE}{\rho dx}$ *and Range R′*

for deuterons and α-particles, all having the same speed as the proton.

Solution. $$\frac{dE}{dx} = \frac{\text{const}}{E^K}, \quad K \simeq 0.8, \quad R' = \rho R$$

From eqs. $\frac{dE}{dx} \,\alpha\, Z^2 f(v)$ and $R \,\alpha\, \frac{m}{Z^2} F(v)$

$$\therefore \quad \frac{dE}{\rho dx} \,\alpha\, Z^2 \qquad R' = \rho R \,\alpha\, \frac{m}{Z^2}$$

$$\frac{-dE}{\rho dx} = Z^2 \times 59 = 59 \text{ for deuterons}$$

$$R' = \rho R = 50 = 100 \text{ for deuterons}$$

$$\frac{-dE}{\rho dx} = 236 \text{ for } \alpha\text{-particles}$$

$$R' = \rho R = 50 \text{ for } \alpha\text{-particles}$$

Example 9.3. *What thickness of concrete (density 2200 kgm^{-3}) is needed to attenuate a collinated beam of 1 MeV γ-rays by a factor of 10^6? The mass attenuation coefficient of concrete $\mu_m = 0.064$ cm^2 g^{-1}.*

Solution. $$I = I_0 \exp.(-N\sigma_x) = I_0 \exp(-\mu x) \quad \mu = N\sigma$$

$$\mu_m = \frac{\mu}{\rho}$$

$$\mu = \rho\,\mu_m = 0.141 \text{ cm}^{-1}$$

$$e^{-\mu x} = 10\text{-}6$$

$$\therefore \quad x = 98 \text{ cm.}$$

Example 9.4. *Calculate the average no. of collisions needed for neutrons of initial energy 2 MeV to be thermalised (average energy = 0.025 eV) when scattered by beryllium (A = 9). In this case how many collisions would be needed to reduce the energy to about half its initial value?*

Solution. $$\xi = 1 + \frac{(A-1)^2}{2A} \ln\left(\frac{A-1}{A+1}\right)$$

Approximately $$\xi = \frac{2}{A} - \frac{4}{3A^2}$$

and $$n = \frac{1}{\xi} \ln \frac{E_0}{E_n'}$$

$$\therefore \quad \xi = \frac{2}{9} - \frac{4}{243} = .207$$

$$n_{\frac{1}{2}} = \frac{1}{.207} \ln \frac{1}{2} = 3.4$$

$$n = \frac{1}{\xi} \ln \frac{2\ MeV}{.025} = 88$$

Example 9.5. *Find the change in the voltage across a G.M tube when 5 MeV α-particles are stopped in Arg/N$_2$ when gas W = ionisation 25 eV capacity of system = 10^{-10} F.*

Solution. No. of ion produced

$$= \frac{5 \times 10^6}{25} = 2{,}00{,}000 \text{ approx.}$$

If the ions are collected by the electrode system having a cap of 10^{-10} F. Change in pot.

$$\Delta V = \frac{2\times10^5\times1.6\times10^{-19}}{10^{-10}} = \frac{Q}{c} = 3.2\times10^{-4} \text{ volts.}$$

EXERCISE IX

1. Describe the Bethe-Block formula in determining the stopping power of the medium. Bring out the dependence of range on the energy.

2. Describe the Bragg curve ? What is the Bragg-Kleeman rule for the Range ?

3. How do electrons loose their energy in a medium ?

4. What are the ways in which a γ-ray loses its energy ? Describe each process.

5. How are γ-rays attenuated ? Obtain an expression for the intensity loss with thickness of the medium.

6. How do neutrons loose their energy in a medium ?

7. Obtain an expression for log energy decrement for neutron moderation.

8. Describe the proportional counter. What is the proportional region ?

9. Describe the Geiger-Mueller counter. Give its characteristic curve. Define : Dead time and recovery time for the counter.

10. Describe the scintillation counter and the Nucleus emulsion technique for counting particles.

11. Describe the Cloud chamber and give its principle. What is a spark Chamber ?

12. The energy of the photon, after being Compton scattered (from an electron at rest) through 60°, is half its initial energy. Calculate its initial energy. [**Ans.** 1.022 MeV]

13. An ionisation chamber is connected to an electrometer of capacity 0.5 pF and voltage sensitivity of 4 divisions per volt. A beam of α-particles causes a deflection of 0.8 divisions. Calculate the no. of ion pairs required and energy of source of α-particles, ion pair requires 35 eV.

[**Ans.** 6.25 × 10^5 ion Pairs, 2.19 MeV]

14. Calculate the no. of collisions needed to thermalise neutrons of initital enegy 2 MeV when scattered by iron ($A = 56$) and lead ($A = 207$). How many collisions are needed to reduce the energy to half its intital value ?

[**Ans.**	ξ	n	$n_{1/2}$
Fe	.035	516	20
pl	.0096	1890	72]

15. For a 20 MeV neutron to be slowed down to an energy of 1.25 eV, how many collisions on an average, will be required when the moderator is Carbon ($A = 12$). [**Ans.** 112.1]

16. Find the mean energy of neutrons in thermal equilibrium with the molecules of a medium at (*a*) 0°C and at (*b*) –150°C.

10

STRUCTURE OF NUCLEI

10.1. Basic Properties

1. Nuclear Size : From scattering experiments Rutherford deduced that the nuclear radius was less than 10^{-14} m. The nuclear radius can also be estimated from Gamow's theory of α-decay. Scattering by fast electrons and neutrons have shown that the nuclear radius can be expressed in terms of mass number A as

$R = R_0 A^{1/3}$ where $R_0 = 1.3 \times 10^{-15}$ m = 1.3 fermi. R_0 depends on the method used to find the nuclear radii and gives slightly smaller value of R_0 from electron scattering than for neutron scattering.

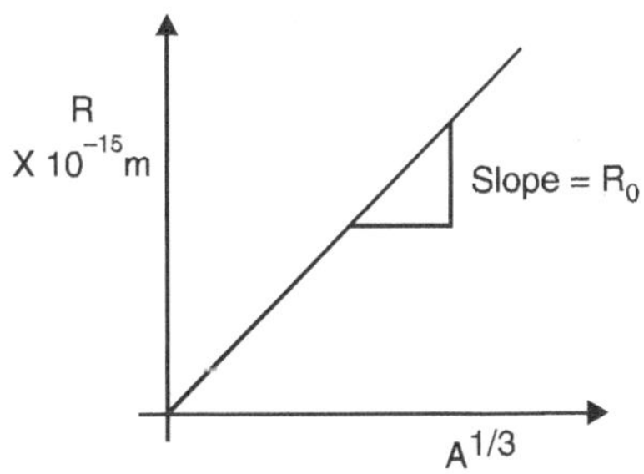

Fig. 10.1.

A distribution of nuclear charge can be obtained from electron diffraction experiments. Fig. 10.2 shows the set up of the experimental apparatus used by Hofstadter and his workers. Electrons are produced in a linear accelerator. The electrons are scattered from a thin target foil placed at the centre of an evacuated scattering chamber. Scattered electrons are detected by the spectrometer which determine their kinetic energy by bending them in a magnetic field. They used 420 MeV electron on a carbon target.

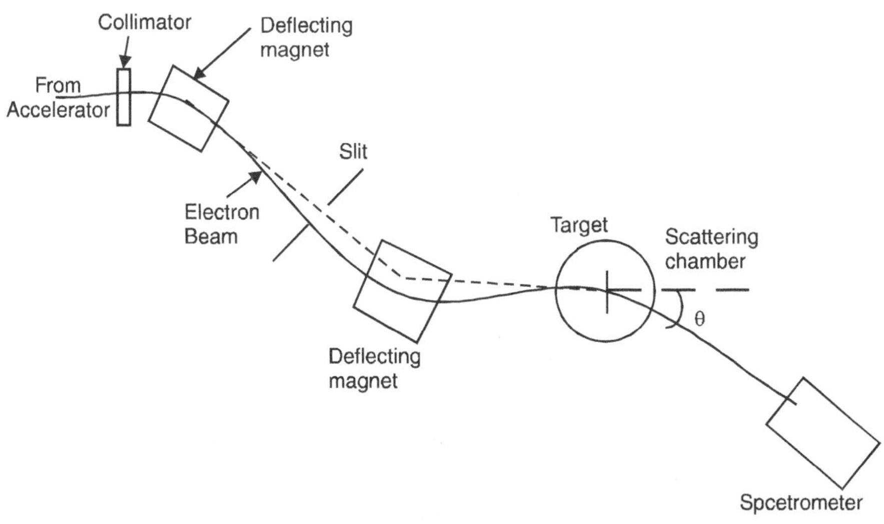

Fig. 10.2.

The variation of the charge density is given by the empirical equations

$$\rho(r) = \frac{\rho_0}{1+e^{(r-a)/b}} \qquad ...(10.1)$$

where $a = 1.07\ A^{1/3} \times 10^{-15}\ \text{m} = 1.07\ A^{1/3}\ \text{F}$

$b = 0.55\ \text{F}$

and ρ_0 = density at the nuclear centre.

The following observations can be made.

1. Charge density the distribution of protons in nucleii is constant in the interior, and falls of to zero at the nuclear surface.

2. The mass density ρ_M which gives the density of all nucleons varies as

$$\rho(r) \propto \frac{Z}{A}\rho_M(r)$$

3. The Radius at which the density has half its interior value is proportional to $A^{1/3}$.

10.2 Spin

Just as there are magnetic moment associated with the angular momentum of the electron, magnetic moment should also be associated with the angular momentum of the nucleus.

The spins of the protons and neutrons measured in experiments have been found to be equal to $\frac{1}{2}$, like the spin of the electron. The **nuclear spin** is equal to the geometric sum of the angular momentum of protons and neutrons. The total angular momentum of each nucleon is the sum of spin and orbital angular momentum. The magnitude of total momentum $\vec{I}$ is given by

$$|\vec{I}| = \sqrt{I(I+1)}\,\hbar$$

I is integer of half integer and called the nuclear spin.

The component of I in any direction is

$$|\vec{I}_z| = M\hbar$$

where M takes on integral values equal to $2I + 1$ from $-I$ to $+I$.

Example for $I = \frac{5}{2}$

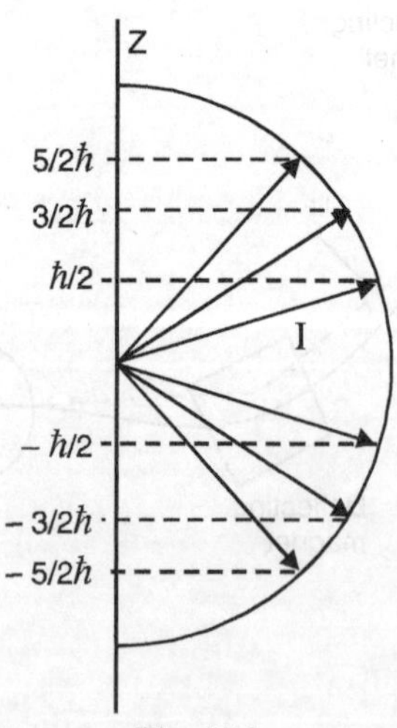

Fig. 10.3

10.3. Nuclear Magnetic Moment

A nucleus with non-zero spin has a magnetic moment μ which describes the interaction of the nucleus with an external magnetic field.

The potential energy of interaction of the magnetic moment μ in a magnetic field B is

$$U = -\mu . \vec{B}$$

Now μ is related to I as

$\mu = g\vec{I}$ where g is called the gyromagnetic ratio. Since μ and I are parallel, the nuclear properties are given by μ.

The nuclear magnetic moments are measured in nuclear magnetic μ_N

where $$\mu_N = \frac{e\hbar}{2m_p} = .505 \times 10^{-26} \text{ Amp} - m^2 \simeq 10^{-3}\ \mu_b$$

$$\mu_b = \text{Bohr Magneton}$$
$$= 3.5 \times 10^{-8} \text{ eV/tesla}$$

Magnetic moment of proton and neutron are :

$$\mu_p = 2.7\ \mu_N$$

and $$\mu_n = -1.91\ \mu_N.$$

10.4. Electric Quadrupole Moment (Q)

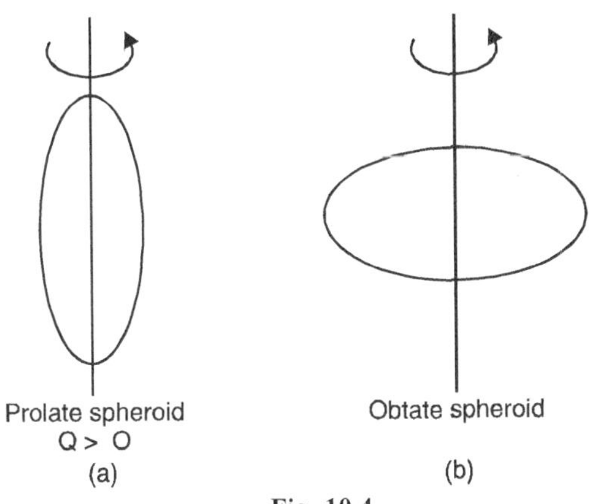

Fig. 10.4.

The observed departure of nuclear charge distribution from spherical symmetry is specified by nuclear electric quadrupole moment Q. As in Fig. 10.4(a) $Q > 0$ the ellipsoidal charge distribution is elongated in the direction of its symmetry axis; with the elongation increasing as Q is more + ve. In Fig. (b) $Q < 0$ the elliposoidal charge distribution is flattened increasing as Q becomes more negative.

The convincing proof of the existence of non-spherical nucleii is obtained by the presence of rotation bands in the spectra of even-even nuclei.

Quadrupole moment equation of a nucleus is defined as

$$Q = Z\int (3Z^2 - r^2)\,\rho\left(\vec{r}\right) dr$$

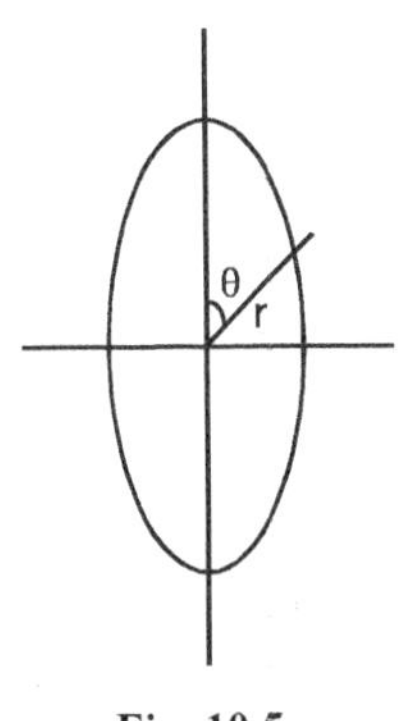

Fig. 10.5

where $\rho(\vec{r})$ is the density of nuclear matter at a point r defined in Fig. (10.5). Ze is the nuclear charge Q has dimensions of area and is given in m^2 or barns (10^{-24} cm^2) or F^2 (fermi^2).

Experimentally it has been found that quadrupole moment of some nuclei (z = 2, 8, 20, 50, 82, etc.) is zero (spherical nuclei). The quadrupole moment of all other nuclei is non-zero *ie.,* there charge distribution is not spherical, but is elliptical.

10.5. Atomic Mass Unit and Binding Energy

Atomic mass unit (amu) is the common unit of mass and energy in nuclear physics. Previous to 1960, the isotope of oxygen $_8O^{16}$ was taken as one-sixteenth of the mass of one atom of oxygen. But in 1960, the *General Assembly of the International Union of Pure and Applied Physics adopted carbon $_6C^{12}$ as the reference for the atomic mass unit.* Now one atomic mass unit (1 amu) is taken as one-twelfth of the mass of carbon atom $_6C^{12}$. The masses of protons, neutrons etc. are expressed in amu Carbon of atomic weight 12 and atomic number 6 has a mass equal to 12 amu.

Accordingly, 1 amu = 1.66038×10^{-27} kg

The mass of all particles and atoms have been found in terms of amu.

The mass of proton = 1.007277 amu

The mass of neutron = 1.008665 amu

It is found that the mass of a nucleus is less than the sum of the masses of the constituent particles in the free state. According to Einstein, the decrease in mass is due to the release of energy when the particles combine to form a nucleus. The energy released is given by the relation $E = mc^2$, where m is the decrease in mass and c the velocity of light.

The amount of energy represents the binding energy of the nucleus. If the binding energy is large, the nucleus is stable.

If a nucleus is to be broken into a constitutent particles, an energy equal to or more than the binding energy must be supplied to the nucleus.

It is convenient to calculate energy in joule or electron volt, when a mass equal to 1 amu is converted into energy.

$$E = mc^2$$

$$m = 1 \text{ amu} = 1.66 \times 10^{-27} \text{ kg}$$

$$c = 3 \times 10^8 \text{ m/second.}$$

$$\therefore \quad E = (1.66 \times 10^{-27})(3 \times 10^8)^2$$

$$= 1.49 \times 10^{-10} \text{ Joules}$$

But 1 eV of energy = 1.6×10^{-19} Joule

$$\therefore \quad E = \frac{1.49 \times 10^{-10}}{1.60 \times 10^{-19}}$$

$$E = 931 \times 10^6 \text{ eV}$$

$$E = 931 \text{ MeV} \qquad ...(10.2)$$

The relation (10.2) represents the conversion of amu into MeV and *vice-versa.* Thus as nucleus of mass M is amu in equivalent to $931 \times M$ million electron volt energy. The energy equivalent of an electron can also be calculated in the same way.

Mass of an electron = 5.483×10^{-4} amu

Equivalent energy of an electron at rest

$$= 5.483 \times 10^{-4} \times 931 \text{ MeV} = 0.511 \text{ MeV}$$

The binding energy of different nuclei can be calculated. Take the case of lithium of A = 7 and Z = 3. Its nucleus contains three protons and 4 neutrons.

The mass of the lithium nucleus = 7.016005 amu

The mass of three protons = 3×1.007277 amu

= 3.021831 amu

The mass of four neutrons = 4×1.008665 amu

= 4.03466 amu

Total initial mass of the three protons and four neutrons

= 3.021831 + 4.034660

= 7.056491 amu

As the mass of lithium nucleus is less than the total initial mass of 3 protons and 4 neutrons decrease in mass

= 7.056491 – 7.016005 = 0.040486 amu

This shows that in the formation of lithium nucleus the mass has decreased by 0.040486 amu and this decrease in mass is due to the release of energy in the process.

1 amu = 931 MeV

Binding energy of lithium nucleus = 0.040486×931

= 37.7 MeV

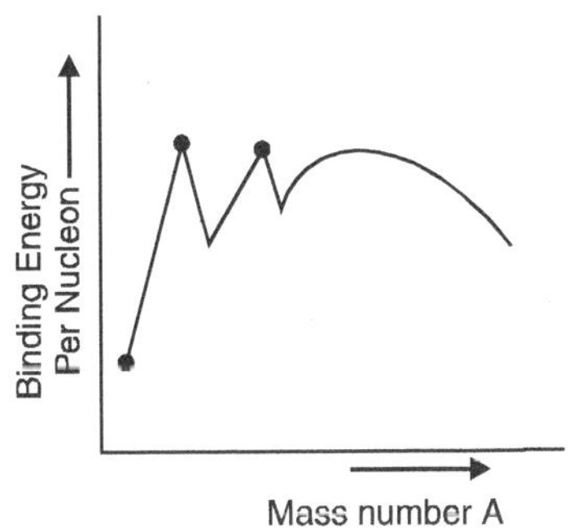

Fig. 10.6

In order to break the lithium nucleus, an energy more than 37.7 MeV must be supplied. Therefore, the lithium nucleus is very stable in structure. In any attempt to break the nucleus, energy must be supplied by some external source.

In the case of lithium, there are 7 particles. Hence the binding energy nucleon = $\frac{37.7}{7}$ = 5.4 MeV. The binding energy per nucleon for most of the nuclei has been calculated in the same way as discussed above. A graph between binding energy and atomic number A is shown in Fig. 10.6. It is found that ${}_2He^4$, ${}_6C^{12}$ and ${}_8O^{16}$ are stable and show peak values. All other elements with A less than twenty have low values. The nuclei of intermediate values of A are most stable.

10.6. Mass Defect and Packing Fraction

The mass defect of the atom is the difference of its atomic mass and its mass number. For carbon ${}_6C^{12}$ the mass defect is zero because its mass $M = 12$ amu and $A = 12$

$\therefore$ Mass defect $\Delta = M - A$

A mass defect gives the deviation of the atomic mass from the whole number A. The packing fraction p is the mass defect per elementary particle in the nucleus.

$$p = \frac{\Delta}{A} = \frac{M - A}{A}$$

A graph between packing fraction p and mass number A is shown in Fig. 10.7. The graph shows interesting results.

(1) The packing fraction is positive for elements having mass number below 20.

(2) The packing fraction is negative for elements having mass number between 20 and 200.

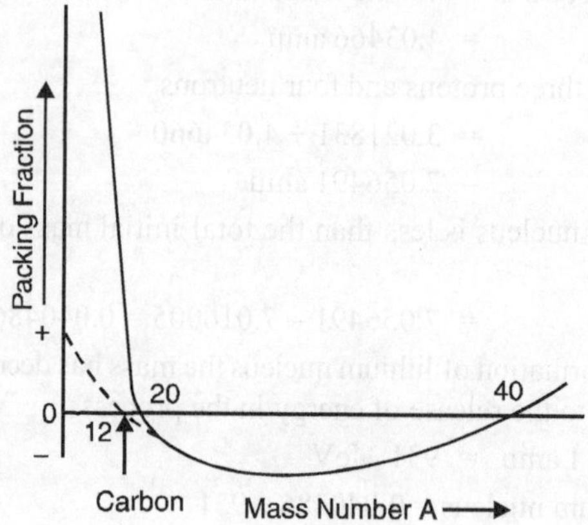

Fig. 10.7

(3) The packing fraction is positive for elements having mass number beyond 200.

The packing fraction is zero for carbon $A = 12$ and $Z = 6$. This does not mean that its binding energy is zero. The mass defect and packing fraction only show their relations with respect to carbon. A nucleus having positive packing fraction has its mass M greater than its mass number A. It means that the loss of mass due to binding energy arrangement in this nucleus is less than for carbon (It is common to represent packing fraction in terms of $p \times 10^4$).

Example 10.1. *Give (i) energy equivalence in MeV and (ii) mass equivalence in electron mass unit for a gamma ray of wavelength 4.5×10^{-11} cm.*

Sol. (*i*) Here, $\lambda = 4.5 \times 10^{-11}$ cm

or $\lambda = 4.5 \times 10^{-13}$ m

$$h = 6.624 \times 10^{-34} \text{ J-s}$$

$$c = 3 \times 10^8 \text{ m/s}$$

$$U = h\nu = hc/\lambda$$

$$U = \frac{6.624 \times 10^{-34} \times 3 \times 10^8}{4.5 \times 10^{-13}}$$

$$U = \mathbf{4.4 \times 10^{-13} \ J}$$

$$U = \frac{4.4 \times 10^{-13}}{1.6 \times 10^{-19}} \text{ eV}$$

$$U = 2.75 \times 10^6 \text{ eV}$$

$$U = \mathbf{2.75 \ MeV}$$

(*ii*) Equivalent mass

$$m = \frac{U}{c^2}$$

$$= \frac{4.4 \times 10^{-13}}{(3 \times 10^8)^2}$$

$$= 4.89 \times 10^{-30} \text{ kg}$$

Mass of electron, $m_e = 9.1 \times 10^{-31}$ kg

$$\therefore \quad \frac{m}{m_e} = \frac{4.89 \times 10^{-30}}{9.1 \times 10^{-31}}$$

$$m = \mathbf{5.37\ m_e}$$

Example 10.2. *Two isotopes of oxygen having nuclear masses 15.990523 amu and 17.994768 amu calculate the binding energy per nucleon in the two cases in MeV. What do you expect the relative abundance of the two isotopes ?* [*I.A.S. 1990*]

Sol. *Mass of Proton* = 1.007276 amu

Mass of neutron = 1.008665 amu

(1) For isotope $_8O^{16}$

$$\begin{aligned}
\text{Number of protons} &= 8 \\
\text{Number of neutrons} &= 8 \\
m_p &= 1.007276 \\
m_n &= 1.008665 \\
\text{Binding energy} &= 8[m_p + m_n] - 15.990523 \\
&= 8[1.007276 + 1.008665] - 15.990523 \\
&= 16.127528 - 15.990523 \\
&= 0.137005 \text{ amu} \\
&= 0.137005 \times 931 \text{ MeV} \\
&= 127.55 \text{ MeV}
\end{aligned}$$

As there are 16 nucleons, binding energy per nucleon

$$= \frac{127.55}{16} = \mathbf{7.972\ MeV}$$

(*ii*) For isotope $_8O^{18}$

$$\begin{aligned}
\text{Number of protons} &= 8 \\
\text{Number of neutrons} &= 10 \\
\text{Binding energy} &= [8m_p + 10m_n] - 17.994768 \\
&= [8 \times 1.007276 + 10 \times 1.008665] - 17.99476 \\
&= 8.058208 + 10.086650 - 17.994768 \\
&= 0.15009 \text{ amu} = 0.15009 \times 931 \text{ MeV} = 139.73 \text{ MeV.}
\end{aligned}$$

As there are 18 nucleons,
Binding energy per nucleon

$$= \frac{139.73}{18} = \mathbf{7.762\ MeV.}$$

Example 10.3. *Find the Binding energy per nuclear when (i) two deuterons combine to form an alpha particle (ii) two neutrons and two protons combine to form an alpha particle.*

Solution:

1. B.E. for α-particle = $[2 \times M(_1H^2) - M(_2He^4)] \times 931$ MeV

$= [2 \times 2 \times 2.013554 - 4.002604] \times 931$

$= 22.8132$ MeV

B.E./nucleon = 5.703MeV

2. B.E. for α-particle

$= [2 \times M_p + 2\,M_n - M(_2He^4)] \times 931$ MeV

$= [2 \times 1.00\,7825 + 2 \times 1.008665 - 4.002604] \times 931$

$= 28.28$ MeV

B.E./Nucleon = 28.28/4 = 7.07 MeV

Though α-particle in both the cases has got the same constituents, yet the mass defects are different due to the combination of different particles.

10.7. Deuteron Binding Energy

The deuteron $_1H^2$ (or the deuterium nucleus) is the simplest system in the ground state consists of proton and neutron bound by a nuclear force.

A method to finding the Binding energy is to observe the smallest energy γ-ray required to break up the deuteron into neutron and proton.

Experimentally the threshold energy

$$n + p \rightarrow d + \gamma$$

deuteron recoil energy is 1.3 keV. ∴ B.E. = 2.22 MeV

Also

$$\begin{aligned} \Delta m &= m_p + m_n - M_n \\ &= (1.007276 + 1.008665 - 2.013553) \text{ amu} \\ &= 0.002388 \text{ amu} \\ &= 2.22 \text{ MeV} \end{aligned}$$

Nuclear spin = I = 1 Nuclear Parity-even

Ground state of the deuteron is characterised by the following quantities :

$$\mu = \text{Magntic dipole moment} = +0.857\ \mu_n$$

$$Q = \text{Quadrupole moment} = +\ 2.7 \times 10^{-31}\ \text{m}^2$$

Charge distribution at half value radius a = 2.1 F.

The value of Q shows that the charge distribution is not spherically symmetric. ∴ The nuclear potential *i.e.*, force between two nucleons is not spherically symmetric. But the departure is not large.

Deuetron is a system in which the nucleons are bound in a $^3S_1$ state.

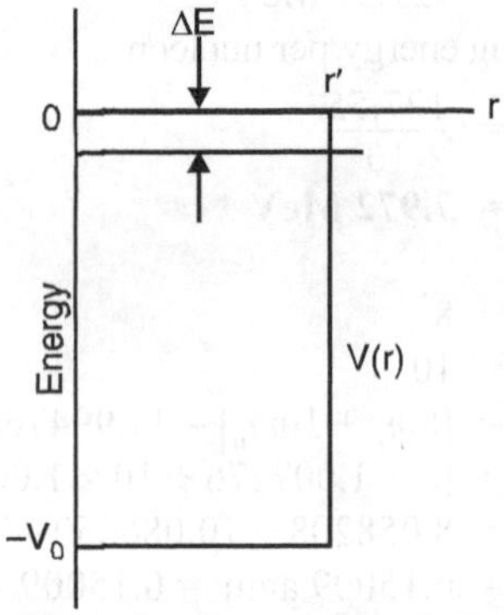

Fig. 10.8. A square well potential of radius r and depth V_0, ΔE is the ground state eigenvalue. For the deuteron this state is the only bound state of the potential.

If we assume $V(r)$ the potential as square well then r' = $2F$ and depth V_0 = 40 MeV refer Fig. 10.8.

Now, the spin of neutrons and protons are parallel in $^3S_1$ state of the deuetron. There are no bound states with spins anti-parallel, hence there is no $^1S_0$ state.

The nucleon potential is spin dependent being weaker when the two nucleons interact with anti-parallel spins therefore no $^1S_0$ states. Additional evidence for the spin dependence of nucleon potential is the absence of bound state for a system of two protons or of two neutrons, where the spins are anti-parallel. In this case the potential is too weak to lead to binding.

10.8. *n-p* Scattering

Their is evidence for the spin dependence of nuclear potential from scattering of unbound neutrons from protons. The total cross-section σ is proportional to the total probability that a neutron is scattered by a proton. The cross-section is made up of neutron-proton interaction of $^1S_0$ and $^3S_1$ states. If spins of neutrons in the incident beam and protons in the target are random then all four spin states of the two nucleon are equally probable.

There are three $^3S_1$ states (triplet states) when spins are parallel *i.e.*, $-\hbar$, O, $+\hbar$. The nucleons can also interact with $^1S_0$ state *i.e.*, anti-parallel. In this case the total spin has only one component zero.

∴ We have 3 : 1 ratio of $^3S_1$ and $^1S_0$ interactions. Hence the relative strengths cannot be had from the total cross-section.

Ortho-hydrogen has total spin = 1 where as para hydrogen has spin = *O*. If we consider very low neutron scattering, the $\lambda_{\text{De-Broglie}}$ is much larger than the distance between the protons in the Hydrogen molecule. Therefore in one interacting scattering of the neutrons from the two protons is coherent *i.e.*, amplitudes add.

Since ortho and para have different mixtures of $^3S_1$ and $^1S_0$ states, the strengths of the two spin states can be compared. This shows that the singlet potential is 40% weaker than the triplet states potential *i.e.*, square well is less deep by 40%. ∴ Low energy gives data of the extra state of the deuetron. The spin dependence is also confirmed by low energy *p-p* reaction.

Low energy nucleon potential does not give any information of the form of the necteon potential. ∴ only high energy nucleons have enough resolving power to be effective as probes for studying the nature of the nucleon potential.

10.9. *p-p* Scattering

p-p (proton-proton) scattering is more difficult to analyze than n-p scattering due to :

(1) At long ranges the Coulomb repulsion is dominant over the purely nuclear forces and there is a large amount of Rutherford-type scattering at low angles and low energies.'

(2) There is no distinction between the scattered and the scattering particle. In fact due to the two identical particles introduces a wave-mechanical effect like interference of waves.

According to Paul's principle both protons cannot interact in the same state which is allowed in *n-p* system.

In *S*-state (l = 0) only possible *p-p* interaction is one in which the spins are opposite *i.e.*, $^1S_0$ whereas for *n-p* both $^1S_0$ and $^3S_0$ are possible.

At low energies there is an appreciable contribution due to $^1S_0$ and the effective well depth is nearly the same as for *n-p* scattering in the same state. This result gives support to the charge-independence theory (wherethere *n-n*, *n-p* or *p-p*) of nuclear forces are the same when coulomb effects are ignored.

10.10. Nuclear Force

The following featurs emerges from scattering experiments :

(1) Short Range : The nuclear force between two nucleons is strongly attractive for a very short range. In early scattering experiments there were deviations from the Rutherford scattering cross-section when the surfaces of the colliding nuclei were within a few fermis of one another. This short-range property was confirmed by studies of nucleon-nucleon scattering.

(2) Repulsive Core : At high (scattering) energies the effective inter-nucleon force also has a repulsive nature, which is dominant at very short distances (0.5 F). The potential energy vs distance is shown in Fig. 10.9.

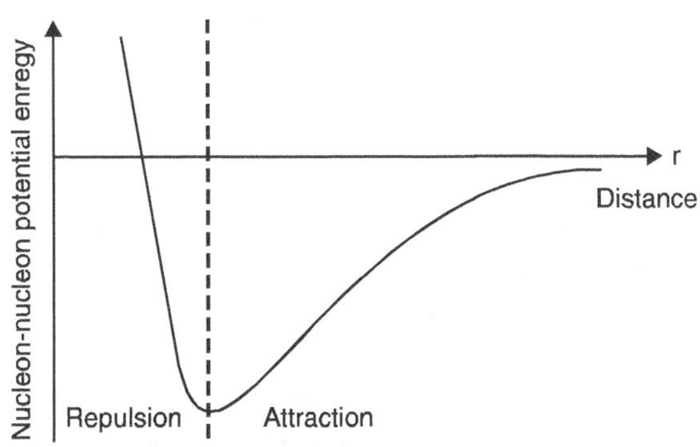

Fig. 10.9

The P.E. is a minimum where the attractive and repulsive forces are equal and opposite. This is the equilibirum distance and therefore, a nucleon inside a nucleus tends to maintain an average separation with its neighbours. This leads to the concept of the **saturation of the nuclear force** *i.e.*, the nuclear B.E. of a nucleon approaches a saturation value independent of mass number A. The situation is similar to a liquid drop, were the density and B.E. of a molecule are independent of the size of a drop.

(3) Charge dependence : The nuclear force is charge symmetric *i.e.*, the force between two neutrons is the same as that between two protons (in the same states). This is indicated by the fact that for light nucleii ($A \leq 40$) the Coulomb force is small compared to the nuclear B.E., stable ones are those for which $N \simeq Z$. This would not be so if *n-n* and *p-p* forces were significantly different.

The nuclear force is almost **charge independent** *i.e.*, same for *n-p*, *n-n* and *p-p*. For two neutrons or two protons to occupy the same spatial state the spins must be opposite. $\therefore$ The net spin is *O* for the lowest diproton or dineutron state.

The deuteron only has a bound state. This means that the average force between a neutron and a proton inside the nucleus is greater than than that between two identical nucleons. There a factor of two difference.

(4) Spin Dependence : The nucleon-nucleon force is different for the two nucleons in the $S = 0$ and $S = 1$ state and hence it is spin dependent. The spin-orbit force varies as the scalar product $\vec{L} \cdot \vec{S}$. The force is attractive if $\vec{S}$ and $\vec{L}$ are parallel and repulsive if anti-parallel. If a nucleon is will inside the nucleus, it has other nucleons with spin up and an equal number with spin down and therefore some form of spin dependence averages out. This is not so everywhere and the spin-orbit force leads to an important contribution to the force.

10.11. Nuclear Force and Pions

Yukawa proposed that a nucleon frequently emits a particle with an appreciable rest mass, called the π **meson** or **pion**. The particle hovers near the nucleon in the π-meson field as it is called for a very short time, and is then absorbed by the nucleon. During the process the nucleon has its normal rest mass, and while the process takes place there is a violation of mass-energy conservation. This is because there is more rest mass than bfore the π meson is emitted or absorbed. The π-meson cannot permanently escape the nucleon and then the mass-energy would be permanently violated. Such a pion is called a **virtual** particle because of its short existence and limited by its violation of mass-energy conservation.

For two nucleons close enough for their meson fields to overlap a π meason may leave one field and join another. In the interaction momentum carried by the π-meson is tranferred from one field to another. This is effectively like force transfer. Thus the exchange of a virtual pion-between two nucleons leads to the nucleon force acting between them.

Experimental evidence for the exchange of pions between two interacting nucleons is found in *n-p* scattering. The C-S implies that for half the scatterings the neutron changes into a proton and the proton changes to a neutron.

This is given by the set of reactions

$$n \rightarrow p + \pi^- \quad \text{then} \quad \pi^- + p \rightarrow n.$$

The neutron emits a negatively charged π^- meson into its field and becomes a proton. The π^- meson joins the field of the proton and is absorbed giving a neutron. The scattering process can also be taking place by the following set of reactions.

$$p \rightarrow n + \pi^+ \text{ then } \pi^+ + n \rightarrow p.$$

In this case a π^+ meson is emitted and then absorbed to give back the proton.

Similarly $$p \rightarrow n + \pi^0 \text{ then } \pi^0 + p \rightarrow p$$

and $$p \rightarrow p + \pi^0 \text{ then } \pi^0 + n \rightarrow n$$

can occur. The neutral π^0 meson transfers momentum, but no charge between the interacting nucleons.

An isolated proton is surrounded by a meson field which sometimes contains a π^0 meson and sometimes a π^+

$$p \to p + \pi^0 \text{ or } \quad p \to n + \pi^+$$

The nucleon of course must absorb the meson it emits in a very short time.

Similarly an isolated neutron contains a π^0 or a π^- meson

$$n \to n + \pi^0 \quad \text{or} \quad n \to \; p + \pi^-$$

But the proton field cannot contain a π^- meson and the neutron field cannot contain a π^+ meson. Electron scattering expts have given direct experimental verification.

10.12. Geiger-Nuttal Law

The relation between the range of an α-particle and the average life period of the radioactive element has been studied by Geiger and Nuttal. The reciprocal of the average life period is known as the transformation constant λ. According to Geiger Nuttal Law :

$$\log R = A \log \lambda + B \qquad ...(10.3)$$

where R is the range of the α-particle and λ is the transformation constant. A and B are constants. A graph between R and λ is a straight line. The law is useful to find the transformation constant of certain radioactive elements which cannot be easily determined by direct methods. In this way the average life periods of certain radioactive elements have been determined.

Example 10.1. *The decay constants of two α-emitters of uranium-series are 4.80×10^{-18} per second and 4.225×10^3 per second. If the ranges of α's are 4.19 cm and 7.86 cm respectively, calculate the range of α's from the third member of this series with a decay constant of 3.786×10^{-3} per second.* [GNDU, 1992]

Sol. Here,

$$\lambda_1 = 4.8 \times 10^{-18}/\text{s}$$
$$\lambda_2 = 4.225 \times 10^3/\text{s}$$
$$\lambda_3 = 3.786 \times 10^{-3}/\text{s}$$

According to Geiger Nuttal law

$$\log R = A \log \lambda + B$$
$$R_1 = 4.19 \text{ cm}$$
$$R_2 = 7.86 \text{ cm}$$
$$R_3 = ?$$
$$\log 4.19 = A \log 4.8 \times 10^{-18} + B \qquad ...(i)$$
$$\log 7.86 = A \log 4.225 \times 10^3 + B \qquad ...(ii)$$

Subtracting (*i*) from (*ii*) and arranging

$$\log_e \left(\frac{7.86}{4.19}\right) = A\left[\log_e \left(\frac{4.225 \times 10^3}{4.8 \times 10^{-18}}\right)\right]$$

or

$$0.629 = 48.227$$

∴

$$A = \frac{0.629}{48.227} = \mathbf{0.013}$$

From equation (*ii*)

$$\log_e 7.86 = 0.013 \log_e 4.225 \times 10^3 + B$$
$$2.0618 = 0.013 \times 8.3488 + B$$
$$2.0618 = 0.1085 + B$$
$$B = \mathbf{1.9533}$$

Now,

$$\log R_3 = \text{A} \log \lambda_3 + B$$
$$\log R_3 = 0.013 \log 3.786 \times 10^{-3} + 1.9533$$
$$\log R_3 = -0.725 + 1.9533$$

$$\log_e R_3 = 1.2283$$
$$R_3 = \mathbf{7.270\ cm.}$$

10.13. Alpha-Spectra and Fine Structure

Most α-particles have one associated α-energy. This is due to α-emission between two definite nuclear energy states.

But some α-emitters have a fine structure. Example of Bi^{212} is shown in fig. Bi decays to Tl^{208} (Thallium) by 5 groups of α-energies : 5.603 MeV, 5.622, 5.765, 6.044 and 6.086 MeV.

α-particles exhibit a fine structure in their energies. Like electrons, nuclei also have discrete spectra of excited states. Hence α-decay leaves the nucleus in an excited state or ground state.

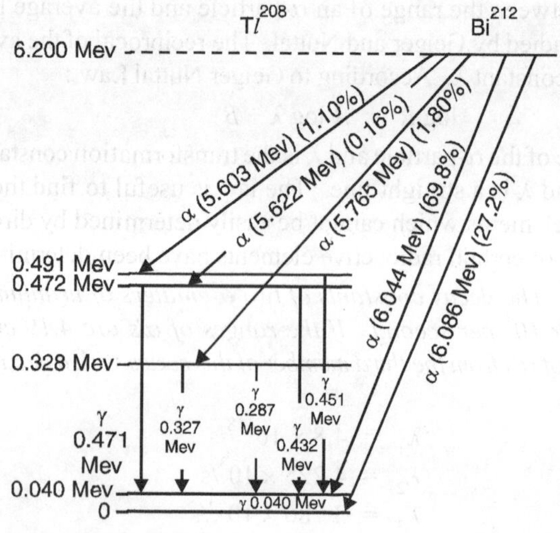

Fig. 10.10. Energy-level for the alpha-particle decay of bismuth-212 to thallium-208.

It has been observed that nuclides ThC′ and RaC′ formed by β-decay from ThC and RaC, respectively,

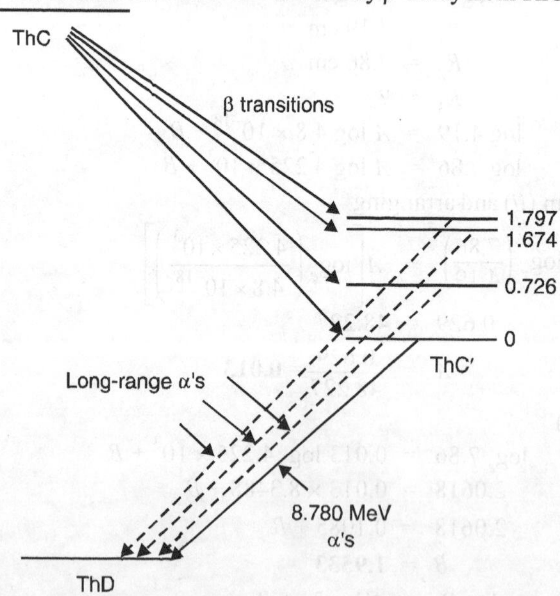

Fig. 10.11. The occurrence of high-energy groups in the alpha-particles spectrum of ThC′ is due to direct decay from excited states of ThC′ to the ground of ThD, these states being fromed in the beta decay of ThC

ocassionally emit alpha particles of phenomenally **long range.** Mean energy of α-particles from ThC′ is 8.78 MeV, there are also weak long range groups at 9.49, 10.42 and 10.54 MeV. In this case the Q value is lower than other Q values showing that the weak groups are not due to excitation of residual ThD. Instead we assume that ThC′ nuclei are formed from ThC in several excited states some are 0.726, 1.674 and 1.797 MeV above the ground state. These normally proceed to ground state by γ emission, but the probability of α-decay is not negligible. Ref. Fig. 10.11.

10.14 Alpha Decay

α-decay can be represented by equation

$$^{A}_{z}X \rightarrow \; ^{A-4}_{z-2}Y + ^{2}_{4}\text{He} + Q$$

Q is the disintegration energy of the reaction.

Let M be the mass of the parent nuclei at rest. m mass of the α-particle which has a velocity v which is emitted.

M_d is the mass of the recoiling nucleus with velocity V_d.

∴ By conservation of momentum

$$M_d V_d = mv \qquad \text{...(10.3)}$$

Now,
$$Q = \text{Final Kinetic Energy} - \text{Initial Kinetic Energy}$$

$$= \frac{1}{2}MdV_d^2 + \frac{1}{2}mv^2 \quad \text{Kinetic Energy of Parent} = 0$$

From (i), we get
$$= \frac{1}{2}M_d\left(\frac{mv}{M_d}\right)^2 + \frac{1}{2}mv^2$$

$$= \frac{1}{2}mv^2\left[\frac{mv}{M_d} + 1\right], \quad \text{Substituting } \frac{1}{2}mv^2 = K$$

$$\therefore \quad Q - K\left[\frac{m}{M_d} + 1\right] \qquad \text{...(10.4)}$$

Now
$$\frac{m}{M_d} = \frac{4}{A-4}, \; A \text{ is the mass number of parent nucleus}$$

$$\therefore \quad Q = \frac{A}{A-4}K \qquad \text{...(10.5}$$

Since A is large we have $Q = K$ *i.e.*, α-particle takes away most of the energy in the disintegration.

10.15. Gamow's Theory of α-decay

Alpha Decay : Tunnel Theory of Alpha Decay

α-particles incident on a heavy Uranium nucleus experience a coulomb barrier about 20 MeV. In spite of this α-particles are emitted from Uranium with energies less than 5 MeV, apparently this seems to violate energy conservation in the vicinity of the barrier.

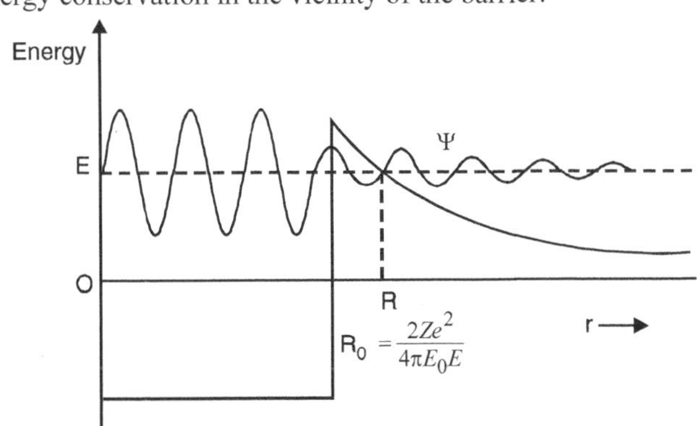

Fig. 10.12

α-decay theory was given by Gamow, Gurney and E. Condon. In dealing with a finite potential well problem in Quantum Mechanics the transmission probability T for a particle to pass through the barrier is given by

$$T = e^{-2KL} \qquad ...(10.6)$$

$$L = \text{Width of the barrier and } K = \frac{\sqrt{2m(U-E)}}{\hbar}$$

$$U = \text{barrier height and } E = \text{incident energy of } \alpha\text{-particle of mass } m.$$

Now an α-particle inside a nucleus is faced with a barrier of varying height as seen in fig. 10.12. Using the equation (1) for T is the case of the α-particle.

We see that $\quad l_n T = -2KL$

$$\therefore \quad \ln T = -2\int_0^L K(r)dr = -2\int_{R_0}^{R} K(r)dr \qquad ...(10.7)$$

R_0 = Radius of the nucleus; R is the distance from centre at which $U = E$.

Now, kinetic energy $E >$ Potential energy U for $r > R$

∴ Beyond R the α-particle escapes the nucleus.

The Potential energy of α-particle at distance r from a nucleus of charge Ze is

$$U = \frac{2Ze^2}{4\pi \in_0 r}$$

Ze = nuclear charge menicus the alpha particle of charge 2e.

Z = atomic number.

$$\therefore \quad K = \sqrt{\frac{2m(U-E)}{\hbar}} = \left(\frac{2m}{\hbar^2}\right)^{\frac{1}{2}}\left(\frac{2ze^2}{4\pi \in_0 r} - E\right)^{\frac{1}{2}}$$

$$\because \quad U = E \text{ when } r = R$$

$$E = \frac{2Ze^2}{4\pi \in_0 R}$$

and K can be written as

$$K = \left(\frac{2mE}{\hbar^2}\right)^{\frac{1}{2}}\left(\frac{R}{r} - 1\right)^{\frac{1}{2}}$$

Substitute in (*ii*)

$$\therefore \quad l_n T = -2\int_{R_0}^{R} K(r)dr$$

$$= -2\left(\frac{2mE}{\hbar^2}\right)^{\frac{1}{2}} \int_{R_0}^{R}\left(\frac{R}{r} - 1\right)^{2} dr$$

$$= -2\left(\frac{2mE}{\hbar^2}\right)^{\frac{1}{2}} R\left[\cos^{-1}\left(\frac{R_0}{R}\right)^{\frac{1}{2}} - \left(\frac{R_0}{R}\right)^{\frac{1}{2}}\left(1 - \frac{R_0}{R}\right)^{\frac{1}{2}}\right]$$

Since the potential barrier is wide $R >> R_0$

and $\quad \cos^{-1}\left(\frac{R_0}{R}\right)^{\frac{1}{2}} \simeq \frac{\pi}{2} - \left(\frac{R_0}{R}\right)^{\frac{1}{2}}$

$$\left(1-\frac{R_0}{R}\right)^{\frac{1}{2}} \simeq 1$$

$$\therefore \qquad l_n T = -2\left(\frac{2mE}{\hbar^2}\right)^{\frac{1}{2}} R\left[\frac{\pi}{2}-2\left(\frac{R_0}{R}\right)^{\frac{1}{2}}\right]$$

Now $$E = \frac{2Ze^2}{4\pi\in_0 R} \qquad \therefore R = \frac{2Ze^2}{4\pi\in_0 R}$$

$$\therefore \qquad l_n T = \frac{4e}{\hbar}\left(\frac{m}{\pi\in_0}\right)^{\frac{1}{2}} Z^{1/2}R_0^{\frac{1}{2}} - \frac{e^2}{\hbar\in_0}\left(\frac{m}{2}\right)^{\frac{1}{2}} ZE^{-1/2}$$

Calculating the constant we get

$$l_n T = 2.97\, Z^{1/2} R_0^{\ 1/2} - 3.95\, ZE^{-1/2}$$

E = α-particle K.E. in MeV

R_0 = Nuclear radius in Fermi (10^{-15}m)

Z = Atomic number of nucleus minus α-particle

Now since $\log_{10} A = \log_{10} e\, l_n(A) = 0.4343\, l_n A$

$$\therefore \qquad \log_{10} T = 1.29\, Z^{1/2} R_0^{\ 1/2} - 1.72\, ZE^{-1/2}$$

Now decay constant λ is given by

$$\lambda = vT = \frac{v}{2R_0}T$$

v = velocity of α-particle taking log on both sides.

$$\log_{10}\lambda = \log_{10}\left(\frac{v}{2R_0}\right) + 1.29\, Z^{1/2} R_0 - 1.72\, ZE^{-1/2}.$$

10.16. Beta Emission

β-particle is identical to an e^- and when it is emitted; the charge on the nucleus increases by one unit ($Z+1$). Similarly when a β^+ particle is emitted the nuclear charge decreases by one unit. β^+ particle is called a positron (e^+). It behaves like a positively charged electron with same mass. β-particles are emitted with a continuous spectra of energies.

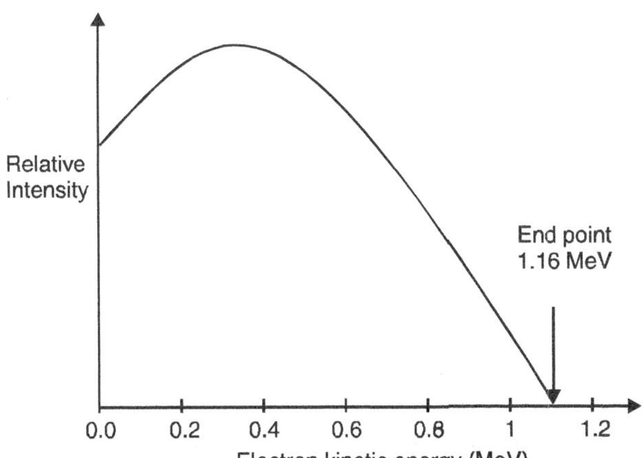

Fig. 10.13

Fig. 10.13 shows energy spectrum of β-particles emitted by ^{210}Bi. The maximum energy (end point) is the energy expected from the mass-energy relation, but the average electron energy is only about a third of this. Energy apparently seems to be missing. Pauli accounted for the apparent-violation of conservation of energy by postulating the existence of a third undetected particle in the decay. Enrico Fermi called it the neutrino ν ('little neutral one'). The neutrino is an uncharged fermion (spin 1/2) with zero or negligible rest mass, which interacts very weakly with matter. It is difficult to detect the neutrino but Reinis and Cowan in a very sophisticated experiment using a nuclear reactor gave conclusive proof of its existence.

Beta decay is the transformation of neutron to a proton or vice-versa, giving a final more stable product. The processes are :

$$n \rightarrow p + e^- + \bar{\nu} \text{ and } p \rightarrow n + e^+ + \nu \qquad \bar{\nu} = \text{anti-neutrion}$$

The leptons (e^-, ν) anti-leptons (e^+, $\bar{\nu}$) are created due to the weak nuclear force. Hence β-decay is relatively slow process compared with strong force transitions or electromagnetic interaction. Since energy has to be conserved, the daughter nucleus must be lighter than the parent.

β^+ decay, for a proton-rich nucleus is electron capture.

i.e. $$p + e^- \rightarrow n + \nu$$

The electron captures an electron from the inner orbit or *k*-shell of the atom *i.e.*, why it is termed K-capture.

10.17. Discrete Spectra

Fig. 10.14 shows β-decay of $_{12}Mg^{27}$ to $_{13}Al^{27}$.

β-decay takes place to either of the two excited states of $_{13}Al^{27}$. The resulting $_{13}Al^{27}$ excited nucleus undergoes one or two γ-rays to reach the ground state. These give rise to discrete β-spectra Cl^{38} gives three groups of electrons. Cu^{64} is a particularly interesting case because it emits both electrons and positrons and also undergoes orbital electron capture.

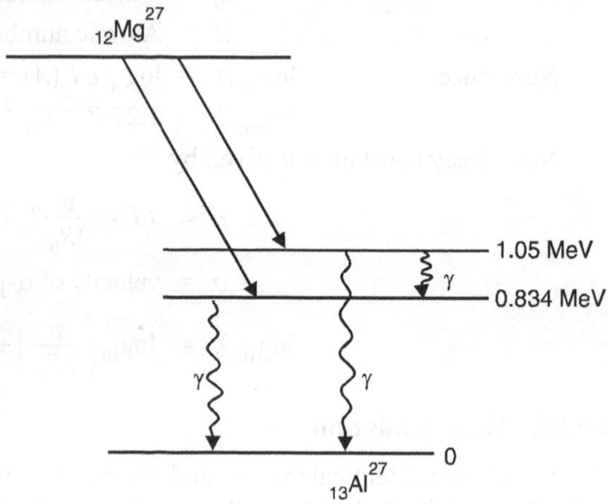

Fig. 10.14.

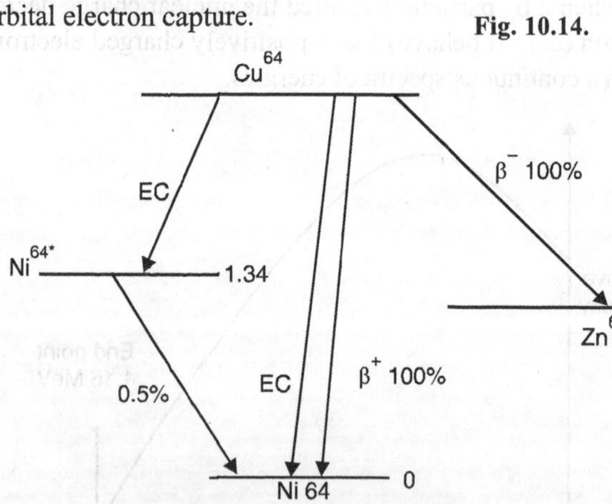

Fig.10.15. Ec = Electron Capture.

Several hundred decay schemes have n7ow been determined, varying in complexity.

Example 10.1. *Calculate the binding energy per nucleon of helium assuming its atomic weight = 4.00260 amu. Assume mass of proton = 1.007895 amu and mass of neutron = 1.008665 amu.*

[*I.A.S.*]

Solution. Mass of 2P + 2N

$$= 2(1.007895 + 1.008665)$$
$$= 4.03312 \text{ amu}$$

$$\text{Atomic wt. of helium} = 4.00260 \text{ amu}$$

Difference of mass

$$\Delta m = 4.03312 - 4.00260$$
$$\Delta m = 0.03052 \text{ amu}$$
$$1 \text{ amu} = 931 \text{ MeV}$$
$$\text{Binding energy} = 0.03052 \times 931$$
$$= 28.414 \text{ MeV}$$

Binding energy per nucleon

$$= \frac{28.414}{4}$$
$$= \mathbf{7.1035\ MeV.}$$

Example 10.2. *Find the amount of energy released when all the nuclei in 1 kg of deuterium fuse by the following reaction.*

$$6\left({}_1\text{H}^2\right) \rightarrow 2\left({}_2\text{He}^4\right) + 2p + 2n + 43 \text{ MeV}$$ **[Osmania, 1992]**

Solution. 2 kg of deuterium contains 6.023×10^{26} atoms

1 kg of deuterium contains $\left(\frac{6.023 \times 10^{26}}{2}\right)$ atoms

Therefore, energy released due to 1 kg of deuterium by fusion is given by

$$U = \left(\frac{43}{6}\right)\left(\frac{6.023 \times 10^{26}}{2}\right) \text{MeV}$$
$$U = \mathbf{2.158 \times 10^{27}\ MeV}$$
$$U = 2.158 \times 10^{27} \times 10^{6} \times 10^{-19} \text{ J}$$
$$U = \mathbf{3.453 \times 10^{14}\ J}$$

Example 10.3. *Calculate the mass of deuterium nucleus if the binding energy per nucleon is 1 MeV.* **[Kanpur, 1990]**

Solution. In deuterium, there is one proton and one proton.

$$1 \text{ MeV} = \frac{1}{931} \text{ amu}$$

$$\text{Binding energy of two nucleons} = \frac{2}{931} = 0.00215 \text{ amu}$$

Mass of deuterium nucleus

$$= \text{mass of proton} + \text{mass of neutron} - \text{binding energy}$$
$$= 1.00758 + 1.00898 - 0.00215$$
$$= 2.01441 \text{ amu}$$
$$= 2.01441 \times 1.66 \times 10^{-27} \text{ kg}$$
$$= \mathbf{3.334 \times 10^{-27}\ kg.}$$

Example 10.4. *Calculate the binding energy per nucleon of* ${}_{28}Ni^{64} = 63.9280$ *amu.*

$$\textit{Mass of Hydrogen atom } (M_H) = 1.007825 \textit{ amu}$$

$$\text{mass of neutron } (M_n) = 1.008665 \text{ amu}$$

Solution. mass of electron $= 0.00055$ *amu* **[Kanpur, 1990]**

$$M_H = \text{mass of proton} + \text{mass of electron}$$

$$\therefore \quad M_p = M_H - M_e = 1.007825 - 0.000550$$

$$M_p = 1.007275 \text{ amu}$$

$$\text{Mass of } _{28}\text{Ni}^{64} \text{ atom} = 63.9280 \text{ amu}$$

$$\text{Mass of } _{28}\text{Ni}^{64} \text{ nucleus} = 63.9280 - 28 \times (0.00055)$$

$$= 63.9280 - 0.01540$$

$$= 63.8126 \text{ amu}$$

$\therefore$ Difference of mass

$$m = (28\, M_p + 36\, M_n) - 63.8126$$

$$m = 28 \times 1.007275 + 36 \times 1.008665 - 63.8126$$

$$m = 64.51564 - 63.8126$$

$$m = 0.70304 \text{ amu}$$

But 1 amu $= 931$ MeV

$$\text{Total Binding Energy} = 0.70304 \times 931$$

$$= \mathbf{654.53\ MeV}$$

There are 64 nucleons,

$\therefore$ Binding energy per nucleon

$$= \frac{654.53}{64}$$

$$= 10.227 \text{ MeV.}$$

Example 10.5. *Estimate the energy released when two deuetron nuclei fuse together to form Helium nucleus, given that binding energies per nucleon of H_2 and He_4 are 1.1 and 7 MeV respectively.* **[Osmania, 1992]**

Solution. $_1H^2 + {_1H^2} = {_2He^4} + E$

Let $x = 1.1$ MeV

and $y = 7$ MeV

2[mass of one proton + mass of one neutron – x]

$$= [\text{mass of 2 proton} + \text{mass of 2 protons} - y] + E$$

$$\therefore \quad E = y - 2x$$

$$= 7 - 2 \times 1.1 = 7 - 2.2$$

$$= \mathbf{4.8\ MeV} = 4.8 \times 10^6 \times 1.6 \times 10^{-19} \text{ J}$$

$$= \mathbf{7.68 \times 10^{-13}\ J}$$

$$\text{Hence energy released} = 4.8 \text{ MeV}$$

$$= \mathbf{7.68 \times 10^{-13}\ J}$$

Example 10.6. *Calculate the binding energy and packing fraction for helium. The atomic masses of proton, neutron and helium are 1.00814 u, 1.00898 u, 4.00387 u and 1 u = 931* MeV. **[Mumbai, 1991]**

Solution. $m_p = 1.00814$ u

$$m_n = 1.00989 \text{ u}$$

$$m_{He} = 4.00387 \text{ u}$$

$$\text{Difference in mass} = 2[m_p + m_n] - m_{He}$$
$$m = 2[1.00814 + 1.00989] - 4.00387$$
$$m = \mathbf{0.03037\ u}$$
$$1\ \text{u} = 931\ \text{MeV}$$
(*i*) $$\text{Binding energy} = 0.03037 \times 931$$
$$= \mathbf{28.275\ MeV}$$

(*ii*) Mass defect,

$$\Delta m = M - A$$
$$M = \text{mass}$$
$$A = \text{mass number}$$

Packing fraction,

$$p = \frac{\Delta m}{A} = \frac{M - A}{A}$$

Here $M = 4.00387$ and $A = 4$

$$p = \frac{4.00387 - 4}{4}$$
$$p = \mathbf{9.675 \times 10^{-4}}$$

Example 10.7. *Estimate the mass of the Yukawa particle.*

Solution. If the range of the nuclear force is approximately $2R$ $= 2.4 \times 10^{-15}$ m. Using Heisenburg's uncertainity principle $\Delta E\ \Delta t \simeq \hbar$. If the particle has a speed of 10^8 m/s time taken is Δt.

$$\therefore \quad S = 10^8\ \Delta t = 2.4 \times 10^{-15}\ \text{m}$$
$$\therefore \quad \Delta t \simeq 2.4 \times 10^{-23}\ \text{sec}$$
$$\Delta E = \frac{4.14 \times 10^{-21}\ \text{MeV}}{2.4 \times 10^{-23}\text{S}} = 173\ \text{MeV}$$

Mass of particle is approximately

$$m = \left(\frac{173}{0.51}\right) me = 340\ me.$$

Example 10.8. *Find the maximum height of the potential barrier for a-penetration through ^{238}U nucleus. The radius of the residual nucleus is 9.3×10^{-13} cm.*

Solution. For $$r = R \quad U = B = \frac{1}{4\pi\epsilon_0}\frac{Zze^2}{R}$$
$$= \frac{92 \times 2 \times (1.6 \times 10^{-19})^2}{4 \times 3.14 \times 8.85 \times 10^{-12} \times 9.3 \times 10^{-15} \times 1.6 \times 10^{-13}}\ \text{MeV}$$
$$\simeq 28\text{MeV}.$$

EXERCISE X

1. Give some basic properties relating to the Nuclear size, Spin, Nuclear Magnetic Moment and Electric Quadrupole Moment.
2. What do you understand by atomic mass unit and the binding energy of the nucleus ? Derive the relation between atomic mass unit and electron volt.
3. Define Mass defect, packing fraction and binding energy.

4. Sketch the curve of the binding energy per nucleon versus mass number. Quantitatively show that there is a possibility of release of energy on fusion of light elements and fission of heavy elements.
5. Give an account of *n-p* and *p-p* scattering.
6. Give the importance of Deutron Binding energy.
7. Explain the features of the Nuclear force.
8. Explain the Yukawa theory for the Nuclear force.
9. What is Geiger Nuttal law ? Explain Alpha spectra and fine structure.
10. Give the theory of α-decay and show that the α-particle takes away most of the energy in the disintegration.
11. Give the Gamow's theory of Alpha decay and obtain the relevant equation.
12. Give an account of Beta Emission and β-ray spectra.
13. Calculate the radius of the ^{227}Thorium nucleus. **[Ans.** 7.32×10^{-15} m]
14. Give the neutrino theory of α-dacay.

11

NUCLEAR REACTIONS

11.1. Nuclear Reactions

In general, a nuclear reaction can be represented by an equation of the form :

$$ {}^{A}_{Z}X + x \longrightarrow {}^{A}_{Z}Y + y \text{ or simply as } {}^{A}X(x, y)^{A'}Y $$

X is the target nucleus which is bombarded by the projectile x. The resulting compound nucleus breaks -up to form a product nucleus Y and particle y almost immediately.

$X(x, x)X$ is an elastic scattering process. The incident particle is deflected while passing a target nucleus, this is scattering

$$ X + a \longrightarrow X^* + a $$

where * asterik indicates that the nucleus is in an excited state. Such a process is called an inelastic scattering. Here the residual nucleus is left in an excited state without any change in its internal structure. The incident particle goes out with lower kinetic energy.

In both elastic and inelastic collisions the incoming and outgoing particles are the same.

Formation of a Compound Nucleus

$$ X + x \longrightarrow Y + y $$

$$ \text{or } X(x, y)\,Y $$

Here, the incoming particle is captured forming a new **compound nucleus**.

The compound nucleus has considerable excess energy and decays with one or more particles.

Conservation laws in Nuclear Reactions

1. **Conservation of Mass number :** *i.e.*, sum of mass numbers of X and x must equal the sum of mass numbers of Y and y. *i.e.*, total of protons and neutrons in a nuclear reaction remains unchanged.

2. **Conservation of Atomic number :** Sum of Atomic numbers of X and x = sum of atomic numbers of Y and y.

The total numbers of protons taking part in a nuclear reaction remains unchanged after the reaction.

3. **Conservation of Energy :** Conservation of energy requires that the total energy *i.e.*, the sum of the rest-mass energy of all the nuclei and their kinetic energy must equal the sum of the rest mass energies and kinetic energies of the products.

4. **Conservation of linear momentum :** Let p_X, p_x, p_Y and p_y be the momentum of the different nuclei taking part in a reaction.

$\therefore$ Law of conservation of linear momentum requires :

$$ p_X + p_x = p_Y + p_y. $$

5. **Conservation of angular momentum :** Let I_X, I_x, I_y and I_y be the spins (total angular momentum) of the nuclei X, x, Y, y.

Then, $$ I_X + I_x + l_x = I_Y + I_y + l_y $$

where l_x and l_y are relative orbital angular momentum of X and x (initial) and of Y and y (final).

6. **Conservation of parity :** In case of strong interaction parity is conserved.

Parity before the reaction is equal to parity after the reaction.

7. **Conservation of isotopic Spin :** Let T_X, T_x, T_Y and T_y be the Isotopic spins of nuclei.

Conservation requires that $\quad T_X + T_x = T_Y + T_y$.

11.2. Q-value of a Nuclear Reaction

The quantity Q is known as the energy balance of a Nuclear reaction. The energy absorbed or released in a nuclear reaction is called the Q-value.

For the Nuclear reaction

$$X\,(x, y)Y$$

i.e.,

$$X + x = Y + y$$

If E_X, E_Y, E_x and E_y be the kinetic energies of X, Y, x and y respectively and if X is initially at rest.

Conservation of Energy requires :

$$M_Xc^2 + (E_x + m_xc^2) = (E_Y + m_Yc^2) + (E_y + m_yc^2)$$

$$\therefore \quad Q = (E_Y + E_y) - (E_X + E_x) = [(M_X + m_x) - (M_Y + m_y)]c^2$$

If Ex = o *i.e.*, target nucleus is at rest

then $\quad Q = (E_Y + E_y) - E_X = [(m_x + M_X) - (m_y + M_Y)]c^2 \qquad \text{...(11.1)}$

Now there are two cases :

1. **Exoergic reaction Q > 0**

If $(M_X + m_x) > (M_Y + m_y)$

i.e., $(E_Y + E_y) > E_x$ $\qquad$ Exothermic

There is a net decrease in mass. Using $E = mc^2$ this energy is converted to the kinetic energy of the product particles.

2. **Endoergic or Endothermic Reaction Q < 0**

$(M_X + m_x) < (M_Y + m_y)$

or $E_Y + E_y < E_x$.

Energy has to be supplied from outside for the reaction to occur called **Threshold energy**.

3. $Q = 0$

gives an Elastic collision. Since no loss of energy only their is change in direction.

$$(M_X + m_x) = (M_Y + m_y)$$

11.3. Nuclear Reaction Kinematics

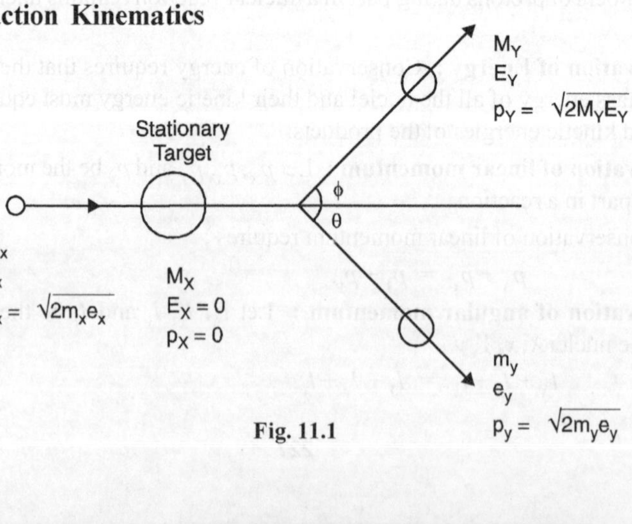

Fig. 11.1

Consider a nuclear reaction in which a bombarding particle strikes a target nucleus at rest. After the collision the product nucleus goes in the direction ϕ and the outgoing particle goes in the direction θ with respect to the incident particle.

Let m_x, M_X, m_y, M_Y be the mass of the perfectly, target nucleus, outgoing particle and the product nucleus respectively. Similarly, let E_x, E_X, E_y, E_Y denote then kinetic energies and $p_x = \sqrt{2m_xE_x}$, $p_X = 0; p_y = \sqrt{2m_yE_y}$ and $p_Y = \sqrt{2m_YE_Y}$ their momentum.

By conservation of linear momenum we have

X-direction $p_y \cos\theta + p_y \cos\phi = p_x$ ...(11.2)

Y-direction $p_y \sin\theta - p_y \sin\phi = o$ ...(11.3)

Eliminate ϕ from (11.2) and (11.3)

Squaring and adding

$$\therefore \quad (p_y \cos\theta - p_x)^2 + (py \sin\theta)^2 = p_Y^2$$

$$p_x^2 + p_y^2 - 2p_xp_y \cos\theta = p_Y^2$$

$$\therefore \quad 2m_xE_x + 2m_yE_y - 2\sqrt{(m_xE_x)(m_yE_y)} \cos\theta = 2M_YE_Y$$

$$\therefore \quad E_Y = \frac{m_x}{M_Y}E_x + \frac{m_y}{M_Y}E_y - \frac{2}{M_Y}\sqrt{m_xm_yE_x\,E_y \cos\theta} \quad ...(11.4)$$

Q-value of the reaction is

$$Q = (E_Y + E_y)\text{-}(E_X - E_x) \qquad \therefore E_X = 0$$

$$= E_Y + E_y - E_x \quad ...(11.5)$$

We have

$$Q = \left[\frac{m_x - M_Y}{M_Y}\right]E_x + \left(\frac{m_y + M_Y}{M_Y}\right)E_y - \frac{2}{M_Y}\sqrt{m_xm_yE_xE_y \cos\theta} \quad ...(11.6)$$

Writing as a quadratic Eq. in $\sqrt{E_y}$ we have

$$E_y - \left[\frac{2\sqrt{m_xm_y\,E_x \cos\theta}}{m_x + M_Y}\right]\sqrt{E_y} - \frac{QM_Y - E_x(M_Y - m_x)}{m_{y+}M_Y} = 0$$

Let $x = \sqrt{E_y}$ $\quad\therefore$ Eq. is of form

$$ax^2 - 2bx - c = 0$$

$$E_y - 2_b\sqrt{E_y} - c = 0$$

where

$$a = 1,\; b = \left[\frac{\sqrt{m_xm_y\,E_x \cos\theta}}{m_x + M_Y}\right]$$

$$C = \frac{QM_Y - E_x(M_Y - m_x)}{m_y + M_Y}$$

Solution is

$$\sqrt{E_y} = x = \frac{-b \pm \sqrt{b^2 - 4ac}}{2a}$$

$$\therefore \quad \sqrt{E_y} = x = \frac{1}{M_Y + m_y}\left[\sqrt{m_xm_y\,E_x \cos\theta}\right] \pm$$

$$\left[m_xm_y\,E_x \cos^2\theta + (M_Y + m_y)\{QM_Y + E_x(M_Y - m_y)\}\right]^{\frac{1}{2}}$$

For threshold energy we have $b^2 - 4ac = 0$

$$\therefore \quad \frac{4}{M_Y^2}\left(m_x M_y\, E_{\min}\right)\cos^2\theta \;=\; 4\left[1+\frac{m_y}{M_Y}\right]\left\{-Q - E_{\min}\left(1-\frac{m_x}{M_Y}\right)\right\}$$

$$\text{This gives } E_{\min} = (-Q)\left[\frac{m_y + M_Y}{m_y + M_Y - m_x}\right] \quad ...(11.7)$$

($\therefore$ At $E_{\min}$ $\theta = 0, \phi = 0$) $E_{\min}$ is known as Threshold energy.

$$(m_x + M_X) - (m_y + M_Y) = Q/c^2$$

$\therefore$ Q/c^2 is much less than the mass of the particles

we have $\quad m_x + M_X = m_y + M_Y$

$\therefore$ Substitute $\quad m_x = m_y + M_Y - m_x$

$$\therefore \quad E_{\min} = (-Q)\left[\frac{m_x + M_X}{M_X}\right]$$

$$E_{\text{Thres}} = -Q\left[1+\frac{m_x}{M_X}\right] \quad ...(11.8)$$

11.4. Channels

Consider a binary reaction in which 'a' particle is incident on a target nucleus A leading to final products b and B.

$$a + A \longrightarrow B + b \text{ or } A\,(a, b)\,B$$

where $(a + A)$ is called the initial (entrance) channel and $(b + B)$ is the final exit channel.

11.5. Compound Nucleus

The characteristic feature of a Nuclear Reaction that involves the formation of a compound nucleus is that the decay of the compound nucleus does not depend on the manner in which it was formed.

$$^{60}\text{Ni} + 4\text{He} \longrightarrow {}^{64}\text{Zn}^* \longrightarrow {}^{63}\text{Zn} + \text{n} + \text{Q}.$$

Compound nucleus

$$^{64}\text{Zn}^* \longrightarrow {}^{62}\text{Cu} + \text{p} + \text{n} + \text{Q}$$

$$^{64}\text{Zn}^* \longrightarrow {}^{62}\text{Zn} + 2\text{n} + \text{Q}.$$

The intermediate nucleus can also be formed by bombarding ^{63}Cu with protons.

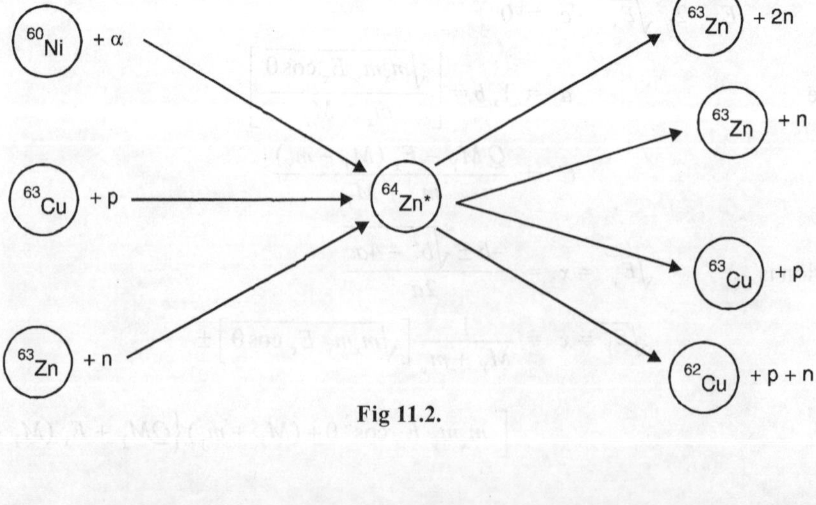

Fig 11.2.

These several decay modes of $^{64}Zn^*$ must have probability independent of the method by which the compound nucleus was formed, provided the nucleus is in the same excited state.

Particles in the initial channel coalesce to form a compound system that stays together long enough for statistical equilibrium to be reached (10^{-16} to 10^{-18} secs). Decay occurs after some time, by chance a nucleon or a number of nucleons acquire enough energy to escape. In general, there may be many possible ways the compound nucleus decay and the final channel is not unique.

If a proton enters the nucleus it will probably collide with nucleon in the nucleus after travelling a short distance. One of the nucleons will be reflected back into the nucleus by the change in nuclear potential (like total reflection). This nucleon will collide with another nucleon, each gives rise to further collisions leading to a cascade. Therefore, many nucleons share the energy in the compound nucleus. At this time no nucleon has enough energy to escape ~ 8 MeV Binding energy. After a time, (10^{-15} secs) a fluctuation in the energy sharing makes energetically possible the escape of a nucleon. Eventually, several nucleons are "evaporated" and their binding energy are largely responsible for removing most of the excitation energy of the compound nucleus. They will almost always be neutrons, since the coulomb barrier acts to retain the protons. When the excitation energy is below the neutron Binding energy, the relatively slow process of γ-decay takes over and allows the system to land up in its ground state.

The lifetime of an excited state is frequently expressed in terms of its width. According to energy time uncertainty Principle, if an average nucleus survives in an excited state only for the life time τ of the state, then its energy in the state can be specified only within an energy range Γ, given approximately by,

$$\Gamma = \frac{\hbar}{\tau}$$

Γ is the full width at half maximum.

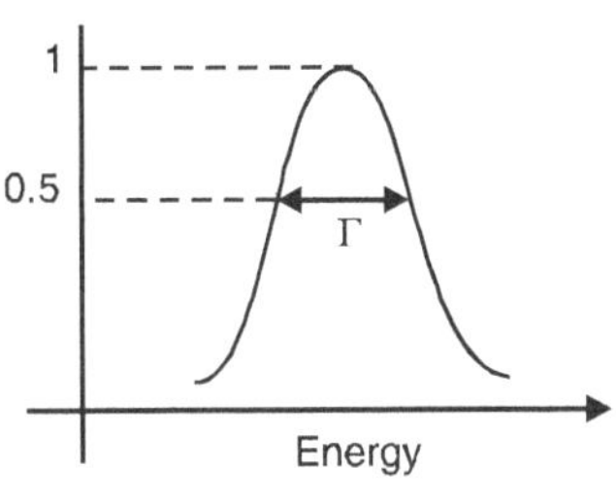

Fig. 11.3

Since it takes a compound nucleus $\tau \sim 10^{-15}$ seconds to decay by neutron emission the width τ of a typical one of its states is

$$\Gamma = \frac{\hbar}{\tau} \sim 10^{-15} \text{ eV sec}/10^{-15} \text{ sec} = 1 \text{ eV}.$$

These narrow states can be observed by measuring as a function of the nucleon energy the probability, or total cross-section to form a compound nucleus.

11.6. Direct Reactions

A direct nuclear reaction occurs when the projectile interacts primarily in the surface of the target nucleus. Energy and material transfer are small and therefore the masses and energies of the outgoing products are related to those of the initial particles.

The time scale of the reaction is rapid (~ 10^{-22} sec) given by the time it takes the projectile to cross the nuclear diameter, and since the projectile–like outgoing particle retains most of the momentum of the initial channel ($a + A$), its angular distribution is generally peaked in the forward direction. Direct Reactions are used to excite and identify nuclear states. They take place at higher energies.

The characteristics of Direct reactions are :

(1) Large number of high energy particles are emitted.

(2) Angular distribution of emitted particle shows a forward peak.

(3) No Resonance.

Simple direct Nuclear Reactions occur when the impact parameter is such that the nuclei brush past each other with little contact.

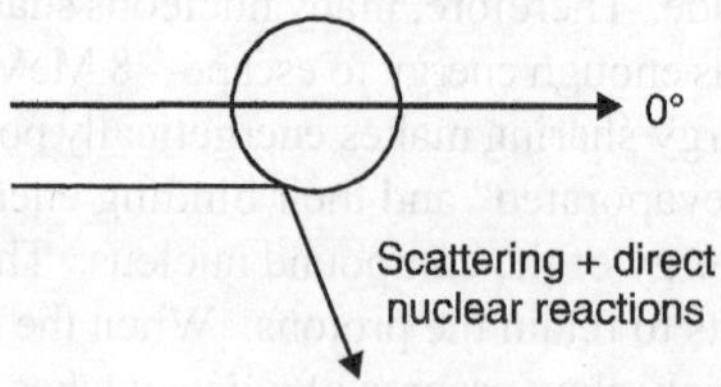

Fig. 11.4.

These reactions do not perturb the path a great deal, and the angular distribution of the outgoing reaction products exhibits a bell-shaped form which peaks at angle θ_g corresponding to grazing collision for $\theta < \theta_g$.

In inelastic scattering the initial and final products are the same. Some of the initial kinetic energy is used to raise the final particles to the excited state (either one or both) which goes to ground state by remission.

A **transfer** reaction is one in which more nucleons are exchanged as the target and projectile pass one another. The nuclei may not be in an excited state. If the transfer of nucleons is from projectile to target it is called a **stripping reaction.** A **Pick up** reaction is the reverse.

Example 1. *Calculate the Q value of the reaction* $^2{}_1H\,(d,n)_2{}^3He$.

Solution.

$$[2M(^2_1H) - M(^3_2He) - m_n]\ u$$
$$= [2 \times 2.0.141 - 3.01603 - 1.008665]u$$
$$= [4.0282 - 3.01603 - 1.008665]u$$
$$= .0035\ u$$

$$\therefore \qquad 1u = 931.5\ \text{MeV}$$
$$Q = 0.00354 \times 931.5 = 3.26\ \text{MeV}.$$

Example 2. *What is the threshold energy for the reaction* $^{13}C\ (n, a)^{10}\ Be$?

Solution. $[13.003355 + 1.008665 - 10.013534 - 4.002603]\ u$

$$= -.004117$$
$$= -3.835\ \text{MeV}$$

–ve sign shows that the reaction is endothermics. The threshold energy in the laboratory is

$$E_{TH} = -Q\left[1 + \frac{m_x}{M_x}\right]$$
$$= +3.835\ [1 + 1/13] = 4.13\ \text{MeV}.$$

Example 3. *Calculate the width Γ for γ-ray emission whose lifetime is* 10^{-9} *secs. What is the width for* $\tau = 10^{-12}$ *sec ?*

Solution. $\tau = \dfrac{\hbar}{\Gamma}$ Γ is in energy units.

$$\tau = \frac{\hbar}{\Gamma} \sim \frac{10^{-15}\text{eV-sec}}{10^{-9}} = 10^{-6}\text{ eV}$$

$$\tau = \frac{\hbar}{\Gamma} = \frac{10^{-15}\text{eV.sec}}{10^{-12}} = \mathbf{10^{-3}eV.}$$

EXERCISE XI

1. What is Q-value of a Nuclear Reaction ? Obtain expressions for exoergic and Endoergic Reactions.
2. Using the conservation of energy and momentum obtain an expression for the threshold energy of a reaction in terms of Q-value.
3. Describe the formation and decay of a compound nucleus. What do you understand by 'channels' ?
4. What are direct reactions ? Explain.
5. The nuclide Li^7 has excited states at

   (*i*) 6.54 MeV with an estimated total width Γ of 1 MeV,

   (*ii*) 7.47 MeV with an estimated total width of 0.14 MeV,

   Find the lifetimes of these states.
6. Calculate the Q-value of the reaction ${}^{10}_{5}B\ (n, \alpha)\ {}^{7}_{3}Li$

   B = 10.0129 *u*

   Li = 7.0160 *u*
7. Calculate the Q-value of the reaction

   ${}^{15}_{8}O\ (d, p)\ {}^{16}_{8}O$

   O^{15} = 15.003064 *u*

   O^{16} = 15.99491 *u*
8. Alpha particles have a kinetic energy of 4 MeV. What is the Q-value of the reaction in the lab system ?

   ${}^{9}_{4}Be\ (\alpha, u){}^{12}_{6}C$

12

NUCLEAR MODELS

12.1. The Liquid Drop Model

The liquid drop model is one in which the nucleus is regarded as a collection of neutrons and protons forming a droplet of incompressible fluid which behaves in some ways like a classical liquid drop. The liquid drop model could account very well for the systematic behaviour of Nuclear Binding energy with mass number.

The liquid drop model is based on two properties :

1. The interior mass densities are approximately the same.
2. Their total Binding energies are approximately proportional to their masses *i.e.*, $\Delta E/A$ = constant (expect for very small A).

For classical drop of various sizes

(1) The interior densities are the same and (2) their heats of vaporization are proportional to their masses. Heat of vaporization is the energy required to disperse the drop and is comparable to the binding energy of the nucleus.

The liquid drop approximates the nucleus to a sphere which has uniform interior density which drops to zero abruptly at the surface.

The mass of the nucleus is not exactly equal to the total mass of the nucleons. The difference is due to the Binding energy of the nucleus.

$$\therefore \quad {}_Z^A\text{M} = ZM_p + (A-Z)M_n - B/c^2$$

According to the liquid Drop Model the total binding energy is due to five effects :

1. Volume : The mass dependence of this term follows directly from the saturation of nuclear forces (described earlier). A nucleon in the nucleus interacts, on an average, with a fixed number of neighbouring nucleons within the short range of the nuclear force. The Binding energy per nucleon will be constant and total contribution to binding energy is proportional to *A*.

$\therefore$ Proportional to nuclear volume.

If a_v is the binding energy per nucleon, then the contribution to the total Binding energy is

$$E_v = a_v A.$$

2. Surface Correction : A nucleon near the surface interacts with fewer nucleons, and is therefore less tightly bound then if it is in the interior. Thus nucleons on the surface have a lesser Binding Energy. The number of such nucleons depends on surface area of the nucleus, which is proportional to R^2 (radius). Thus a term proportional to the surface area (or square of R) must be **subtracted** from the volume term to correct for this and since $R \propto A^{1/3}$ the surface term $\propto A^{2/3}$.

$$\therefore \quad E_s = -a_s A^{2/3}$$

–ve sign shows that the total binding energy gets reduced.

3. Coulomb energy : The protons inside the nucleus, because of their charge also interact with the coulomb force. The electric potential energy of a uniformly charged sphere is $\frac{3}{5}\frac{Q^2}{4\pi\epsilon_0 R}$ where Q = total charge R = radius of the sphere.

Since the nuclear charge is Ze, we get

$$E_c \propto \frac{Z^2e^2}{R} \propto \frac{Z^2e^2}{R_0A^{1/3}} \qquad (\because R = R_0A^{1/3})$$

$$\therefore \qquad E_c = \frac{-a_cZ^2}{A^{\frac{1}{3}}}$$

–ve sign showing energy is reduced.

4. Symmetric term : This term expresses the charge symmetric nature of the nucleon- nucleon forces. In the absence of the coulomb force the most stable nuclei must have equal number of neutrons and protons. This arises from Pauli's Principle. There are two contributions to the symmetry term : – One from the nuclear potential energy and the other from their Kinetic Energy.

The gradual filling of the lowest energy levels favour an equal number of protons and neutrons. The nuclei containing equal number of neutrons and protons are most stable *i.e.*, highest stability of nuclei is for

$$Z = N = A - Z \text{ or } 2Z = A.$$

$\frac{N-Z}{A}$ or $\frac{A-2Z}{A}$ is called the neutron excess ratio.

Neutron excess or proton decrease nuclear stability *i.e.*, Asymmetric Energy term is –ve and proportional to $\frac{(A-2Z)^2}{A}$.

$$\therefore \qquad E_c = -a_a\frac{(A-2Z)^2}{A}$$

$$a_a = \text{constant.}$$

5. Pairing term E_p : The interaction of the nucleons depends on the relative orientation. Binding energy is maximum for even-even nucleon (favoured) and is minimum. The pairing energy is proportional to $A^{-3/4}$.

$$E_p = a_pA^{-3/4} \text{ for} \qquad \text{if } Z \text{ odd, } A - Z = N \text{ odd}$$

$$= 0 \qquad \text{if } Z \text{ even, } A - Z = N \text{ odd}$$

$$\qquad \text{if } Z \text{ odd } A - Z = N \text{ even}$$

$$= -a_pA^{-3/4} \text{ for} \qquad \text{if } Z \text{ even } A - Z = N \text{ even.}$$

i.e. + ve for even – even, –ve for odd – odd, and zero for odd-even.

$$\therefore \qquad B = E_v + E_s + E_c + E_a + E_p \quad \text{(Refer Fig. 12.2)}$$

$$B = a_vA - a_sA^{2/3} - a_cZ^2A^{-1/3} - a_a\frac{(A-2Z)^2}{A} \pm a_pA^{-3/4}$$

Note

$a_s \sim 1.2\,9\,v$

$$a_v = 15.75 \text{ Mev}$$

$$a_s = 17.8 \text{ MeV}$$

$$a_c = 0.711 \text{ MeV}$$

$$a_a = 23.7 \text{ MeV}$$

$$a_p = 11.18 \text{ MeV.}$$

In terms of masses

$$ZM^A = 1.007825Z + 1.008665\,(A-Z) - a_vA + a_sA^{2/3} + a_cZ^2A^{-1/3} + a_a(Z-A/2)^2A^{-1} + \left(\begin{smallmatrix}-1\\0\end{smallmatrix}\right) a_pA^{-1/2}$$

(in unit u)

$a_v = 0.01691$ $a_s = 0.01911$ $a_c = 0.00763$

$a_a = 0.10175$ $a_p = 0.012$

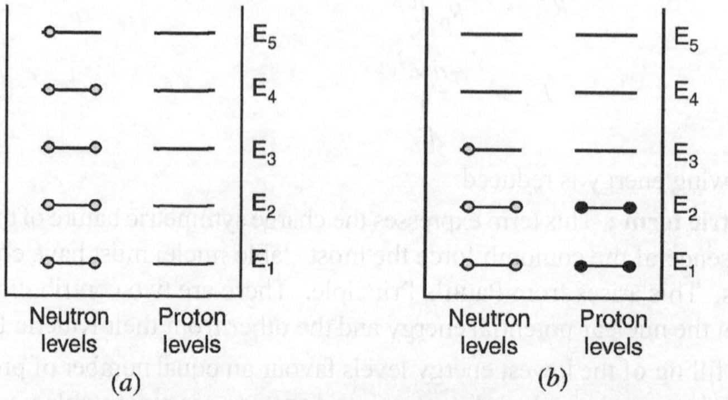

Fig. 12.1

For 9 particles (neutrons) eight occupy the lowest four levels and the 9th goes into energy level E_5. If the nine particles four are protons and five neutrons, only the lowest two levels are fully occupied , the 5th neutron goes into level E_3.

The Weizsaecker formula is

$$BE = a_v A - a_s A^{2/3} - a_c Z^2 A^{-1/3} - a_a (A - 2Z)^2 A^{-1} + a_p A^{-1/2}$$

This is also referred to as Weizsaeckers **Semiempirical mass formula.**

This explains why alpha particle nuclei are especially stable *i.e.*, have low ground-state energies. In three nuclei the total no. of protons and neutrons just fill the states of lowest energy. One more proton or neutron would have to go into a higher energy state. The fact that each energy can accommodate two fermious of opposite spin also explains why, in general nuclei with an even number of protons and even number of neutrons have a greater Binding energy per nucleon than either odd-even, even-odd or odd-odd nuclei.

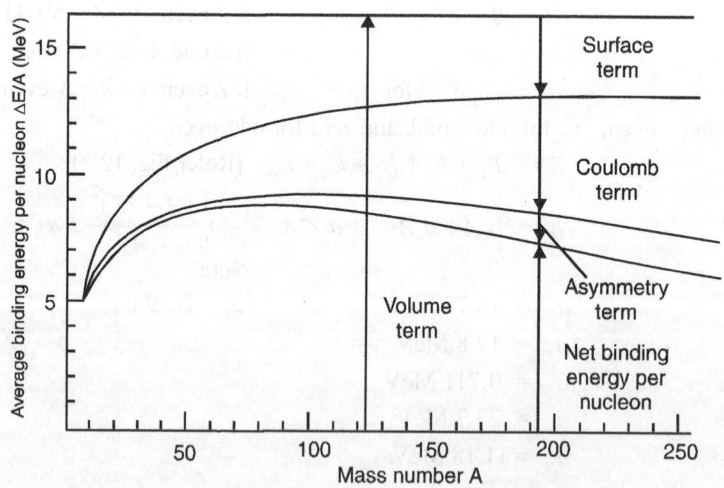

Fig. 12.2

Nuclear Stability Rules

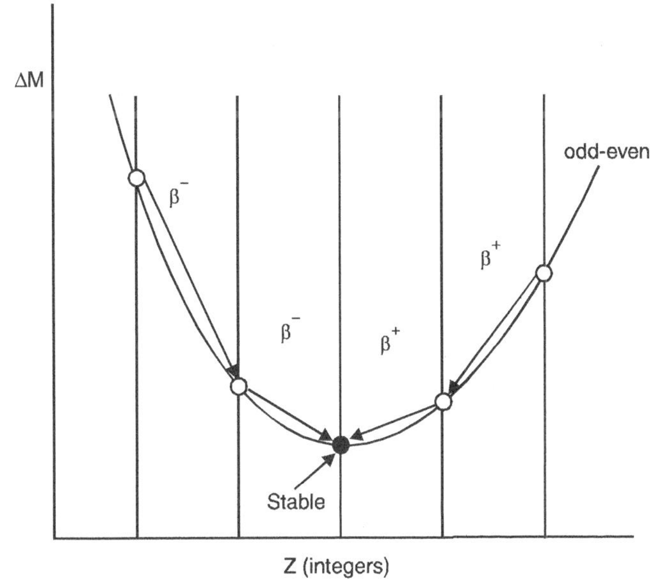

Fig. 12.3

We just consider nuclides with odd A, *i.e.*, those with an odd and an even no. of nucleons, where pairing effects can be neglected

Now by the Semi-Empirical Mass Formula

$$M = 1.007825\,Z + 1.008665\,(A-Z) - A\bar{B}$$

where
$$\bar{B} = 15.2 - 16.0\,A^{-1/3} - 0.667\,Z^2 A^{-4/3} - 22.2\,(A-2Z)^2 A^{-2}$$

We define a "mass excess" ΔM as the difference between the proper atomic mass (M) and mass of A neutrons (in MeV)

$$\therefore \quad \Delta M = (M - 1.008665\,A)c^2$$
$$= -0.78Z - 15.2A + 16.0\,A^{2/3} + 0.667Z^2 A^{-1/3} + 22.2\,(A-2Z)\,A^{-1}$$

In the problem of Beta stability we are concerned with isobaric nuclides, *i.e.* same A. Therefore, Variation with Z of M is required.

$$\therefore \quad \Delta M = \text{const} - 89.67\,Z + Z^2\left(\frac{88.8}{A} + \frac{0.667}{A^{1/3}}\right)$$

This is shown in (Fig.12.3) by a parabolic curve, the minimum occurring at predicted Z value for stability, *i.e.*, $\frac{dM}{dz} = 0$.

$$\therefore \quad Z = \frac{1.01\,A}{2 + 0.015\,A^{2/3}} \qquad \text{(Rule for } Z < 82)$$

Since Z can only take up an integral values, the most stable nuclei for a given A is that with Z closest to the theoretical value. **For even A** there are two groups : Odd-odd (rarely stable) and even-even (commonly stable). Due to pairing nucleons have opposite spins resulting in a net O^+ ground state for Even-even nuclei.

$$\therefore \quad \Delta M = \text{const} - 89.6\,Z + Z^2\left(\frac{88.8}{A} + \frac{0.667}{A^{1/3}}\right) \pm \delta$$

where δ = pairing term + for odd-od –ve even-even.

Plotting ΔM *vs* Z Fig.12.4, and Fig. 12.5 we get two parabolic curves separated by 2δ. Since Z values are alternatively odd and even, the points corresponding to successive nuclei alternate between the two curves. There may be more than one nuclei stable against β-decay in this situation, depending on the position of the theoretical minimum and the steepness of the curves. If minimum occurs for an odd z the curves are not too steep, the nuclei; (Z_0) can decay by either –ve electron or by β^+ emission to nucleides $Z_0 \pm 1$ Fig. 12.5. *Ex.* is Cu^{64}.

On the other hand, it is possible for only one nuclei to be stable as shown in Fig. 12.4 where the curve is very steep. The condition

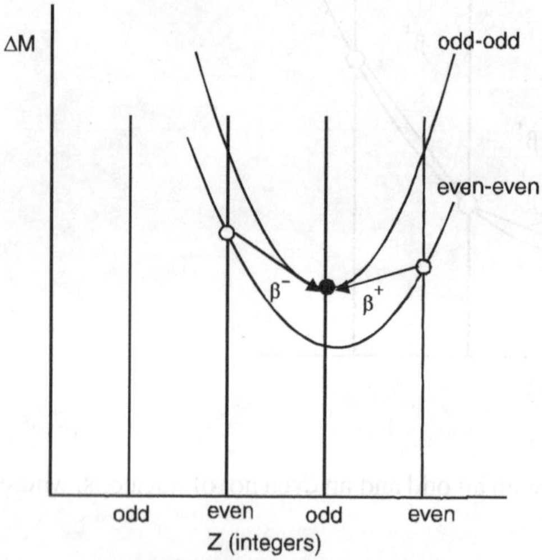

Fig 12.4.

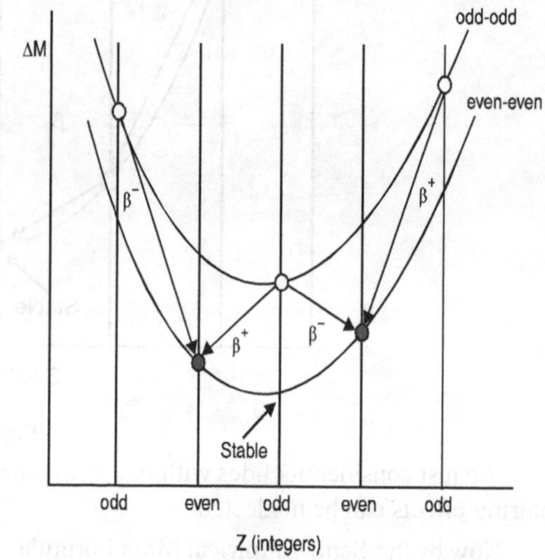

Fig. 12.5.

for this can be obtained if δ is known. Consider light nuclei in which minimum of curve occurs at $Z_0 = A/2$. Increase in ΔM for a change of one unit in Z is

ΔM = const + 22.2 $(A - 2Z)^2 A^{-1}$ approximately with $Z = Z_0 \pm 1$ we find

$$\Delta M = \Delta M_{\min} + \frac{88.8}{A} \text{ MeV}$$

If the increase in ΔM from the minimum exceeds the value of 2δ, it is possible for $Z_0 = A/2$ to be the sole stable (odd-odd) nuclei as shown.

12.2. Shell Model

Experimental studies show that nuclei for which either Z or N equal 2, 8, 20, 28, 82 and 126 are more stable. These are known as magic numbers.

Nuclei for which both Z and N are magic numbers are called doubly magic.

The magic numbers of neutrons and protons have been interpreted as forming closed shells in analogy with the filling of electron shells in atoms. The neutron and proton shells appear to be independent of one another. This is also known as the Independent Particle Model.

Firstly it is necessary to find the neutron and proton energy levels for a potential well of the nucleus. Each nucleon is treated as moving independently in a net nuclear potential well given by the formula :

$$V(r) = \frac{-V_0}{1 + \exp\left(\frac{r - R}{a}\right)} \qquad \text{where } V_0 = 50 \text{ MeV}$$

$R \simeq 1.25 \times A^{1/3}$ fm and $a = 0.524$ fm r = distance measured from the centre of the nucleus.

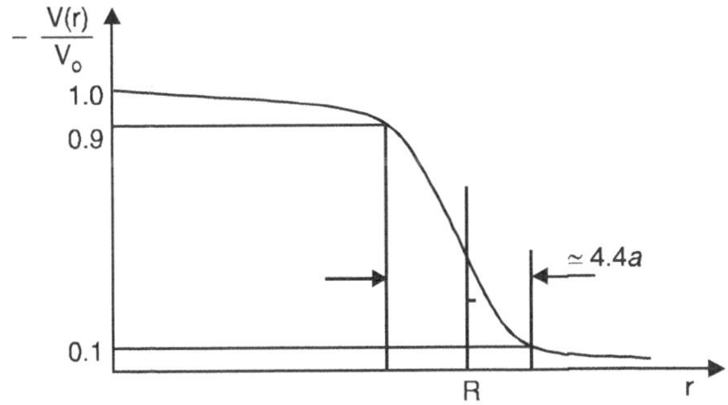

Fig. 12.6.

As in the case of Hartree theory of atoms it is found that the nucleon energy depends on the quantum numbers n and l. The nucleus has no fixed centre of charge unlike the atom and the nucleons are confined to their own orbits. The nucleons do not collide as per Pauli's Principle. Such collisions are allowed only when two nucleons occupy an empty level *i.e.*, energy more than the Fermi level. Since this energy is large only few transfers or collisions take place.

The notation 1s means $n = 1$ and $p = 0$ as usual. The states are labelled s, p, d, f, g for orbital angular momentum quantum number $l = 0, 1, 2, 3, 4$, etc. For each value of l, there are $2l + 1$ substates. The Exclusion Principle limits the number of neutrons or protons occupying a level to $2(2l + 1)$. A magic no. occurs when there is a large gap between the last filled level and the next unoccupied one.

Mayer and independently Jensen introduced the idea of nuclear spin-orbit interactions to account for the proper spacing of levels and the observed shell closings.

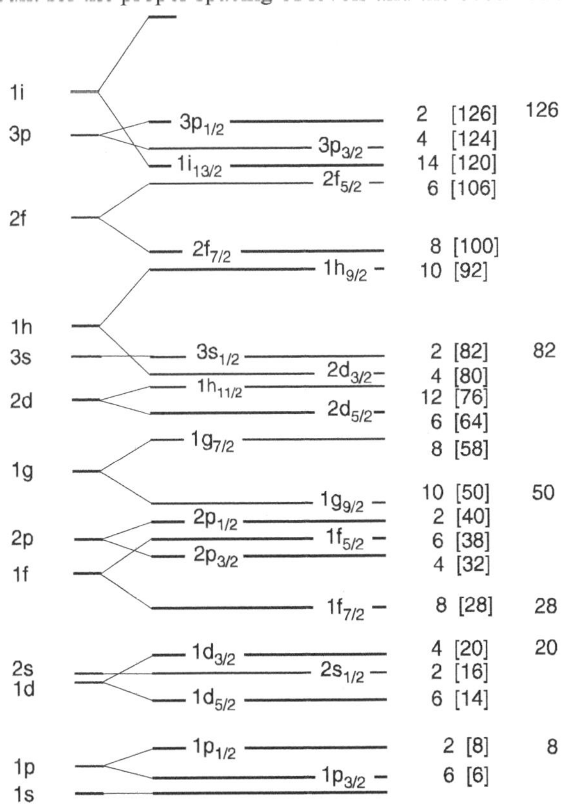

Energy levels of a nucleon in the nuclear potential well according to the shell model. At the left are the levels neglecting spin-orbit coupling. The numbers to the right of each spin-orbit split level gives the number of identical nucleons that level can accommodate. The total number of identical nucleons with energies equal to or less than a given level is shown in brackets. The relatively large separations $1d_{3/2} - 1f_{7/2}$, $1f_{7/2} - 2p_{3/2}$, $1g_{9/2} - 1g_{7/2}$–. $3S_{1/2} - 1h_{9/2}$, and $3p_{1/2} - 1i_{11/2}$ account for the stability of nuclei with a 'magic number of Z or N (*e.g.*, $_{82}$Pb).

Fig. 12.7

There is a strong inverted spin-orbit interaction due to $\vec{S}.\vec{L}$. Strong because the interaction energy is 20 times stronger than atomic spin orbit. Inverted energy because it decreases when $\vec{S}.\vec{L}$ is +ve and increases when $\vec{S}.\vec{L}$ is negative.

Levels are filled as in the case of electrons the radii increases in proportion to the cube root of the number of nucleons.

The total angular momentum has values

$j = l + ½$ and $j = l - ½$: Empirically it is found that the energy level with higher value of j lies below that with smaller j. Therefore the former gives a more tightly bound nucleonic states. the separation between two sub-levels for a given l is large and increases rapidly with $l \geq 4$ the sub-levels lie in different shells. In Fig. 12.7 the split energy levels, are shown. Each level has a capacity $(2j + 1)$. The ordering of l for all the unsplit levels of l are as :

$$s\,p\,d\,s\,f\,p\,g\,d\,s\,h\,f\,p\,i\,g.$$

The shell model does more than predict the magic numbers. It predicts the total angular momentum of all ground states of almost all nuclei.

Yukawa's Pion Field

The nuclear force-field does not extend to infinity but instead, is confined to an extremely small distance around the source. The exact nature of the nuclear force-field was not known earlier than 1935, when the Japanese physicist Hideki Yukawa, gave the new basic idea that paved the way for our present understanding of the nuclear force. He predicted the existence of a new elementary particle (now called a pi-meson) and hypothesized that the two nucleons experience an attractive force at small distances because of the (virtual) exchange between them of a pi-meson. The exchange takes place in a very short time-interval of about 10^{-23} to 10^{-24}S and is a typical of nuclear forces. Since the exchange takes place in a very short time, the uncertainty principle requires that no visible change in masses of the nucleons will be observed. This produces a rapid meson-exchange field between protons and neutrons in which meson is a quantum of nuclear force. [The situation, qualitatively, is analogous to two dogs grappling for control over a bone; one dog grabs the bone from the other and then, the second snatching it back again, the continual exchange of the bone results in each dog being pulled towards the other].

12.3. Nuclear Fission

In 1939, Strassmann and Hahn found, after a number of experiments that uranium ($Z = 92$) when bombarded by neutron was split up into lower elements barium and krypton. In the process, uranium, when bombarded with a neutron, is converted into unstable nucleus which splits up into two nuclei of medium atomic masses, barium $Z = 56$ and krypton $Z = 36$. This process in which a heavy nucleus splits up into two nuclei of nearly comparable masses is called *nuclear fission.*

It is also found that in the process of nuclear fission, a large amount of energy is released. This energy is produced because the original mass of the nucleus is greater than the sum of the products produced after fission and this mass excess appears as energy according to Einstein's equation $E = mc^2$. Bohr and Wheeler observed that slow neutrons are responsible for the fission of uranium $A = 235$ and $Z = 92$, whereas practically no fission was observed with uranium $A = 238$ and $Z = 92$. Neutrons with energies greater than 0.3 MeV are necessary for the fission of uranium $A = 238$. Consider the fission of uranium 235 by thermal neutrons. The reaction can be represented as :

$${}_{92}U^{235} + {}_{0}n^{1} \rightarrow {}_{92}U^{236} \rightarrow {}_{56}Ba^{141} + {}_{56}Kr^{92} + 3\,{}_{0}n^{1} + \text{energy}$$

$$\text{Mass of } {}_{92}U^{235} = 235.045733 \text{ amu}$$

$$\text{Mass of } {}_{0}n^{1} = 1.008665 \text{ amu}$$

$$
\begin{aligned}
\text{Total initial mass} &= 236.044398 \text{ amu} \\
\text{Mass of barium} &= 140.9177 \text{ amu} \\
\text{Mass of Krypton} &= 91.8854 \text{ amu} \\
\text{Mass of 3 neutrons} &= 3 \times 1.008665 \\
&= 3.025995 \text{ amu} \\
\text{Total final mass} &= 235.829095 \\
\text{Difference} &= 236.044398 - 235.829095 \\
&= 0.2153 \text{ amu} \\
\text{Since 1 amu} &= 931 \text{ MeV} \\
\text{Energy released} &= 0.2153 \times 931 \text{ MeV} \\
&= 200.4443 \text{ MeV}
\end{aligned}
$$

Thus, in the process of fission of one nucleus of uranium, about 200 MeV energy is released.

Energy released by 1 gram of uranium. Number of atoms in one gram atom of uranium = 6.025×10^{23}.

Number of atoms in one gram of uranium

$$= \frac{6.025 \times 10^{23}}{235}$$

Energy produced by 1 gm of uranium during fission

$$= \frac{6.025 \times 10^{23}}{235} \times 200 \text{ MeV}$$

$$\therefore \qquad E = 5.128 \times 10^{23} \text{ MeV}$$

$$1 \text{MeV} = 1.6 \times 10^{-13} \text{ Joule}$$

But, $$E = 5.128 \times 10^{23} \times 1.6 \times 10^{-13} \text{ joules}$$

$$1 \text{ kilowatt hour} = 10^3 \times 3600 \text{ joules}$$

$$E = \frac{5.128 \times 10^{23} \times 1.6 \times 10^{-13}}{10^3 \times 3600}$$

$$= 2.26 \times 10^4 \text{ kilowatt hour.}$$

Thus, the disintegration of 1 gram of uranium produces energy = 2.26×10^4 kWt. Due to this reason atomic energy is being used for the generation of electricity.

Fission occurs in the case of nuclei of large Z. Symmetric fission is relatively rare and occurs only once in about 20,000 or so fissions. Coulomb energy is minimized if the nucleus splits into two fission fragments that contain equal number of protons, but usually splitting is not completely symmetrical.

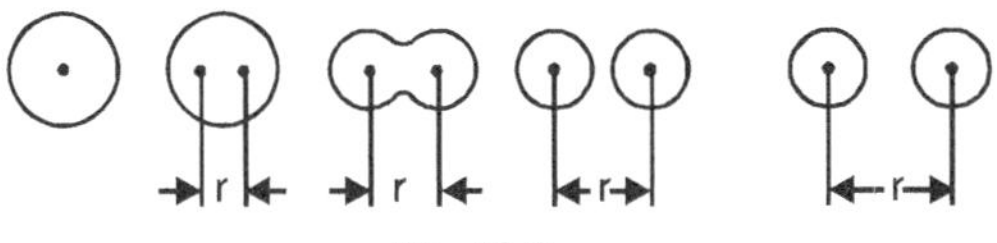

Fig.12.8

Representation of the steps involved in the process of nuclear fission.

Fig. 12.8 shows the elongation of the fissioning nucleus and separation of the two fission fragments.

Fig 12.9 shows the variation of $V(r)$ with distance r. For a very small r there is very little change in the Coulomb repulsive force but the surface area increases rapidly. Now, according to liquid drop model an increase in surface area increases the surface energy. Hence $V(r)$ increases with r. Beyond this as r increases the surface tension effect produces a narrow neck and the nucleus splits. After

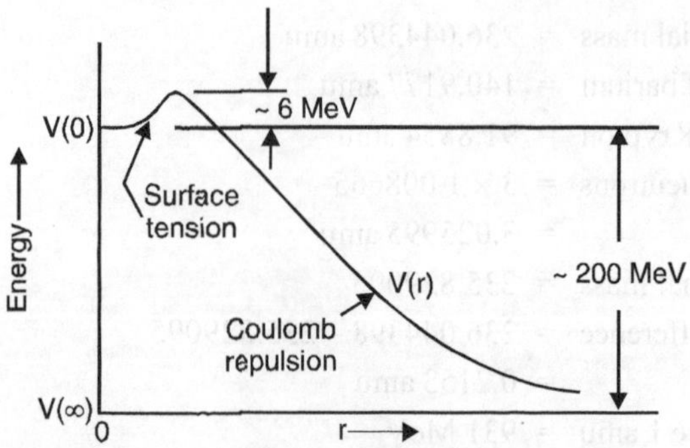

Fig. 12.9. An energy diagram for a fissionable nucleus

splitting surface energy no longer changes with r. $V(r)$ decreases following the Coulomb repulsion of the two fission fragments.

A typical nucleus of large Z has $V_{(r)\max}$ of 6 MeV above $V(0)$. $V(0) = 200$ MeV from $V(\infty)$. Thus we see that nuclei are normally stable to decay by fission as they are at energy $E = V(0)$, at the bottom of the potential $V(r)$. The probability of barrier penetration is very small.

The process which is important is by induced fission. The extra energy is obtained by the nucleus capturing a low-energy neutron. Since the binding energy of the last neutron in a nucleus of large Z is about 6 MeV, the nucleus receives enough energy to go to the top of the fission barrier. The incoming neutron gives it about 6.5 MeV, so that fission takes place even if the neutron brings no Kinetic energy.

The possibility of using fission to produce power in a chain reaction is because of the fact that two or three neutrons are emitted in each fission process.

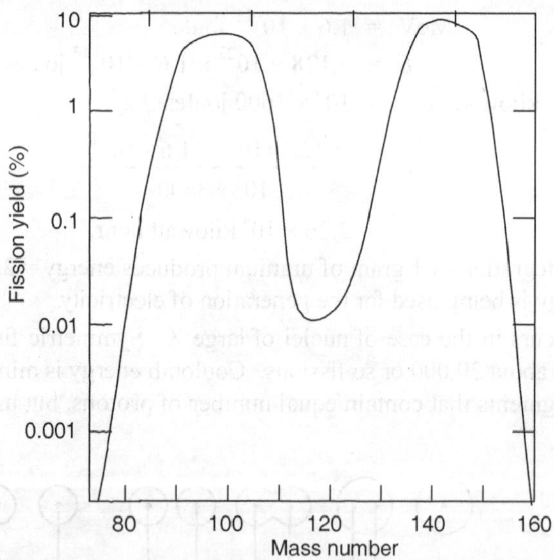

Fig. 12.10. shows the asymmetry of the fission fragments as function of A.

Fragment mass numbers vary between 70 and 160 with the most probable values being about 96 and 135. The fragments have a neutron to proton ratio of over 1.5 and thus decay to stable nuclei. These are mainly by a series of β^- emissions, although in some cases by neutron emission. The radioactive decay of fragments gives the second release of energy due to fission, though it is much less than prompt energy.

12.4. Fission by Photons

Baldwin and Koch in 1945 used high energy photons produced by the betatron (X-ray machine) to induce the fission of uranium. Uranium oxide coated on a cylinder was bombard by photons of energies ranging from 8 MeV to 16 MeV. The fission fragments were collected on a piece of paper. It was estimated that the photons of energy less than 7 MeV cannot disintegrate the nucleus and thus the threshold energy for the fission of uranium by photons is 7 MeV. Fission of the nuclei of thorium and transuranic elements has been produced by high energy photons.

12.5. Chain Reaction

The tremendous amount of energy released during nuclear fission and the release of 2.5 neutrons on the average has made it possible for the fission process to continue. Neutral uranium consists of 99.28% of uranium 238 and 0.72% of uranium 235. As most of the mass of neutral uranium consists of U 238 the neutrons releases during nuclear fission will try to bombarded the nuclei of U 238 mostly and very few will bombard U 235. It has been found that very few neutrons can cause fission of U 238 but neutrons of all possible energies can cause fission of U 235. Thus chain reaction is not possible in natural uranium.

A chain reaction is possible, if uranium 238 is removed and only uranium 235 or Pu 239 is used. The chain reaction once started continues and a tremendous amount of energy is released because of the fission of all nuclei. In bombs, U 235 and Pu 239 have been used and in fast reactors also, U 235 and Pu 239 used as fuels. In the early reactors, moderators are avoided.

Suppose, a fast neutron bombard 1S U 235. It breaks up into Ba, kr, three neutrons and a large amount of energy is released (Fig. 12.11). These three neutrons will produce the fission of three nuclei of U 235 and the same process is repeated. In this way, the reaction once started, continues and a tremendous amount of energy is released. To control the chain reaction moderators are used. The common moderators in use are (1) graphite, (2) water, (3) heavy water, and (4) beryllium or beryllium oxide. A controlled chain reaction can be stopped whenever required.

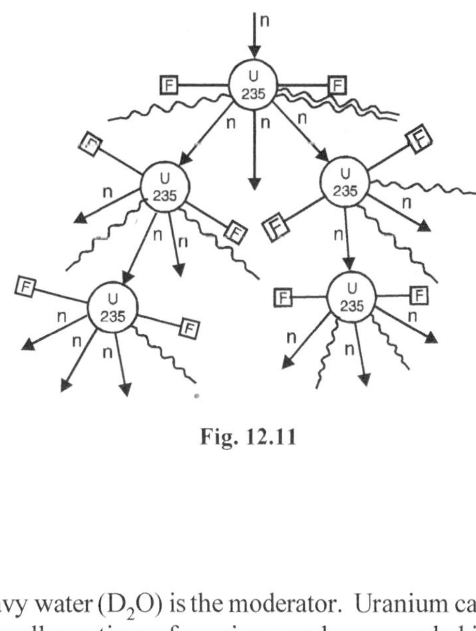

Fig. 12.11

12.6. Nuclear Reactors

There are two types of nuclear reactors :

1. **Homogeneous reactors :** In this reactor, heavy water (D_2O) is the moderator. Uranium can be in the form of a solution of Uranyl sulphate very small practices of uranium can be suspended in D_2O.

2. **Heterogeneous reactors :** In this reactor, graphite is the moderator. Rods of uranium are distributed through graphite in a regular way, forming a lattice.

Nuclear reactors can be classified according to the characteristics of the chain-reacting system. The following are the essential features of the nuclear reactors.

1. The Fuel : The common fuels are :

(*a*) Natural uranium, containing 99.28% of U238 and 0.72% of U235

(*b*) U 235

(*c*) Pu 239

(*d*) U 233

2. **Neutrons :** Neutrons used are :
   (*a*) Low energy or thermal neutrons
   (*b*) Intermediate energy neutrons
   (*c*) High energy neutrons
3. **Moderator :** The common moderators are :
   (*a*) Heavy water D_2O
   (*b*) Graphite
   (*c*) Beryllium oxide or beryllium
   (*d*) Water
4. **Coolant :** The energy produced in the reactor must be removed by circulating a coolant through the reactor. These are :
   (*a*) Water or other suitable liquids
   (*b*) Air, CO_2 or He
   (*c*) Liquid metal
5. **Nature of Assembly :**
   (*a*) Heterogeneous
   (*b*) Homogeneous
6. **Control Rods :** These are made of *cadmium.*

The nuclear reactor at Trombay in India is a heavy water uranium heterogeneous reactor. The nuclear reactor at Chalk-River, Canada, is a natural uranium, thermal neutrons, heterogeneous, heavy water moderated, ordinary water coolant reactor. In most successful reactors, liquid metal is used as the coolant, which removes the energy produced in a short time. Cadmium rods are used in the reactors to control the nuclear reactions. In power reactors the coolant, through a head exchanger, supplies steam to operate the turbines and electrical generators.

Breeder reactors have been built up in Britain ad Scotland. In this reactor, uranium 238 is converted into fissionable material plutonium 239. Uranium 235 is fissionable and plutonium is also fissionable. Plutonium is produced by the resonance capture of neutrons by uranium 238. A large scale breeder reactor in Scotland has a liquid metal of potassium-sodium alloy, as the coolant. In such reactors more of the fissionable material would be produced than consumed.

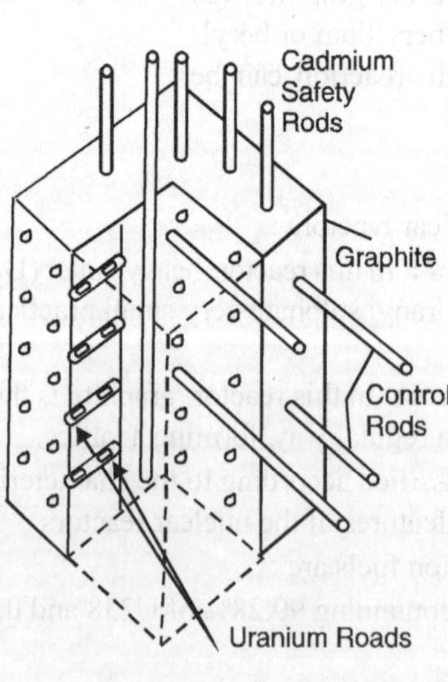

Fig. 12.12

A typical type of a nuclear reactor is shown in Fig. 12.12. The uranium rods 10cm in length and

2.75 cm in diameter are inserted in an air tight aluminium cylinder. This prevents the oxidation of uranium. These rods are fixed in graphite blocks such that the distance between two rods is about 20 cm. Air is used as the coolant which keeps the uranium rods at a temperature below 250°C. The control rods are made of steel coated with boron and have high absorption power for neutrons. For safety, cadmium rods are used and can be inserted whenever the fission is to be controlled or stopped.

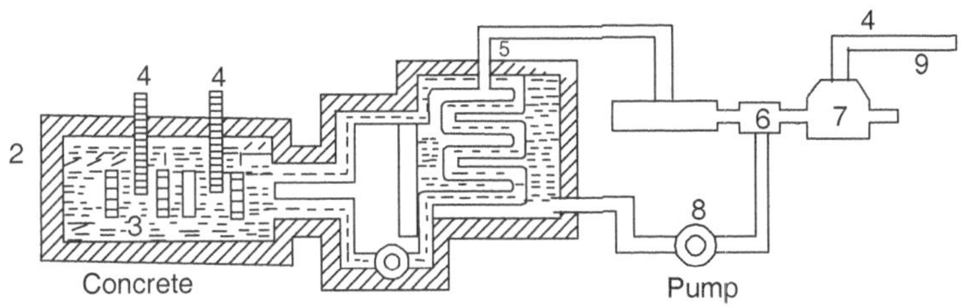

Fig. 12.13.

Power reactor with heat exchanger : 1. Fuel elements (uranium or enriched uranium). 2. Moderator (graphite block). 3. Coolant (water, gas or liquid alloys). 4. Control rods (boron, steel or cadmium). 5. Heat exchanger. 6. Turbines. 7. Generator. 8. Pump. 9. Power lines (Fig. 12.13).

The heat produced by the nuclear disintegration is extracted by following carbon through a pile. Water boils to generate necessary steam to drive the turbines.

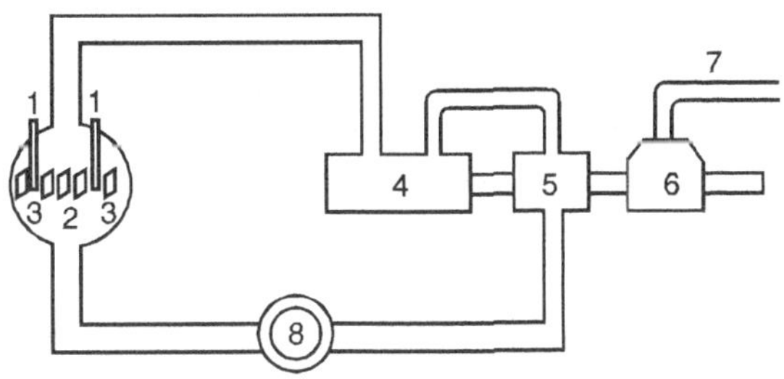

1. Boron Steel or cadmium rods. 2. Water
3. Uranium or platinum. 4. Steam. 5. Turbine.
6. Generator. 7. Power line. 8. Pump

Fig. 12.14.

The breeder reactor is important because it uses uranium 238 which is easily obtainable as a fuel.

In practice, all the U 238 cannot be burnt in the first time and it is necessary to extract Pu from partly spent fuel element and put back in the reactor.

Cadmium has the property of hungrily absorbing slow neutrons thus, removing them from circulation.

It is essential that the nuclear reactor should be shielded by thick concrete walls on all sides. The walls are usually 2 metre thick and are made of concrete. The energy produced during fission is removed by circulating the coolant and the coolant gives out energy in the heat exchangers (Fig. 12.13 and 12. 14). When neutrons

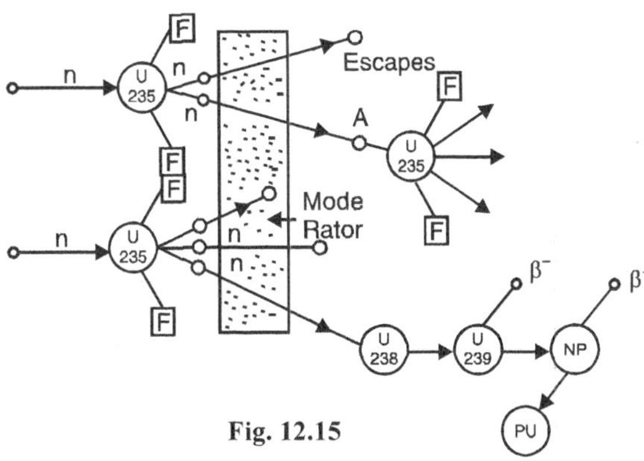

Fig. 12.15

bombard the nuclei of uranium 235, on an average 2.5 neutrons are produced in each fission. Some of the neutrons after passing through the moderator escape (*i.e.*, *A*) while others again bombard U 235 and fission takes place (Fig. 12.15). Some of the neutrons (*B*) are absorbed in the moderator. The neutrons like *C* after passing through the moderator cannot break uranium 238 and U 238 is converted into Pu 239. Stable elements may be inserted in the reactor to become radioactive by neutron absorption. In this way cobalt 60 and iodine 131 are produced for use in industry and medicine.

Four Factor Formula

Neutron balance is important in a thermal reactor. Starting with U^{235} fission by thermal neutrons. Let v be the fast neutrons emitted in this process. Some of these neutrons have energies greater than fission threshold for U^{238}. Hence they can cause fission of U^{238} which leads to a corresponding increase in the number of fast neutrons. Therefore, the no. of fast neutrons increases from υ to $\upsilon\in$ where $\in$ is called the fast fission factor. It is very close to unity in a homogeneous reactor and can be as high as 1.1 in heterogeneous reactors. On the other hand in several natural uranium, graphite moderated reactors $\in$ is 1.03.

Now, from a finite size reactor core some fast neutrons can leak out before slowing down to thermal energies. Let l_f be the leakage factor for fast neutrons *i.e.*, to say l_f escapes before being slowed down and $v\in l_f$ neutrons are lost. The remaining $\upsilon\in(1-l_f)$ are slowed down by colliding with the moderator atoms, but in the process some of them may be captured by U^{238} to form U^{239}. These decay to Np^{239} and finally to Pu^{239} at resonant energies. Out of $\upsilon\in(1-l_f)$ neutrons which start the slowing down process, a fraction $v\in(1-l_f)p$ escape resonance capture and remaining $v\in(1-l_f)(1-p)$ captured go to form Pu^{239}. p is called the resonance escape probability. The neutrons escaping resonance absorption are slowed down to thermal energies. Some of these thermal neutrons may again leak out of the core. So, if the fraction leaking out is denoted by l_t (leakage factor for thermal neutrons), the no. of thermal neutrons per fission of U^{235}, that escape is equal to $v\in(1-l_f)\,p\,l_t$ and the remaining *i.e.*, $v\in(1-l_f)\,p\,(1-l_t)$, a fraction f is absorbed in uranium. The fraction $1-f$ is absorbed in other materials and are lost. At this stage the number of neutrons still available for containing the chain reaction is $v\in(1-l_f)\,p\,(1-l_t)\,f$. f is termed as the thermal utilisation factor.

It should be noted that not all the neutrons absorbed in the fuel cause fission of U^{235}, some are absorbed by U^{238} to form Pu^{239}. Some others are absorbed in U^{235} to form U^{236}. If σ_a is the cross-section of thermal neutrons absorbed in uranium and σ_f that of neutions causing fission of U^{235} then σ_f/σ_a denote the fraction of thermal neutrons which are absorbed in uranium that cause fission of U^{235}.

Now, the number of second generation fission in U^{235} per fission of U^{235} by a first generation neutron is called the *reproduction factor* or *multiplication factor* K, and is given as

$$K = \upsilon\in(1-l_f)\,p\,(1-l_t)\,f\left(\frac{\sigma_f}{\sigma_a}\right)$$

The quantity $\upsilon\left(\frac{\sigma_f}{\sigma_a}\right)$ represents the no. of fast fission neutrons produced per thermal neutron absorbed in uranium, it is denoted by η

$\therefore$ we have $K = \eta\in p\,f\,(1-l_f)(1-l_t)$

For a core of very large size, the leakage of fast and slow neutrons can be disregarded. Hence, the multiplication factor K_∞ (for very large) reactor is given by

$$\boxed{K_\infty = \eta\in p\,f}$$

This is known as the *four factor formula*

For an infinite system $K = 1$ for self sustained chain reaction (*i.e.,* each neutron just replaces the previous one). The system is said to be *critical.* If $K > 1$, the system is *supercritical* and for $K < 1$ it is *subcritical.*

In the very special case where the fuel consists of only U^{235} and no U^{238} both $\in$ and p are practically unity. $\therefore\ K_\infty = \eta f.$

For a reactor of finite dimensions, the *effective* multiplication factor is less than K_∞

$$\therefore \qquad K_{eff} = K_\infty (1 - l_f)(1 - l_t).$$

The magnitude of K_{eff} gives the speed with which the neutrons build up and also the rate at which fission occurs.

For $K_{eff.} > 1$, we can progressively reduce the reactor size, as a result the neutron loss increases through leakage from the assembly. The reactor size for which $K_{eff.} = 1$ is known as the *critical size.*

12.7. Nuclear Fusion

In this process, small nuclei are fused together to form a single heavy nucleus. The mass of the single nucleus formed is less than the initial mass. The difference in mass is converted into energy, according to Einstein equation $E = mc^2$. Consider a single helium nucleus formed by the fusion of two deuterium nuclei.

$${}_1H^2 + {}_1H^2 = {}_2He^4$$

The initial mass of two deuterium atoms = 2 × 2.01471

$$\begin{aligned} &= 4.02942 \text{ amu} \\ \text{Mass of helium atom} &= 4.00388 \text{ amu} \\ \text{Decrease in mass} &= 4.02942 - 4.00388 \\ &= 0.02554 \text{ amu} \\ \text{But, 1 amu} &= 931 \text{ MeV energy} \\ \therefore \qquad \text{Energy released} &= 0.02554 \times 931 \\ &= 24 \text{ MeV} \end{aligned}$$

Thus, the energy released in fusion in 24 MeV which is much less than the energy produced in the fission of uranium which is about 200 MeV. But energy released per unit mass during fusion of the light nuclei is much greater as compared to the fission of uranium.

The process of fusion appears to be simple. But it is not easy to fuse the light nuclei into a single nucleus. The fusion process can be carried out only at a extremely high temperatures. It is calculated that the fusion is possible at a temperature of the order of 10^7 to 10^8 °C. Therefore, before fusion takes place, the light nuclei must have their temperatures is raised by several million degree Celsius. These types of reactions are called *thermonuclear reactions.* This high temperature can be attained during the explosive type of fission process and also in the intense electrical discharges through gases.

A hydrogen bomb has been made in which fusion of light nuclei takes place and tremendous amount of energy is released. In this bomb *Fission-Fusion* process is applied. The central part contains a fission bomb containing uranium 235 or plutonium 239.

The plutonium core is surrounded by the fusible materials. The outer part also consists of uranium or plutonium where fission takes place due to the release of neutrons during fusion.

There is no limit to the size of the bomb because there is no necessity for critical mass in this process.

The materials used during a fusion process are deuterium and tritium. The reactions are :

$${}_1H^2 + {}_1H^2 \longrightarrow {}_2He^3 + {}_0H^1 + 4.04 \text{ MeV}$$

When two deuterium nuclei combine to form a helium nucleus of $A = 3$ and $Z = 2$ and a proton, energy released per fusion is 4.04 MeV. If instead of proton, a neutron is released.

$$_1H^2 + {}_1H^2 \longrightarrow {}_2He^3 + {}_0n^1 + 3.26 \text{ MeV}$$

The energy released per fusion is 3.25 MeV. Fusion reaction based on deuteron-deuteron and deuteron-triton are known as 'Wet hydrogen bomb' reactions. The deuteron-triton reaction is

$$_1H^2 + {}_1H^2 \longrightarrow {}_2He^4 + {}_0n^1 + 17.6 \text{ MeV}$$

In this case, energy released per fusion is 17.6 MeV. The neutron released, further fuses with lithium, according to the reaction $_3Li^6 + {}_0n^1 \rightarrow + {}_1H^3 + {}_2He^4 + 4.7$ MeV and energy released is 4.7 MeV.

Moreover, ordinary walls cannot suffice for the fusion reactions because they would vaporise under bombardment of the high energy particles and these would cool below the fusion temperature. Thus in a controlled type of fusion, there must be suitable arrangement. At the high temperature required for fusion, the electrons are stepped out of the atoms. This results in a fourth state of matter, a fully ionised gas known as *plasma*. This plasma consists of nuclei of light elements *e.g.*, deuterons, protons, tritons, and electrons moving in all directions in rapid motion. As no material wall can contain such a plasma at a very high temperature, a type of magnetic bottle *i.e.*, a suitably defined magnetic field can hold such matter. In such a process, plasma is compressed and a very high temperature suitable for nuclear fusion is produced.

12.8. Stellar Burning

Energy production in stars *p–p* reaction Hydrogen burning

When a cloud of interstellar gas contracts to form a star, its temperature increases as the gravitational potential energy is converted to kinetic energy. The interior heats upto 10^7 K. At this stage the first thermonuclear reactions take place converting H_2 to He. This sequence of reactions is known as proton-proton chain.

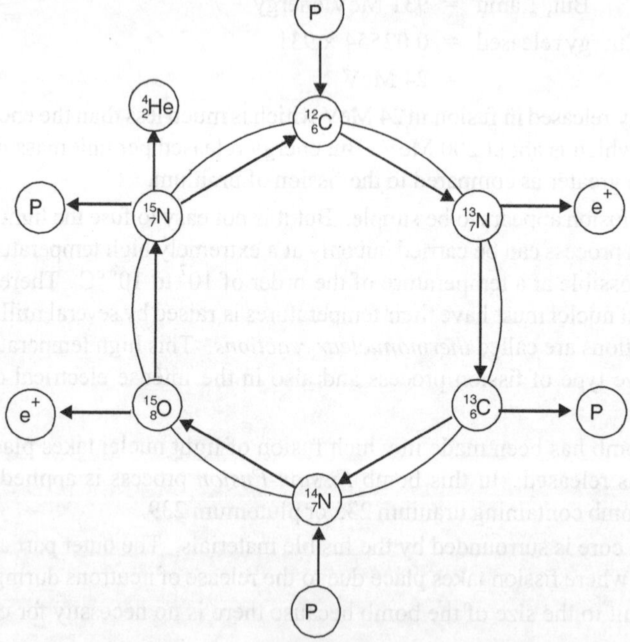

C.N.O. Cycle

Fig. 12.16. 12.16 (C.N.O. cycle). The carbon cycle also involves the combination of four hydrogen nuclei to form a helium nucleus with the evolution of energy. The $^{12}_6C$ nucleus is unchanged by the series of reactions. This cycle occurs in s tars hotter than the sun.

The first hydrogen burning reaction in a star is the conversion of two protons into a deuteron.

$$\therefore \qquad p+p \longrightarrow d+e^{+}+\nu$$

β-decay takes place and the cross-section for the reaction is very low.

The deuteron so formed rapidly undergoes the reaction.

$$d+p \longrightarrow {}^{3}\text{He}+\gamma$$

No d-d reaction occur as this is quite unlikely. Since the average lifetime 10^{18} sec. For a proton before fusion. The mean life time of deuteron is 2 secs and therefore it is 10^{18} times more probable that a deuteron will collide with a proton than with another deuteron.

Reactions between ^{3}He nucleus and a proton are not possible since ^{4}Li is unbound.

∴ ^{3}He reacts with another ^{3}He.

$$\therefore \qquad {}^{3}\text{He}+{}^{3}\text{He} \longrightarrow \alpha+2p.$$

The overall effect of *p–p* reaction is that four protons are converted into an α- particle and an energy released equal to B.E. of α less twice neutron-proton mass diff. or 26.7 MeV.

Tracks of heavy nuclei up to Z = 26 were found. Protons, α-particle, C, N, O, Mg, Si and Fe are found in the primary cosmi rays.

There are different theories for the origin of cosmic rays. But no theory gives a satisfactory answer. The study of upper atmosphere has become important due to launching of stellites and tempt by mans to reach the moon. The theory that cosmic rays originate from the sun or the novae or from supernova has been found to show contradictory results. It is still an installed question.

12.9. Stellar Energy-Carbon Nitrogen Cycle

It is established that the stellar energy is due to the process of fusion of light nuclei into a heavy nucleus. The temperature of the interior of the sun is about 2×10^{7} °C and the temperature of some stars is of the order of 10^{8} °C. Bethe has given a detailed theory of nuclear reactions in the sun and the stars. The theory involves the synthesis of protons to give helium. Carbon acts as a catalyst in the nuclear reaction. All the light elements like hydrogen and helium are in a state of plasma at such high temperature and all the extra-nuclear electrons are stripped off the atoms. The energy produced in fusion is responsible for the maintenance of the high temperature of the stars and also for the emission of energy by radiations

The *carbon-nitrogen cycle* is responsible for nuclear fusion of protons and a tremendous amount of energy is produced during the process. Protons interact with carbon nuclei in the interior of the stars according to the relation.

$${}_{6}\text{C}^{12} + {}_{1}\text{H}^{1} \longrightarrow {}_{7}\text{N}^{13} + \text{E}_{1}$$

E_1 is the amount of energy released. Nitrogen A = 13 and Z = 7 decays by emitting a positron

$${}_{7}\text{N}^{13} \longrightarrow {}_{6}\text{C}^{13} + {}_{+1}e^{0}$$

${}_{6}\text{C}^{13}$ further captures a proton and more energy is released according to reaction

$${}_{6}\text{C}^{13} + {}_{1}\text{H}^{1} \longrightarrow {}_{7}\text{N}^{14} + E_{2}$$

${}_{7}\text{N}^{14}$ further captures a proton according to the reaction

$${}_{7}\text{N}^{14} + {}_{1}\text{H}^{1} \longrightarrow {}_{8}\text{O}^{15} + E_{3}$$

and energy is released.

${}_{8}\text{O}^{15}$ is radioactive and gives out positron.

$${}_{8}\text{O}^{15} \longrightarrow {}_{7}\text{N}^{15} + {}_{+1}e^{0}$$

${}_{7}\text{N}^{15}$ further captures a proton according to the reaction

$${}_{7}\text{N}^{15} + {}_{1}\text{H}^{1} \longrightarrow {}_{6}\text{C}^{12} + {}_{2}\text{He}^{4}$$

and $_6C^{12}$ is again produced. Therefore, in the process 4 protons combine to produce the helium nucleus and a tremendous amount of energy is released. Carbon acts as a catalyst. The energy released during each fusion is 28 MeV.

The sun is the most carefully studied star and the rate of emission of energy is about 4×10^{16} J/s. Hydrogen and helium together form about 90% by weight of the sun's matter. Thus the energy must be released in the nuclear process concerning hydrogen and helium. The reactions which are possible when the sun was formed involve proton-proton reaction and is important at lower temperature.

$$_1H^1 + {}_1H^1 \longrightarrow {}_1H^2 + E$$
$$_1H^1 + {}_1H^2 \longrightarrow {}_2He^3 + E$$
$$_2He^3 + {}_2He^3 \longrightarrow {}_2He^4 + 2\,{}_1H^1 + E$$

The total amount of energy produced when 4 protons combine to form a helium nucleus, is 26.2 MeV.

The other possible reactions in the present state of sun, with its central portion at a higher temperature are

$$_1H^1 + {}_1H^1 \longrightarrow {}_1H^2 + E$$
$$_1H^1 + {}_1H^2 \longrightarrow {}_2He^3 + E$$
$$_2He^3 + {}_2He^4 \longrightarrow {}_4Be^7 + E$$
$$_4Be^7 + {}_1H^1 \longrightarrow {}_5B^8 + E$$
$$_5B^8 \longrightarrow {}_4Be^8 + {}_{+}e^0$$
$$_4Be^8 \longrightarrow {}_2He^4 + {}_2He^4$$

The carbon-nitrogen cycle is less possible in the case of sun. The carbon-nitrogen cycle is only possible in the case of stars of the main sequence which are at higher temperatures than the sun. It has been estimated in the case of the sun, that one gram of its mass contains about 2×10^{23} protons. The energy released by the fusion of 2×10^{23} protons is about 2×10^{11} J. If the sun were to continue to radiate at its present rate, it would take billions of years to exhaust its supply of protons.

Example 12.1. *A nuclear reactor using $_{92}U^{235}$ is to operate at a power level of 250 megawatts. If the energy released per fission of $_{92}U^{235}$ is 200 MeV, calculate the rate of consumption of $_{92}U^{235}$ per year; assuming that there are no losses and that the mass of atom is equal to the sum of masses of nucleus where each nucleon has a mass equal to 1* a.m.u. *[Delhi]*

Solution. Output power $= 250 \times 10^6$ watts $= 2.5 \times 10^8$ J/s

Energy released per fission $= 200$ MeV

$= 200 \times 10^6$ eV

$= 200 \times 10^6 \times 1.6 \times 10^{-19}$ J

$= 3.2 \times 10^{-11}$ J

Number of fission per second

$$= \frac{2.5 \times 10^8}{3.2 \times 10^{-11}}$$

$= 1.87 \times 10^{18}$

Mass of each atom of $_{92}U^{235}$

$= 235 \times 1.6 \times 10^{-27}$ kg

$= 3.76 \times 10^{-25}$ kg

Consumption per second

$= 7.81 \times 10^{18} \times 3.76 \times 10^{-25}$ kg

$= 2.936 \times 10^{-6}$ kg

Consumption per year

$$= 2.936 \times 10^6 \times 86400 \times 365$$
$$= \mathbf{92.59\ kg.}$$

Example 12. 2. *A reactions developing energy at the rate of 32 × 10⁶ watts. How many atoms of U 235 undergo fission per second ? Assume that on the average, an energy of 200 MeV is released per fission.* *[Delhi]*

Solution. Energy released per second

$$E_1 = 32 \times 10^6 \text{ joules}$$

Energy released per fission

$$E_2 = 200 \text{ Mev} = 200 \times 10^6 \text{ eV}$$
$$= 200 \times 10^6 \times 1.6 \times 10^{-19} \text{ joule}$$

Number of atoms undergoing fission per second,

$$N = \frac{E_1}{E_2} = \frac{3.2 \times 10^6}{200 \times 10^6 \times 1.6 \times 10^{-19}}$$
$$N = \mathbf{10^{18}}$$

Example 12.3. *What is the minimum energy required to break a $_2He^4$ nucleus into free protons and neutrons ? The masses of hydrogen atom, a neutron and the $_2He^4$ atom in amu are 1.007825, 1.008665 and 4.002603 respectively.* *[Delhi]*

Solution. Difference in mass

$$m = 2 \times 1.007825 + 2 \times 1.008665 - 4.002603$$
$$m = 2.01565 + 2.01730 - 4.002603$$
$$m = 4.032980 - 4.002603 = 0.030377 \text{ amu}$$
$$\text{But 1 amu} = 931 \text{ MeV}$$

∴ Energy required,

$$E = 0.030377 \times 931 \text{ MeV}$$
$$\boldsymbol{E} = \mathbf{28.28\ MeV}$$

Example 12.4. *What will be the energy released if two deuterium nuclei fuse into an α-particle ? Given*

Solution.

$$M_H = 2.014102 \text{ amu}$$
$$M_{He} = 4.002603 \text{ amu}$$
$$1 \text{ amu} = 931.5 \text{ MeV.}$$ *[Delhi]*

$$\text{Difference in mass} = 2M_H - M_{He}$$
$$= [2 \times 2.014102 - 4.002603] \text{ amu}$$
$$= [4.028204 - 4.002603] = 0.025601 \text{ amu}$$
$$\text{But 1 amu} = 931.5 \text{ MeV}$$

Energy released

$$U = 0.025601 \times 931.5$$
$$U = \mathbf{23.847\ MeV}$$

Example 12.5. *Calculate the energy in Joules required to break one gram mole of helium into its constituents particles.* *[I.A.S.]*

Solution. Mass of proton = 1.007825 amu

Mass of neutron = 1.008665 amu

Mass of helium atom = 4.002603 amu

Avogadro's Number $N = 6.06 \times 10^{23}$ per gram mole

Difference in mass

$$m = 2 \times 1.007825 + 2 \times 1.008665 - 4.002603$$
$$m = 2.01565 + 2.01733 - 4.002603$$
$$m = 0.030377 \text{ amu}$$
$$1 \text{ amu} = 931 \text{ MeV}$$

Energy required to break 1 atom of helium,

$$E_1 = 0.030377 \times 931 = 28.28 \text{ MeV}$$

Energy required to break 1 gram mole of helium,

$$E = NE_1$$
$$E = 6.06 \times 10^{23} \times 28.28$$
$$E = \mathbf{1.714 \times 10^{25} \text{ MeV.}}$$

Example 12.6. *A proton of energy 0.45* MeV *collides with a lithium nucleus resulting in production of 2α-particles. Calculate the energy liberated in the process.*

Given *mass of* ${}_1H^1$ = *1.007825 amu*

mass of ${}_3Li^7$ = *7.016005 amu*

mass of ${}_2He^4$ = *4.002604 amu* [*I.A.S.*]

$${}_3Li^7 + {}_1H^1 \rightarrow 2\,{}_2He^4 + E$$

Solution. Differences of mass,

$$\Delta m = 7.016005 + 1.007825 - 2 \times 4.002604$$
$$\Delta m = 8.023830 - 8.005208$$
$$\Delta m = 0.018622 \text{ amu}$$
$$1 \text{ amu} = 931 \text{ MeV}$$

Energy released,

$$U = 0.018622 \times 931 \text{ MeV}$$
$$U = \mathbf{17.34} \text{ MeV.}$$

Example 12.7. *Calculate the kinetic energy of the neutrons from the reaction*

$$d + H^3 = He^4 + {}_0n^1$$

assuming the deuteron bombarding energy to be negligible compared to the binding energies. Given binding energies of the nuclei

Solution. $d = 2.2$ MeH, $H^3 = 8.5$ Mev, $He^4 = 28.3$ MeV. [*I.A.S.*]

Kinetic energy of the neutron

$$= 28.3 - 2.2 - 8.5$$
$$= 28.3 - 10.7$$
$$= \mathbf{17.6 \text{ MeV}}$$

Example 12.8. *In the nuclear reaction*

${}_{92}U^{235} + {}_0n^1 \rightarrow {}_{92}U^{236} \rightarrow {}_{56}Ba^{141} + {}_{36}Kr^{92} + 3\,{}_0n^1 + Q$ *the average value of Q is found to be 180 MeV. Compute the consumption rate of* ${}_{92}U^{235}$ *per year if power generation is to be an average 100 MW (take efficiency = 100%)* [*I.A.S.*]

Solution. Power = 100, MW = 10^8 watts = 10^8 J/s

Energy required in one year

$$U = 10^8 \times 86400 \times 365 \text{ J}$$

$$U = \mathbf{3.1536 \times 10^{15}\ J.}$$

Energy produced by one atom fission of uranium

$$U_1 = 180 \text{ MeV}$$
$$= 180 \times 10^6 \times 1.6 \times 10^{-19} \text{ J}$$
$$= \mathbf{2.88 \times 10^{-11}\ J}$$

$$\text{Number of atoms required} = \frac{U}{U_1}$$

$$= \frac{3.1536 \times 10^{15}}{2.88 \times 10^{-11}}$$
$$= 1.095 \times 10^{26} \text{ atoms}$$

Now 235 kg of uranium contains 6.023×10^{26} atoms.

∴ Mass of uranium 235 required

$$= \frac{1.095 \times 10^{26} \times 235}{6.023 \times 10^{26}}$$
$$= \mathbf{42.7237\ kg/year.}$$

Example 12.9. *The minimum energy required for the disintegration of the nucleus of stable isotope of hydrogen is 2.21* MeV. *Calculate the mass of this nucleus in* amu. [*I.A.S. 1991*]

$$\text{Mass of neutron} = 1.008665 \text{ amu}$$
$$\text{Mass of proton} = 1.007276 \text{ amu}$$
$$1 \text{ amu} = 931 \text{ MeV.}$$

Solution. Stable isotope of hydrogen is deuteron which is the nucleus of heavy hydrogen. It has one proton and one neutron.

$$\text{Mass of proton} = 1.007276 \text{ amu}$$
$$\text{Mass of neutron} = 1.008665 \text{ amu}$$
$$\text{Binding energy} = 2.21 \text{ MeV.}$$
$$E = \frac{2.21}{931} \text{ amu}$$
$$E = 0.002374 \text{ amu}$$
$$\text{Mass of the nucleus} = (m_p + m_n) - \text{E}$$
$$= (1.007276 + 1.008665) - 0.002374$$
$$= \mathbf{2.013567\ amu.}$$

Example 12.10. *Assuming the energy released per fission of U^{235} is 200 MeV, calculate the rate in kg per year at which fission should occur in a nuclear reactor in order to operate at a power level of 1 watt. Mass of hydrogen atom= 1.67×10^{-27}* kg] [*I.A.S.*]

Solution.

$$\text{Power} = 1 \text{ watt} = 1 \text{ J/s}$$

Energy needed in one year

$$U = 1 \times 3600 \times 24 \times 365 \text{ J}$$
$$U = 3.1536 \times 10^7 \text{ J}$$

Energy release per fission

$$= 200 \text{ MeV}$$
$$= 2 \times 10^8 \text{ eV}$$

$$= 2 \times 10^8 \times 1.6 \times 10^{-19}\ \text{J}$$
$$= 3.2 \times 10^{-11}\ \text{J}$$

Number of atoms undergoing fission per year

$$= \frac{3.1536 \times 10^7}{3.2 \times 10^{-11}}$$
$$= 9.855 \times 10^{17}$$

1 kg of U^{235} contains 6.023×10^{26} atoms.

Mass of uranium required

$$m = \left[\frac{9.855 \times 10^{17}}{6.023 \times 10^{26}}\right] \text{kg}$$
$$\mathbf{m = 1.636 \times 10^{-9}\ kg.}$$

Example 12.11. *Fission of one atom of* U^{235} *releases 200* MeV *energy. Calculate the amount of energy released by* 10^{-3} *kg of Uranium.* *[Delhi, 1990]*

Solution.

$$\text{Number of atoms in 235 grams of uranium} = 6.023 \times 10^{23}$$
$$\text{Number of atoms in } 235 \times 10^{-3} \text{ kg of uranium} = 6.023 \times 10^{23}$$
$$\text{Number of atoms of } 10^{-3} \text{ kg of Uranium} = \frac{6.023 \times 10^{23}}{235}$$
$$\text{Energy produced by fission of one atom} = 200 \text{ MeV}$$

Energy produced by 10^{-3} kg of Uranium,

$$U = \frac{6.023 \times 10^{23}}{235} \times 200$$
$$U = \mathbf{5.126 \times 10^{23}\ MeV.}$$

Example 12.12. *Deduce the minimum energy of gamma -ray photon* (*in* MeV) *which can cause electron-positron pair production.* *[Delhi, 1990]*

Solution. Here,

Mass of electron = mass of positron = 9.1×1^{-31} kg

Energy of γ – ray photon = U

$$U = 2mc^2$$

Here $c = 3 \times 10^8$ m/s

$\therefore$ $U = 2 \times 9.1 \times 10^{-31} \times (3 \times 10^8)^2$

$$U = 1.638 \times 10^{-13}\ \text{J}$$

But $1\ \text{eV} = 1.6 \times 10^{-19}\ \text{J}$

and $1\ \text{MeV} = 1.6 \times 10^{-13}\ \text{J}$

$$U = \frac{1.638 \times 10^{13}}{1.6 \times 10^{-13}}\ \text{MeV}$$
$$U = \mathbf{1.023\ MeV}$$

Example 12.13. *When an aom of* U^{235} *undergoes fission in a reactor 200* MeV *energy is liberated. Suppose that the power out is 800 Mw and reactor is 25% efficient. (i) How many uranium atom does it consume in one days ? (ii) What mass of uranium does it consume in one day ?* *[Mumbai, 1991]*

Solution. (*i*) Power out $= 800 \times 10^6 = 8 \times 10^8$ W $= 8 \times 10^8$ J/s

Energy required in one day

$$E_1 = 8 \times 10^8 \times 86400 \text{ J}$$

$$E_1 = \mathbf{7.712 \times 10^{13}\ J} \quad ...(i)$$

Let number of uranium atoms used be N per day.

Energy produced by N atoms $= N \times 200 \times 10^6 \times 1.6 \times 10^{-19}$ J

$= (3.2 \times 10^{-13}\ N)$ J

Efficiency $= 25\%$

Useful energy required produced in one day $= \mathbf{3.2 \times 10^{-13} N \times 0.25}$

$$E_2 = 8 \times 10^{-14} N$$

$$E_2 = E_1$$

$$8 \times 10^{-14} N = 7.712 \times 10^{13}$$

$$N = \mathbf{9.64 \times 10^{26}\ atoms.}$$

(*ii*) Avogardo's Number $= 6.02 \times 10^{26}$ /kg mole

Atomic wt. $= 235$

mass, $$m = \frac{9.64 \times 10^{26} \times 235}{6.02 \times 10^{26}}$$

$$m = \mathbf{376.31\ kg.}$$

Example 12.14. *A city requires 100 megawatts of electrical power on the average. If this is to be supplied by a nuclear reactor of efficiency 20%. Using* U^{235} *as a nuclear fuel, calculate the amount of fuel required for one days operation. Given that energy released per fission of* U^{235} *nuclide is 200* MeV. [*Calicut, 1992*]

Solution. Energy required per day

$$E_1 = 100 \times 10^6 \times 86400 \text{ J}$$

$$E_1 = 8.64 \times 10^{12} \text{ J} \quad ...(i)$$

Suppose amount of fuel required $= m$ kg

Number of atoms $= \dfrac{m \times 6.023 \times 10^{26}}{235}$

Energy released per fission $= 200$ MeV

$= 200 \times 10^6 \times 1.6 \times 10^{-19}$ J

$= 3.2 \times 10^{-11}$ J

Total energy released $= \left[\dfrac{m \times 6.023 \times 10^{26} \times 3.2 \times 10^{-11}}{235}\right]$ J

Efficiency $= 20\%$

Useful energy $= E_2 = \left[\dfrac{0.20 \times m \times 6.023 \times 10^{26} \times 3.2 \times 10^{-11}}{235}\right]$ J

$$E_2 = [16.403 \times 10^{12}\ m] \text{ J} \quad ...(ii)$$

$$E_1 = E_2$$

$$\therefore \quad 8.64 \times 10^{12} = 16.403 \times 10^{12} \text{ m}$$

$$m = \frac{8.64}{16.403}$$

$$m = \mathbf{0.5267\ kg/day}$$

Example 12.15. *Use the Weizsaecker formula to determine the binding energy of* $^{56}_{26}$Fe *and compare the result with the experimental value.*

Solution. We have

$$BE = 15.7 \times 56 - 17.8\,(56)^{2/3} - 0.711\,(26)^2\,(56)^{-1/3} - 23.7\,(56-52)^2/56 + 11.18\,(56)^{-1/2}$$

$$= 487.75 \text{ MeV}$$

The atomic mass of ^{56}Fe is 55.939395. Hence the binding energy is

$$BE = (26 \times 1.007825 + 30 \times 1.008665 - 55.939395)\,(931.50) = 488.11 \text{ Mev}$$

The result of the semi empirical formula agrees with the experimental value within 0.1 percent.

EXERCISE

1. Discuss any type of nuclear reactor for the production of electric power. What is the action of coolants and control rods in such a reactor ? *[Delhi]*
2. What is the difference between nuclear fission and nuclear fusion ? How is the principle of nuclear fusion applied in the case of hydrogen bomb ?
3. What is the source of stellar energy ? What are the important reactions in the sun ?
4. What is nuclear fission ? Discuss its application in nuclear reactor. *[Delhi]*
5. Work out the theory of the liquid drop model of the nucleus. How can you explain nuclear fission on the above model ? *[Delhi (Hons)]*
6. A nuclear reactor has a power of 1.28×10^8 watts. How many atoms of U_{235} would undergo fission per second ? Assume that on the average an energy of 200 MeV is released per fission.
7. On the basis of the liquid drop model, develop the semi - empirical mass formula by considering the various factors which affect the binding energy of the nucleus.
8. Discuss the Nuclear stability rules for β-decay.
9. Discuss the shell model of the nucleus. What is spin-orbit interaction ? Explain its use.
10. How is energy released in Nuclear fusion ? Discuss.
11. Explain p-p reaction in stars.
12. Explain the Carbon-Nitrogen cycle.

APPENDIX A

Zeeman Effect

A spectral line emitted by the excited atoms is split up into a doublet or a triplet when the emitting atoms are placed in a magnetic field. This effect of the splitting of a spectral line under the action of a magnetic field is known as *normal Zeeman effect*.

To produce Zeeman effect, the source of light such as a sodium lamp or a mercury arc or gas discharge in a Geissler tube is placed between the poles of a powerful electro-magnet [Fig.1.(a)]. The light coming from the source is examined by means of a spectroscope of high resolving power. In order to view the light parallel to the magnetic field, a hold is drilled in one of the pole-pieces along the axis of the magnet.

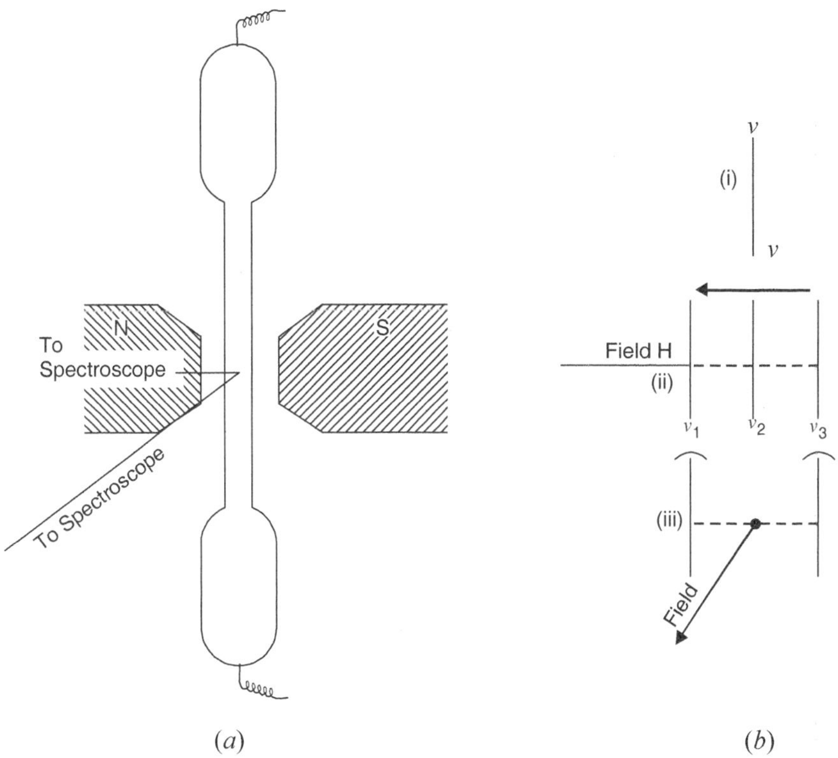

Fig.1. Normal Zeeman Effect (*i*) No field (*ii*) Perpendicular to the field (*iii*) Parallel to the field

When no magnetic field is applied, the spectroscope is focussed on one of the lines in the spectrum of the source of light. When a magnetic field is applied, it is observed :

1. That when the light is viewed in a direction perpendicular to the magnetic field, three component lines are observed. One of the lines is in the same position as the original line and the other two lines are in the two sides of the original line. The outer two lines, when observed by means of a

nicol prism as an analyser, are polarized at right angles to the undisplaced line. This effect is known as *transverse Zeeman effect*.

2. When the light is viewed in a direction parallel to the direction of the field, there is no line in the position of the original line, only two outer lines are present. These lines are found to be circularly polarized in opposite directions. The effect is known as *longitudinal Zeeman effect.*

The normal Zeeman effect is explained on Lorentz electron theory. Consider an electron moving in a circular orbit of radius r with a velocity υ (Fig. 2). The centripetal force is, $F = \frac{m\upsilon^2}{r}$. If an external magnetic field is applied, an additional force acts which is directed perpendicular to the direction of motion of the electron. This force is also perpendicular to the direction of the magnetic field and is along the radius. When this force acts inwards along the radius, the velocity of the electron increases. When this force acts outwards along the radius, the velocity of the electron decreases. Suppose, this force due to the magnetic field = F_1 and let the velocity of the electron be increased to υ_1 by application of the magnetic field. Then, $R_1 = Be\upsilon_1$. Suppose, this force is directed towards the centre, total force along the radius.

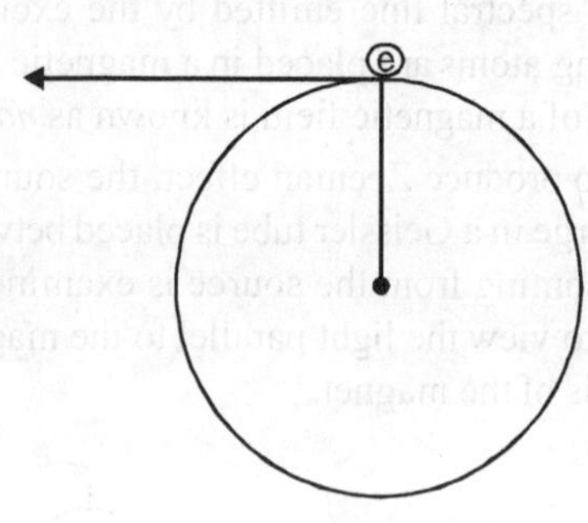

Fig. 2.

$$= F + F_1 = \frac{m\upsilon^2}{r} + Be\upsilon_1$$

$$\text{Total force} = \frac{m\upsilon_1^2}{r}$$

$$\therefore \quad \frac{m\upsilon_1^2}{r} = \frac{m\upsilon^2}{r} + Be\upsilon_1 \qquad ...(i)$$

But, $\upsilon = r\omega$ and $\upsilon_1 = r\omega_1$

where ω and ω_1 are the respective angular velocities.

From equation (*i*)

$$\frac{mr^2\omega_1^2}{r} = \frac{mr^2\omega^2}{r} + Ber\omega_1 \qquad ...(ii)$$

$$mr\omega_1^2 - mr\omega^2 = Ber\omega_1$$

$$\omega_1^2 - \omega^2 = \frac{eB\omega_1}{m}$$

$$(\omega_1 - \omega)(\omega_1 + \omega) = \frac{eB\omega_1}{m}$$

$$\omega_1 - \omega = \frac{eB\omega_1}{m(\omega_1 + \omega)}$$

$(\omega_1 + \omega)$ is approximately equal to $2\omega_1$

$$\therefore \quad \omega_1 - \omega = \frac{eB\omega_1}{m\omega_1} = \frac{eB}{2m}$$

$$\therefore \quad \omega_1 = \omega + \frac{eB}{2m}$$

If ν_1 and v are the frequencies, then, $\omega_1 = 2\pi v_1$, $\omega = 2\pi$ frequency v

$$2\pi v_1 = 2\pi v + \frac{eB}{2m}$$

$$v_1 = v + \frac{eB}{4\pi m} \qquad ...(iii)$$

When the electron moves in the opposite direction, the magnetic field produces a force in the opposite direction and the velocity decreases to v_2. In that case

$$\text{Total force} = F - F_2 = \frac{mv^2}{r} - Be\upsilon_2$$

$$\therefore \quad \frac{m\nu_2^2}{r} = \frac{mv^2}{r} - Be\upsilon_2$$

$$\text{But,} \quad \upsilon_2 = r\omega_2 \text{ and } \upsilon = r\omega$$

$$\therefore \quad \frac{mr^2\omega_2^2}{r} = \frac{mr^2\omega^2}{r} - Ber\omega_2$$

$$mr\omega_2^2 - mr\omega^2 = -Ber\omega_2$$

$$\omega_2^2 - \omega^2 = -\frac{eB\omega_2}{m}$$

$$(\omega_2 - \omega) = -\frac{eB\omega_2}{m(\omega_2 + \omega)}$$

$$\omega_2 - \omega = -\frac{eB}{2m} \qquad (\therefore \omega_2 + \omega = 2\omega_2 \text{ approx.})$$

$$2\pi v_2 - 2\pi v = -\frac{eB}{2m}$$

$$\therefore \quad v_2 = \frac{e\,B}{4\,\pi\,m} \qquad ...(iv)$$

From (*iii*) and (*iv*) we get

$$\Delta\, v \text{ (in each case)} = \frac{eB}{4\pi m} \qquad ...(v)$$

This shows the two lines are displaced equally on the two sides of the original line,

$$v_1 - v_2 = \frac{eB}{2\pi m} \qquad ...(vi)$$

$$\text{In general} \quad \Delta v = \frac{eB}{4\pi m} \qquad ...(vii)$$

Alternative Treatment. Consider an electron moving in a circular orbit of radius r with a velocity υ.

The centripetal force $F = \frac{m\upsilon^2}{r} = mr\omega^2$. If an external field B is applied, the change in the force,

$$dF = \pm Be\upsilon = \pm Ber\omega \qquad ...(i)$$

$$F = mr\omega^2$$

$$\text{Differentiating,} \quad dF = 2mr\omega d\omega \qquad ...(ii)$$

Equating (*i*) and (*ii*)

$$\pm Ber\omega = 2mr\omega d\omega$$

$$\frac{e}{m} = \pm\frac{2d\omega}{B} \qquad ...(iii)$$

Also $$\omega = 2\pi\nu = \frac{2\pi c}{\lambda}$$

Differentiating $$d\omega = \frac{-2\pi c}{\lambda^2}d\lambda$$

Substituting the value of $d\omega$ in equation (*iii*)

$$\frac{e}{m} = \pm\frac{4\pi c d\lambda}{\lambda^2 B} \qquad ...(iv)$$

$$d\lambda = \pm\frac{eB\lambda^2}{4\pi mc} \qquad ...(v)$$

$$\lambda_1 = \lambda+\frac{eB\lambda^2}{4\pi mc}$$

$$\lambda_2 = \lambda-\frac{eB\lambda^2}{4\pi mc}$$

$$\lambda_1-\lambda_2 = \frac{eH\lambda^2}{2\pi mc} \qquad ...(vi)$$

$$\nu = \frac{c}{\lambda}$$

Differentiating, $$d\nu = -\frac{c\,d\lambda}{\lambda^2}$$

Substituting the value of $d\lambda$ from equation (v)

$$d\nu = \pm\frac{eB}{4\pi m} \qquad ...(vii)$$

$$\nu_1 = \nu+\frac{eB}{4\pi m}$$

$$\nu_2 = \nu-\frac{eB}{4\pi m}$$

$$\nu_1-\nu_2 = \frac{eB}{2\pi m} \qquad ...(viii)$$

The quantity $\frac{eB}{2\pi m}$ is known as ***normal Zeeman separation***. Knowing the values of *n* and *B*, $\frac{e}{m}$ can be calculated. The value of $\frac{e}{m}$ calculated from the measurements of Zeeman effect = 1.757 X 10^{11} C/kg and it is in agreement with the value of $\frac{e}{m}$ of the electron obtained from Thomson's experiment.

This experiment established that the electron in the atom is responsible for the emission of spectral lines.

Example 1. *What is Bohr magneton ? Calculate its value.* [*Delhi*]

Solution. Bohr magneton

$$\beta = \frac{eh}{4\pi m}$$

Here $$e = 1.6\times10^{-19}\text{ C}$$

$$h = 6.624 \times 10^{-34} \text{ J-s}$$

$$m = 9.1 \times 10^{-31} \text{ kg}$$

$$\beta = \frac{1.6 \times 10^{-19} \times 6.624 \times 10^{-34}}{4 \times 3.142 \times 9.1 \times 10^{-31}}$$

$$\beta = \mathbf{9.27 \times 10^{-24}} \textbf{ SI units.}$$

Example 2. *The experimental value of Bohr magneton is 9.21 × 10^{-24} SI units and Planck's constant h = 6.6 × 10^{-34} joule-second. Calculate the value of $\frac{e}{m}$ of an electron.* [*Gujarat*]

Solution. Here

$$\beta = 9.21 \times 10^{-24} \text{ SI units}$$

$$h = 6.6 \times 10^{-34} \text{ Joule second}$$

$$\beta = \frac{eh}{4\pi m}$$

or

$$\frac{e}{m} = \frac{9\pi\beta}{h} = \frac{4 \times 22 \times 9.21 \times 10^{-24}}{7 \times 6.6 \times 10^{-34}}$$

$$= \mathbf{1.755 \times 10^{11}} \textbf{ C/Kg.}$$

Example 3. *Calculate the wavelength separation between the two component lines which are observed in the normal Zeeman effect. The magnetic field used is 0.4 weber/m^2, the specific charge = 1.76 × 10^{11}* coulombs/*kg and λ = 6000* Å. [*Delhi*]

Solution. Here

$$\lambda = 6000\text{Å} = 6000 \times 10^{-10} \text{ m}$$

or

$$\lambda = 6 \times 10^{-7} \text{ m}$$

APPENDIX B
STERN AND GERLACH EXPERIMENT

Experimental Confirmation of the Vector Model-stern and Gerlach Experiment

Experimental study of the behaviour of atoms in non-homogeneous magnetic fields, initiated by Stern and Gerlach in 1921 and since then elaborated in many directions by different workers, gave a very direct and most convincing confirmation of the essential features of the vector model, viz. spatial quantisation electron and quantised atomic magnetic moment.

Principle. The magnetic moments due to the orbital and spin motions of the electrons in an atom may be estimated as follows :

Considering the simple case of an electron moving in a central orbit, such as an ellipse, the current i due to the motion of the electron round the orbit is given by $i = e/T$, where e is the charge and T the periodic time.

Applying Ampere's theorem, this current gives rise to a magnetic moment μ_1 given by

$$\mu_l = (e/T)\text{A} \qquad \text{...}(i)$$

where A is the area enclosed by the orbit.

Since the areal velocity of motion in a central orbit is $\frac{1}{2}r^2\ (d\theta/dt)$

$$\text{A} = \int_0^T \frac{1}{2}r^2\left(\frac{d\theta}{dt}\right)dt$$

now the angular momentum p_i is given by

$p_l = mr^2\ (d\theta/dt)$, so that $\frac{1}{2}r^2\ (d\theta/dt) = p_{\frac{1}{2}m}$

Hence $$A = \int_0^T \left(\frac{p_l}{2m}\right)dt$$

Since p_i and m are constants.

$$A = \frac{p_l T}{2m}$$

Substituting this value of A in equation (i)

$$\mu_l = \frac{e}{T}\cdot\frac{p_l T}{2m} = \frac{e}{2m}\,p_i \qquad \text{...}(ii)$$

As p_1, according to the quantum theory, is equal to $1(h/2\pi)$.

$$\mu_l = \frac{eh}{4\pi m}\,l \qquad \text{...}(iii)$$

Hence the orbital magnetic moment μ_l is directly proportional to the orbital quantum number l.

if $l = 0$, $\mu_l = 0$; if $l = l$, $\mu_l = \frac{eh}{4\pi m}$ etc.

The quantity $eh/4\pi m$ represents the smallest unit, in terms of which the magnitude of atomic and sub-atomic magnetic moments are measure. It is known as the **Bohr magneton**, whose value can be readily computed from the known value of h and e/m; it is found to 9.253×10^{-24} **SI units**.

For various reasons, both theoretical and experimental, it has been found necessary to attribute *one Bohr magneton to the magnetic moment due to electron spin.*

The atom, with its magnetic moment arising from the orbital and spin motions of the electrons in it, may be regarded as an elementary magnet, whose dimensions, through small, are yet finite. If this atomic magnet be placed in a magnetic field, it will be acted upon by the field. But the nature of this action will depend on the nature of the field.

Prof. Stern

If the field is homogeneous, the atomic magnet will experience merely a couple which will rotate its axis into the direction of the field, since the external field acts upon its poles with equal and opposite forces. *If the atomic magnet moves in such a field* in a direction normal to the field, although its magnetic axis will be rotated into the direction of the field, it *will trace a straight line path* without any deviation, *i.e.*, without any transitional displacement.

On the other hand, *if the field is non-homogeneous,* the forces on the two poles will not be equal, which would result in a translatory displacement of the atom as a whole over and above the rotation of its axis into the direction of the field. *If then, the atomic magnet flies across such a non-homogeneous filed* normal to the field direction, *it will be deviated away from its rectilinear path.*

An expression for the amount of the deviation thus produced may be obtained as follows :

Let the magnetic field be non-homogeneous along the X-direction, so that the field gradient is dB/dX and is positive (Fig.1).

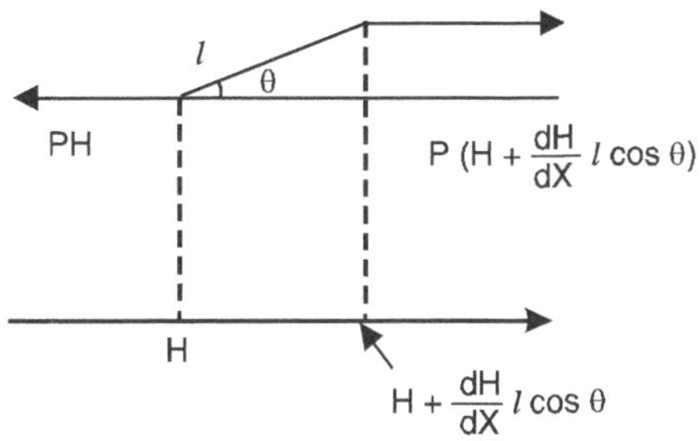

Fig.1.

Let the atomic magnet, whose pole strength is P, length l and magnetic moment M *be placed* in the non-homogeneous field with its axis inclined at an angle θ to the field direction.

If the field strength at one of the poles be B then at the other it will be

$$B + (dB/dX)\, l \cos\theta$$

Hence the force on the two poles of the atomic magnet due to the field are pB and $p\,\{\boldsymbol{B} + (\boldsymbol{dB/dX})\, l \cos\theta\}$ respectively. This means that, over and above the equal and opposite forces pH constituting the rotating couple, there is an extra force equal to $p.\,(dB/dX)\, l \cos\theta$ acting on one of the poles alone; it is this force which displaces the atom as a whole; calling this translatory force F_x, we have

$$F_x = pl\left(\frac{dB}{dX}\right)\cos\theta = M\cos\theta\left(\frac{dB}{dX}\right) \qquad ...(iv)$$

Supposing that the atomic magnet flies across the non-homogeneous field at right angles to the lines of force, it will be displaced from its straight path in the field direction. To find the amount of this displacement, let υ be the velocity of the atomic magnet of the mass m as it enters the field, L the length of its path in the field and t the time of flight through the field.

The acceleration α_x impaired to the atom along the field direction by the translatory force F_x is given by F_x/m.

The displacement D_x of the atom along the field direction at the end of time t is given by F_x/m.

The displacement D_x of the atom along the field direction at the end of time t is given by

$$D_x = \frac{1}{2}\alpha_{x.}t^2 = \frac{1}{2}(F_x/m)(L/\upsilon)^2$$

Since $t = (L/\upsilon)$, the velocity of the atom perpendicular to the field being unaffected by the force in the field direction.

Substituting the value of F_x from (*iv*)

$$D_x = \frac{1}{2}\cdot\frac{M\cos\theta}{m}\cdot\frac{dB}{dX}\cdot\left(\frac{L}{\upsilon}\right)^2$$

If μ be the resolved component of the magnetic moment in the field direction, then

$$D_x = \frac{1}{2}\cdot\frac{\mu}{m}\cdot\frac{dB}{dX}\cdot\left(\frac{L}{\upsilon}\right)^2 \qquad ...(v)$$

From this relation we see that if D_x, m, dB/dX, L and υ are known μ can be calculated. If $dB/dX = 0$, *i.e.*, if the field is made homogeneous, $D_x = 0$, *i.e.*, there is no deviation.

Since $D_x \propto dB/dX$, the amount of deviation is determined by the degree of inhomogeneity of the field. In fact, to produce even a small measurable displacement, the inhomogeneity must be so marked that the final changes even within the small length (of the order of 10^{-8} cm) of the atomic magnet. Stern and Gerlach succeeded in producing a sufficient degree of non-homogeneity by a suitable construction of the pole-pieces of a magnet, one piece being shaped as a knife-edge, while the other hand a flat face provided with a groove, as shown in the figure below. The magnetic lines of force, in consequence, crowd together at the knife-edge so that the field strength is considerably greater there that at the other pole -piece.

Apparatus : The experimental arrangement devised by Stern and Gerlach is shown diagrammatically in Fig. 2. A sample K of the substance to be examined is heated in an electric oven O and, in consequence, sends out atomic rays in all directions with a velocity corresponding to the temperature of vaporisation. By the use of slits S_1 and S_2 a sharp linear beam of atoms is obtained, which then passes between the specially shaped pole-pieces of the electromagnet MM. The magnetic field is

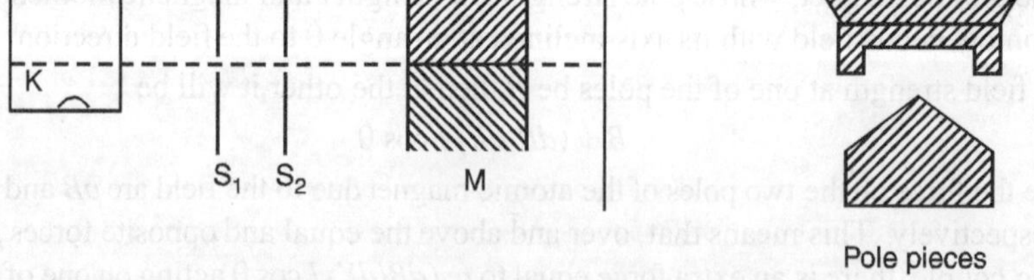

Fig.2. Apparatus used in the Stern and Gerlach experiment.

made as intense and non-homogeneous as possible, its lines of force being perpendicular to the direction of the beam. Finally the beam is made to strike a suitably prepared plate P, placed normal to the initial direction of the beam. The entire apparatus is in evacuated chamber. The arrangement, though simple in design, involves a very delicate technique in which a large number of details have to be carefully attended to. The different parts of the apparatus are quite small. In some cases, the pole-piece is about 5 cm long and the recording plate 3 mm square. The distance between the knife-edge and the plane of the slotted pole-piece is about 1 mm.

Experimental procedure : In performing the experiment, the following precautions have to be taken :

(*a*) As the deflections obtained are small, the different parts of the apparatus must be carefully and accurately aligned.

(*b*) The oven must be capable of withstanding a high temperature and this must be under delicate control as regards constancy over long periods.

(*c*) The velocity of the atoms depends on the oven temperature which must be measured thermoelectrically or optically. For atoms in thermal equilibrium, the kinetic theory of gases gives the relation

$$\frac{1}{2}m\upsilon^2 = \frac{3}{2}kT \text{ or } \upsilon = \sqrt{\frac{3kT}{m}}$$

where m and υ are the mass and velocity of the atom, k the Boltzmann's constant and T the absolute temperature. Some preliminary experiments, however, showed that the mean velocity of the atoms issuing through slits as somewhat greater than this and hence the average value is to be taken as

$$\upsilon = \sqrt{\frac{3.5\ KT}{m}}.$$

(*d*) In order to prevent collisions of the atoms in the beam with the atoms and molecules of any residual gas, the chamber must be exhausted for a long time, liquid air and charcoal being used.

(*e*) Accurate measurement of dB/dX is not easy. It is obtained by measuring the repulsion, at various points between the pole-pieces, of a thin bismuth wire mounted parallel to the edge of the knife-edge pole-pieces. A correction is always found necessary for the variation in the value of dB/dX along the path of the deflected beam, as the beam is nearer the pole-piece at entrance into the field then at exist.

(*f*) Another problem is the development of the traces to render them visible. Receiving plates of different kinds are used for the study of different atoms.

On account of these several technical difficulties to be overcome it is not surprising that of the many experiments begun, few were carried to a successful conclusion.

Silver was the first to be investigated very completely, many series of experiments being made. With other elements, it was then possible to test whether the apparatus was functioning properly by interpolating experiments with silver.

Meissner and Scheffers improved the technique in the following manner which enabled the deviation D_x to be measured very accurately. Working with alkali metals, they caused the deflected beam to impinge on a heated tungsten filament instead of the receiving plate. The atoms of the alkali metals, on striking the hot filament, lost their electrons and were liberated as positive ions. The ionization current thereby produced was measured in different positions and form the position of the maximum current the magnetic deviation was accurately found.

Phipps and Taylor successfully carried out the difficult but important experimental determination of the magnetic moment of the hydrogen atom. The hydrogen atoms were produced in a long discharge tube under appropriate conditions or by dissociating the molecules by means of a hot fragment. A stream containing a fairly high proportion of atomic hydrogen was passed through three special slits in a glass tube between which two high vacuum pumps were operating so that the emergent beam contained effectively free atoms. This atomic beam after passing through a powerful non-uniform magnetic field was received on a plate coated with molybdenum trioxide. The atoms reduce the oxide and so produce a visible trace. A sharply defined blue line against a white background was the result in the absence of the field, while when the field was on, the beam was separated into two, and two traces were obtained. For determining the exact positions of the deviated lines microphotographs were taken. From the deviation thus measured the magnetic moment of the hydrogen atom was calculated to be one Bohr magneton within the limits of experimental error.

Results : In the case of silver, with the field off, a fairly sharp line was traced by the atomic beam on the receiving plate. On establishing the non- homogeneous field a double trace was obtained, one on either side of the original trace. The traces obtained were somewhat as shown in Fig.3. The irregularity in the right hand traces is due to the irregularity of the magnetic field near the knife-edge pole-piece. The convergence at the top and bottom arises from the fact that the field gradient decreases transversely. The traces actually obtained are much smaller than what is shown here, but they can be magnified for easy measurement. A certain diffusiveness is unavoidable owing to the velocity distribution (Maxwellian) of the atoms, but the magnetic splitting of the beam is relatively great enough, that the diffusiveness becomes of secondary importance. The mean deviation D_x of the beam caused by the field can be measured from the traces with fair accuracy. Thus D_x being known, and the other quantities such as dB/dX, L and u being readily measured, μ can be calculated. In the case of silver, it was found to be equal to one Bohr magneton. The results, obtained with other elemented experimented upon, may be summarised as follows :

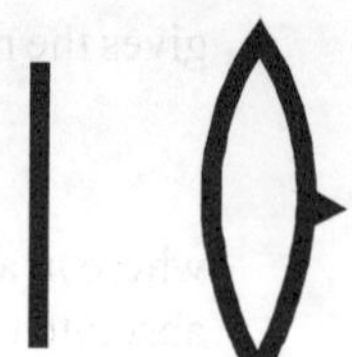

(*a*) (*b*)

Fig. 3

Cu, Au, H, Li, Na, K : all gave double traces, the separation leading to the value of μ = ± 1 Bohr magneton, like silver.

Zn, Cd, Hg, Sn, Pb : no effect was produced by the field. Hence $m = 0$.

Tl (*Thallium*) : The trace was divided into two, there being a definite absence of undeviated atoms. The deviation gave the value of

$$\mu = +1\frac{1}{3} \text{ Bohr magneton.}$$

Sb, Bi : Sb showed only an undeviated trace. The *Bi* trace obtained by Stern and Gerlach was remarkable in showing a one-sided continuous broadening indicating not only undeviated but also strongly attracted carriers. But latter, a symmetrical splitting was obtained by Leu who showed that the distribution of density was consistent with the presence of atoms having values for in the ratio 1 : 3.

Ni, Co, Fe : Nickel showed curious results in that there clearly defined traces appeared, one of them in the undeviated position itself, which meant that some of the atoms have no magnetic moment (μ = 0). For the deviated traces μ was somewhat greater than one magneton. In addition to the three pronounced traces, there were indications of the presence of the atoms might having a value of μ greater than six. In the case of cobalt, a value μ = 6 could just be established.

Interpretation of results

It is important to consider, in the first place, whether the beam consists of single atoms or molecules, since the experiments directly give the resolved moments of the carriers, whether atoms or molecules. At the temperature employed, there is little doubt that the vapours of Cu, Ag, Au, Sn, PB, and *Tl* are monoatomic, so that the values of μ deduced for them do correspond to atomic moments. Ni, Co, Fe also probably monoatomic. But with *Bi* and *Sb* there is a large proportion of molecules in the beam.

Considering only the sure cases, it is possible to show how the results obtained offer a beautiful confirmation of the fundamental postulates of the vector model :

(*a*) **Spatial quantisation :** *The classical theory* lays no restriction on the orientation of the magnet with respect to the field direction, all values from 0° to 180° for the angle θ in the relation $\mu = M \cos \theta$ being permissible. Hence the displacement of the atoms in the non-homogeneous field should cover a continuous range, so that a *continuous diffuse band* should cover a continuous range, so that a *continuous diffuse band* should be obtained on the receiving plate. Further, on account of the Maxwellian distribution of velocities of the atoms in the beam, the densest part of the trace should be in the centre, since the number of atoms whose axes make angles between θ and (θ + dθ) with the field is merely proportional to sin θ and hence a large fraction of the atoms would be oriented merely transversely to the field and so would suffer very small deflections.

On the other hand, *according to the spatial quantisation theory,* not all settings are possible for the atomic magnet with respect to the field direction, but only a certain discrete number. If the total angular momentum number of the atom is J, the permitted orientations are (2J + 1). Hence on the plate, instead of obtaining a continuous band (2J + 1), distinct traces should be obtained. Further, if the atom under test belongs to the *one electron system* and is in the *ground state,* one can have higher values of J. Since although L = 0, S = ½ = J so that (2J+1) = 2. In such a case, therefore, only two orientations are permitted and hence two traces should be obtained on the plate. In the case of many *electron system,* the atoms in the *ground state* can have higher values of J, since although L= 0, S can assume values O, 1/2, 3/2, 2. etc., and J = S. The number of permitted orientations may be greater than two so that more complex setting of the beam could be expected. *e.g.,* for atoms of $^6S_{5/2}$ term J being equal to 5/2, six traces should be obtained on the plate.

All these conclusions of the spatial quantisation theory are exactly verified by the results of the Stern and Gerlach experiment. Thus

(*i*) *H, Li, Na, K, Cu, Ag,* all belonging to the *first column of the periodic table,* have in the ground state the spectral term $^2S_{½}$, which means that 2 J + 1 = 2. They all give only two traces as expected from theory. As soon as the atoms of these elements enter magnetic field their magnetic axes orient themselves in the two permitted directions parallel and antiparallel to the field direction, no intermediate position being allowed. Further, in this case it can be shown that μ = ±1 Bohr magneton, actually obtained from experimental data.

(*ii*) *Zn, Cd, Hg* belonging to the *second column,* have two *s* electrons outside closed shells, and their normal state, as indicated by their spectra, is $^1S_0$. This means that J = 0 and hence the atom as whole has no resultant angular momentum and no magnetic moment. That is why no effect is produced on the atoms of these elements by the field.

(*iii*) *Tl* belonging to the third column, has one *p* electron in addition to completed groups. This gives a $^2P_{½}$ ground term *i.e.*, L= 1, J= ½, 2J + 1 = 2. Hence it gives rise to two traces and calculations show that *m* = 0, which agrees with the experimental results.

(*iv*) *Sn, Pb* belonging to the *fourth column* have two *p* electrons not forming a completed group, but the total magnetic moment vanishes in their normal $^3P_0$ state. This is why the atoms of these elements are undeviated by the field in the experiment.

(*b*) **Electron spin :** Considering the well established case of silver, experiment clearly shows that the atomic beam of silver is split up by the magnetic filed into two parts of approximately equal intensity. Further the value of the magnetic moment obtained is ± 1 Bohr magneton. From theory it is known that for the normal state of the silver atom L = O. Hence, if the spin of the electron does not exist, S would be zero, so that J also would be equal to zero and there would be no splitting of the atomic beam into two components. If, on the contrary, the existence of electron spin is admitted and the value of 1/2 be given to S in this case, there is perfect agreement between theory and experiment.

(*c*) **Atomic and quantum nature of magnetism :** The idea of atomic nature of magnetism goes back to the days of weber. The view that *paramagnetic* substances have a permanent molecular magnetic moment, while *diamagnetic* substances possess no such moment is a long established fact in the physical theory of magnetism. Weber was the first to develop this idea on the molecular hypothesis of Ampere. It was next rendered certain by Langevin's treatment of paramagnetism based on the kinetic theory of gases and of diamagnetism on the basis of the classical electron theory. Weiss, in 1911, from experimental data then available, concluded that there was a fundamental unit of magnetic moment, of which all atomic and molecular moments were multiples. This unit was called after him the **Weiss magneton**, whose value was estimated to be 1123.5 gauss cm. per gram -atom or per gram-molecule, which gives 1.85×10^{-21} gauss-cm. per atom of the molecule. With the advent of the Quantum theory, another fundamental unit, known as the **Bohr magneton**, came to be recognised. Its value was found to be 5,590 gauss-cm. per gram -atom or 9.21×10^{-21} gauss - cm. per atom, which is therefore, almost five times as large as the older unit.

The Stern and Gerlach experiment not only confirms the general conclusions of the classical theory about para and diamagnetism, but goes much further and establishes the quantum nature and atomic origin of magnetism. For, in the first place, according to the results of the experiments, substances which are found to be diamagnetic have no atomic magnetic moment ($\mu = 0$) *e.g.*, Zn, Cd, *Hg*, *Pb*, *Sn*, etc. Substances which are paramagnetic are made up of atoms with one valency electron in the ground state, thereby giving rise to a magnetic moment of 1 Bohr magneton due to the spin of the single free electron : *e.g.*, *H*, *Li*, the alkalies, *Cu*, *Ag*, *Au*,. Ferromagnetic substances (*Fe*, *Ni*, *Co*) are made upon atoms with intermediate incomplete electronic shells and have a large value for the atomic moment.

Secondly, the results establish also that the true fundamental unit is Bohr magneton and not the Weiss magneton. The latter based on, the classical theory, according to which all possible orientations of the atomic axis with respect to the field are permissible, has only an apparent existence. The former is the one given by the experiment of Stern and Gerlach and conforms to the quantum theory; it alone, therefore, has a real substance.

There is no doubt that the vector atom model is a superb and many sided conception, which enables one to solve most of the important and nitrate problems concerning atomic structure and spectral phenomena.

Example 1. *Calculate the value of Bohr magneton.*

Solution. Given,

$$h = 6.6 \times 10^{-34} \text{ joule -second}$$

$$m = 9.1 \times 10^{-31} \text{ kg}$$

$$e = 1.6 \times 10^{-19} \text{ coulomb}$$

Bohr magneton,

$$B = \frac{eh}{4\pi m}$$

or

$$B = \frac{1.6 \times 10^{-19} \times 6.6 \times 10^{-34}}{4\pi \times 9.1 \times 10^{-31}}$$

$$B = \mathbf{9.253 \times 10^{-24}} \textbf{ SI units.}$$

Example 2. *The experimental value of Bohr magneton is* 9.21×10^{-24} SI *units and Planck's constant* $h = 6.6 \times 10^{-34}$ Joule-second.

Calculate the value of $\frac{e}{m}$ *of an electron.* [*Calicut, 1992*]

Solution. Here

$$\beta = 9.21 \times 10^{-24} \text{ SI units}$$

$$h = 6.6 \times 10^{-34} \text{ joule-second}$$

$$B = \frac{eh}{4\pi m}$$

or

$$\frac{e}{m} = \frac{4\pi B}{h}$$

$$\frac{e}{m} = \frac{4 \times 22 \times 9.21 \times 10^{-24}}{7 \times 6.6 \times 10^{-34}}$$

$$\frac{e}{m} = \mathbf{1.755 \times 10^{11}} \text{ C/kg}$$

Wave Mechanical Atom Model

The great drawback of the vector atom model is that it does not present itself as a single unified system, containing the necessary theorietical justification of its principles and postulates. These are introduced somewhat piecemeal, partly on empirical data and partly analogy. No adequate reasons are given for the use of electron spin was half-integer value; the assumption of discrete energy states are emission of radiation when the atom passes from one of these states another are arbitrary as in older models. The model therefore lacks sound theoretical foundation. The most recent theory, known as the wave *mechanical atom model*, supplies it his information and presents a picture of the peripheral electronics structure of the atom, the best that can be in the present state of our knowledge.

A detailed study of this model is beyond the scope of this book masterly treatment by Dirac of the relativistic theory of the electron based on wave mechanics offers, almost without any special effort theorietical justification of the postulates and conclusions of the vector model. We shall limit ourselves to a brief survey of the essential feature of this new conception, indicating the advances made over the older models.

Acciording to wave mechanics, in order to have a complete picture of the electron, we should consider it under the double aspect of particle a wave. The change of fidning a moving elecron at a about this exact position. The older models, which place.

APPENDIX C

APPROXIMATE VALUES OF PHYSICAL CONSTANTS IN SI UNITS

Charge of electron	=	1.602×10^{-19} C
Mass (rest) of electron	=	9.1×10^{-31} kg
Mass (rest) of proton	=	1.6725×10^{-27}
Mass (rest) of Neutron	=	1.6748×10^{-27}kg
Speed of light	=	3×10^{8} m/s
Planck's constant	=	6.624×10^{-34} J-s
e/m of electron	=	1.7589×10^{11} C/kg
First Bohr radius	=	5.29×10^{-11} m
Bohr magneston	=	9.27×10^{-29} J/T
Electron volt	=	1.6×10^{-19} J
Boltzmann's constant	=	1.38×10^{-23} J/K
Universal gas constant	=	8.31 J/K -mole
Rydberg constant	=	1.097×10^{7} m^{-1}
Permittivity of free space	=	8.85×10^{-12} F/m
Permeability of free space	=	$4\pi \times 10^{-7}$ H/m = 1.26×10^{-3} H/m
Avogardro's number	=	6.02×10^{26} per kg mole
Gravitational constant	=	6.67×10^{-11} N-m^2/kg^2
Nuclear Magneton	=	5.05×10^{-27} J/T
1 u	=	931.5 MeV
Fermi (1 fm)	=	10^{-15} m
Barn '1*b* (C.S)'	=	10^{-28} m^2
m_n	=	1.008665 u
m_H	=	1.007825 u
$\hbar$	=	1.055×10^{-34} J.S

Bibliography

1. Quantum Physics of Atoms, Molecules, Solids, Nuclear Particles by *Robert Eisberg & Robert Resniek.*
2. Nuclear Physics : J*ohn Lilley*
3. Modern Physics by *Ritchmeya, Kennard and Coopa.*
4. Atomic and Nuclear Physics by *H. Semat.*
5. Fundamentals of Molecular Spectroscopy by *Banwell.*
6. Nuclear Physics by *Kaplan I.*
7. Perspectives of Modern Physics by *Beiser.*
8. Modern Physics 6th Ed. by Arthur Beiser.
9. Relativity by *Robert Resnik.*
10. Atomic & Nuclear Physics by A.B. Gupta & Dipak Ghosh.
11. Atomic & Nuclear Phy. by *Derek. L. Lwisey.*
12. Atomic & Nuclear Phy. *T.A. Littlefield & N Thorley*
13. Modern Phy. by *Frank J. Blatt.*
14. Nuclear Phy. by *S.B. Patel.*
15. Atomic & N. Phys by *S.N. Ghoshal.*
16. Modern Phy. *Murgeshan*
17. Atomic Phy. *J. B. Rajan*
18. Atomic & Molecular Phy. *Raj Kumar*
19. Modern Phy. by *R.B. Singh*
20. Nuclear Phy. by *D.C. Tayalo.*
21. Modern Phy. by *H.S. Mani & C. K. Mehta.*

Appendix D

Franck and Hertz Experiment

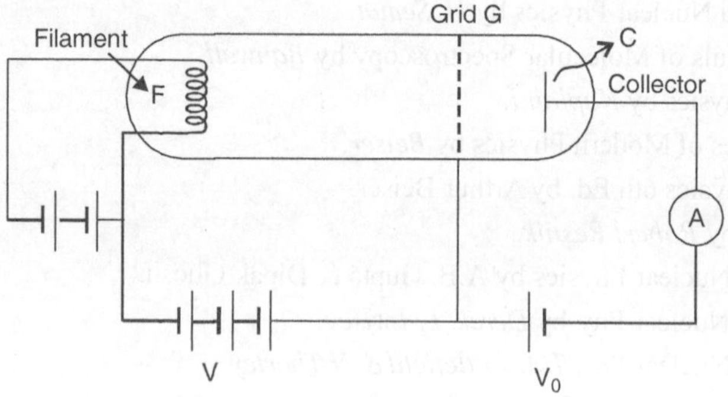

Fig. D-1

The Franck and Hertz experiment gives conclusive evidence of energy levels in an atom and provides confirmation of Bohr's Theory.

J. Franck and G.Hertz established the existence of energy levels in the mercury atom.

The apparatus consists of three electrodes in a glass tube as shown in fig. D-1. A small quantity of mercury was present in the tube and its concentration was controlled by the temperature of the tube (about 150°C).

This was achieved by placing the tube in a small oven. Note that the distance between the filament and the grid G was much greater than the distance between the grid and the collecter plate C.

Electrons from the filament F are accelerated through a potential difference V towards the grid. beyond which there was a small reverse voltage V_0 applied between the grid and the plate.

On increasing the positive potential of the grid, the electrons gain energy as per the equation $1/2\ mv^2 = eV$. These accelerated electrons soon have sufficient energy to reach the collector plate overcoming the retarding potential.

At first the collision of the electrons with the mercury atoms are elastic. But as the energy is increased a critical value is reached, at this stage the collisions became inelastic and the energy of the incoming electron is transferred to the mercury atom.

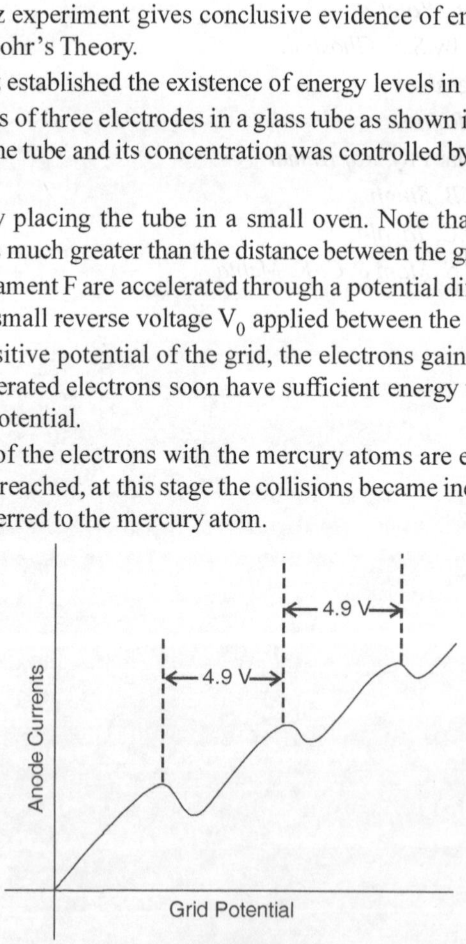

Fig. D-2

This occurs when the energy of the incoming electron is sufficient to raise the electron in the mercury atom from the lowest orbit to the next orbit in mercury atom. The incoming electron having lost its energy is not capable of reducing the collector plate due to the retarding potential. Hence the anode current shows a sudden drop when the grid potential reaches this critical value. As the potential is increased further the current also increases as the electrons again acquire enough energy to reach the plate. A second critical potential is reached when the electrons having excited an atom loses its energy. This is shown by the second peak in the anode current as illustrated in Fig. D-2. A number of such peaks can be obtained. The potential difference between successive peaks for mercury is 4.9V, the value of the first peak is ignored due to contact potentials in the circuit.

Mercury has a spectral line,. 2536Å which corresponds to 4.9eV. Franck and Hertz found a line nearly this value. This and the fact that the energy difference is 4.9V between consecutive dips confirms the existence of discrete energy levels in the mercury atom.

Franck and Hertz were awarded the Noble Prize in Physics (1925) for this work.

Correction for the finite Mass of the Nucleus

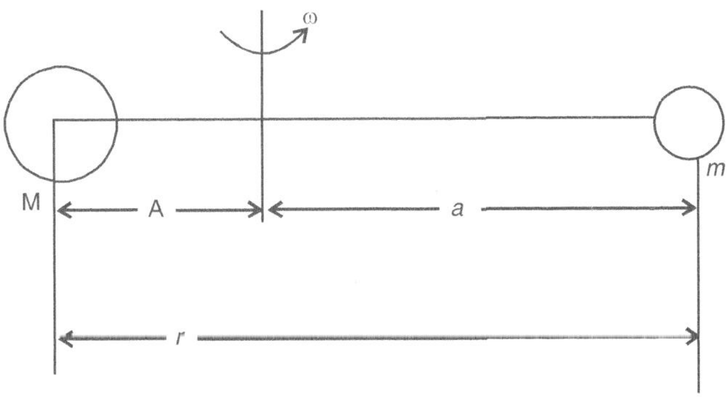

Fig. D-3.

In the Bohr's theory the nucleus was fixed and the electron was assumed to move round in a circle. But this assumption cannot be correct unless the nucleus is infinitely massive.

For a nucleus of mass M both the nucleus and the electron will rotate the common centre of mass of the two body system. Let the angular velocity of the system be ω.

In fig. D-3. , the electron and the nucleus are shown to rotate about the point-O, the centre of mass. Let *a* be the distance of the electron from the centre of mass and A the distance of the nucleus from the centre of mass. The distance from the electron to the nucleus is *r* as shown in fig. D-3.

$$\therefore \quad r = a + A \qquad \text{D-1}$$

According to the definition of centre of mass

$$am = AM \qquad \text{D-2}$$

Eliminating A from Eqs. D-1 and D-2 we have

$$a = \frac{r.M}{M+m} \qquad \text{D-3}$$

or Eliminating *a* from the equations, we get.

$$A = \frac{r.M}{M+m} \qquad \text{D-4}$$

If *v* and *V* be the speed of the electron and the nucleus respectively, then we have

$$\left.\begin{aligned} v &= a\omega \\ \text{and} \quad V &= A\omega \end{aligned}\right\} \qquad \text{D-5}$$

The total kinetic energy of the system is the sum of the kinetic energy of the electron and the nucleus.

Hence, $E_k = E_k$ nucleus + E_k electrons

$$= \tfrac{1}{2}MV^2 + \tfrac{1}{2}mv^2$$

Substituting for V and v from D-5 we get

$$E_k = \tfrac{1}{2} MA^2\omega^2 + \tfrac{1}{2} ma^2\omega^2. \quad \text{D-6.}$$

Using the fact that $a = \dfrac{rM}{M+m}$ and $A = \dfrac{rM}{M+m}$ we have

$$E_k = \frac{1}{2} M \frac{r^2 m^2}{(M+m)^2} \omega^2 + \frac{1}{2} m \frac{r^2 M^2}{(M+m)^2}$$

$$\therefore \quad E_k = \frac{1}{2} Mmr^2\omega^2 \left[\frac{M+m}{(M+m)^2}\right]$$

or

$$E_k = \frac{1}{2}\frac{m\,M}{(M+m)} r^2\,\omega^2 \quad \text{...D-7}$$

$$= \frac{1}{2}\mu\, r^2\,\omega^2$$

The form of the equations is similar excepting that the mass is replaced by the reduced mass μ.

i.e.,

$$\mu = \frac{mM}{M+m} = \frac{m}{1 + m/M} \quad \text{...D-8}$$

The reduced mass, therefore, is the product of the masses divided by their sum:
Similarly it can be derived that the potential energy

$$E_p = \frac{-Ze^2}{4\pi \epsilon_0\, r} \quad \text{...D-9}$$

Hence the total energy

$$E = -\frac{1}{2}\mu\,\omega^2\, r^2 - \frac{Ze^2}{8\pi\epsilon_0\, r}$$

According to the 2nd postulate of Bohr, the angular momentum now reads.

$$MA^2\omega + ma^2\omega = n\,\hbar \quad \text{...D-10}$$

Using $a = \dfrac{rM}{(M+m)}$ and $A = \dfrac{rM}{(M+m)}$

we get

$$mr^2\omega = n\,\hbar \quad \text{...D-11}$$

Likewise we can show that

$$r = \frac{\epsilon_0\, n^2 h^2}{\pi\,\mu\, Z\, e^2} \quad \text{...D-12}$$

and hence

$$E = \frac{-\mu e^4 Z^2}{8\,\epsilon_0^2\, n^2\, h^2} \quad \text{...D-13}$$

Using Bohr's third postulate we have

$$\bar{\nu} = \frac{\mu e^4 Z^2}{8\,\epsilon_0^2\, h^2}\left(\frac{1}{n_1^2} - \frac{1}{n_2^2}\right) \quad \text{...D-14}$$

or

$$\bar{\nu} = Z^2 R\left(\frac{1}{n_1^2} - \frac{1}{n_2^2}\right)$$

where R is the Rydberg constant and has a dependence on mass of the nucleus.

R is the value of R when m is equal to μ.

$$\therefore \qquad R = R_{\infty} \frac{1}{1 + m/M} \qquad \text{...D-15}$$

The theoretical expression for the Rydberg's constant becomes $R = 109678\ \text{cm}^{-1}$. This agrees with the experimental result to within six significant figures.